Advanced Classical Electromagnetism

Advanced Classical Electromagnetism

ROBERT M. WALD

PRINCETON UNIVERSITY PRESS
Princeton and Oxford

Published by Princeton University Press
41 William Street, Princeton, New Jersey 08540
99 Banbury Road, Oxford OX2 6JX

press.princeton.edu

Library of Congress Cataloging-in-Publication Data

Names: Wald, Robert M., author.
Title: Advanced classical electromagnetism / Robert M. Wald.
Description: Princeton : Princeton University Press, [2022] | Includes index.
Identifiers: LCCN 2021018752 (print) | LCCN 2021018753 (ebook) | ISBN 9780691220390 (hardback) | ISBN 9780691230252 (ebook)
Subjects: LCSH: Electromagnetism—Textbooks. | BISAC: SCIENCE / Physics / Electromagnetism | SCIENCE / Physics / General
Classification: LCC QC760 .W295 2022 (print) | LCC QC760 (ebook) | DDC 537—dc23
LC record available at https://lccn.loc.gov/2021018752
LC ebook record available at https://lccn.loc.gov/2021018753

British Library Cataloging-in-Publication Data is available

Editorial: Ingrid Gnerlich and Whitney Rauenhorst
Production Editorial: Mark Bellis
Jacket Design: Wanda España
Production: Jacqueline Poirier
Publicity: Matthew Taylor and Charlotte Coyne
Copyeditor: Cyd Westmoreland

Jacket art: Yves Tanguy, *The Storm*, 1926. The Philadelphia Museum of Art: The Louise and Walter Arensberg Collection, 1950, 1950–134–187. Photo: The Philadelphia Museum of Art / Art Resource, NY © Estate of Yves Tanguy / Artists Rights Society (ARS), New York

This book has been composed in Minion Pro & Univers Lt Std

Printed on acid-free paper. ∞

Printed in the United States of America

10 9 8 7 6 5 4 3 2 1

Contents

Preface

This book arose from my teaching the first quarter of the standard graduate course in electromagnetism at the University of Chicago in the winter of 2018. It had been decades since I had previously taught this course, so I approached it with fresh eyes, and it was natural for me to try to rethink how the subject of electromagnetism should be presented at the graduate level. When I did so, it became clear to me that the usual quasi-historical way of presenting the subject promotes some very unhealthy ways of thinking about electromagnetism. Therefore, to avoid starting off on the wrong foot, I decided to spend the first few lectures of the course describing what I now refer to in chapter 1 of this book as "myths" concerning electromagnetism. I found that by starting out in this way, it became much easier to straightforwardly present the subject in a clear and concise manner, without having to make shifts in perspective as the subject is developed. I taught the course again in the following 3 years and provided lecture notes to the class. These lecture notes have now evolved into this book.

The first chapter of this book is thus a quite unconventional introduction to electromagnetism. Instead of beginning with the force between charged particles, discussing how this gives rise to a "field" concept, and so forth, my aim in chapter 1 is to explain to students how they should think about electromagnetism from a modern and mathematically precise perspective. The major points made in this chapter are that (i) the potentials, not the field strengths, are the fundamental dynamical variables in electromagnetism; (ii) the energy and momentum properties of the electromagnetic field are an essential part of the formulation of the theory and cannot properly be derived by "work done" arguments; (iii) electromagnetic fields should not be thought of as being *produced* by charges; and (iv) at a fundamental level, the charged matter in classical electrodynamics must be viewed as continuously distributed rather than consisting of point charges. Many of these points cannot be fully elucidated until the later chapters in the book—particularly chapters 9 and 10—but my intent is to lay out these ideas in a sufficiently clear and explicit way in chapter 1 that I can take these perspectives unapologetically in the remainder of the book.

The topics treated in chapters 2–7 are ones that normally would be covered in any graduate course in electromagnetism. Electrostatics is treated in chapter 2, but starting with Poisson's equation, not Coulomb's law. Dielectric materials in electrostatics are treated in chapter 3, with considerable care given to how the macroscopic averaging is done and to the treatment of energy. Magnetostatics is treated in chapter 4, with a full discussion of the sign difference between magnetostatics and electrostatics in the field interaction energy of a dipole in an external field—and how this relates to the change in the rest mass of a magnet when it is quasi-statically moved in an external magnetic field. Electrodynamics and radiation are discussed in depth in chapter 5. In addition to topics normally found in electromagnetism texts, I derive the initial value formulation for Maxwell's equations in that chapter. Electrodynamics in media is treated in chapter 6, including a discussion of magnetohydrodynamics. The geometric optics approximation to wave dynamics is presented in the first section of chapter 7, followed by a discussion of interference and coherence and an analysis of two problems in diffraction: scattering by a dielectric ball and the propagation of radiation through an aperture.

Special relativity is discussed in chapter 8. Special relativity underlies the formulation of electromagnetic theory, so it really should be presented at the outset of a book on electromagnetism, rather than be relegated to a chapter near the end of the book. However, special relativity remains such an unfamiliar topic for most students that it is not feasible to do this. Many treatments of special relativity focus on the rules for applying Lorentz transformations to quantities, without providing much insight into the underlying geometrical content of the theory. In contrast, it would be natural for a general relativist like me to introduce more mathematical abstraction and geometrical machinery than would be strictly needed to provide a clear description of special relativity. I have put considerable care into writing section 8.1 in such a way that it introduces special relativity in a conceptually clear way without introducing more abstraction than I believe to be essential. This section can be read independently of the rest of the book, and I hope it will provide a useful introduction to special relativity on its own. The formulation of electromagnetism in the framework of special relativity is then given, followed by a discussion of charged particle motion and the radiation from a point charge in arbitrary motion.

Chapter 9 discusses electromagnetism as a gauge theory, thereby bringing the formulation of electromagnetism in this book up to the level of conceptual understanding that was achieved by the mid-twentieth century. There is a considerable gap between how the electromagnetic field and its interaction with charged matter is normally described in courses on classical electromagnetism and how it actually appears as a fundamental interaction of nature in the standard model of particle physics. This chapter should help close this gap.

Finally, the notion of a point charge is discussed in depth in chapter 10. It is shown that a mathematically well-defined limit of a charged body as it shrinks down to zero size can be taken, provided that one also takes the charge and mass of the body to scale to zero proportionally to its size. Lorentz force motion is obtained in this limit. Self-force corrections can then be computed perturbatively in a mathematically rigorous manner. The issue of how to self-consistently describe the motion of a charged body taking the self-force corrections into account—without introducing spurious "runaway" solutions—is addressed in the final section of this chapter.

Throughout this book, I have attempted to formulate all key conceptual ideas and results in the theory of electromagnetism in a clear and concise manner. However, I have not attempted to present an extensive collection of examples or applications. These features of the book account for why it is about one-third the length of some other graduate texts in electromagnetism that have a very similar coverage of topics.

I have attempted to present everything in this book with a high level of mathematical precision. Although I have made an effort to avoid getting sidetracked with unnecessary mathematical detail, I have not knowingly oversimplified any statements in the book and have tried to be careful to insert appropriate caveats when formulas or other results hold only under restricted conditions. In several instances in the early chapters, I have added boxed "side comments" to explain some mathematical points that may be of potential interest and relevance to the reader but are not strictly needed for the discussion.

An extensive collection of problems is provided for chapters 2–8. One purpose of these problems is the usual one of providing students with an opportunity to test their understanding of the basic concepts introduced in the chapter. However, there is an additional important purpose for some of the problems: to present topics that are not essential to the development of the core ideas of the book but are, nevertheless, of considerable interest and importance. Some examples of such topics treated in the problems are hidden momentum, the Hall effect, Gaussian beams, Thompson scattering, optical fibers, Stokes parameters, and Cherenkov radiation. I have written these problems in such a way that the key concepts are explained—and the key results are given—in the statement of the problem. Thus, a reader may find these problems to be a useful introduction to these topics.

The main audience that I have in mind for this book are graduate students in theoretical physics, although I hope that graduate students in experimental physics, undergraduates, and others will also find the book to be of interest. This book is written under the assumption that readers have had an introductory course in electromagnetism and thereby already have some intuition about electric and magnetic fields. I also assume that readers have a solid knowledge of vector calculus, but I do not assume much mathematical background beyond this.

I use SI units throughout the book. Unfortunately, SI units have the highly unpleasant feature of introducing two constants, ϵ_0 and μ_0, that satisfy the relation $\epsilon_0\mu_0 c^2 = 1$, where c is the speed of light. There are good historical reasons that this is the case. It is natural to assign an electric permittivity ϵ and a magnetic permeability μ to many materials, and it therefore was natural to assign corresponding values, ϵ_0 and μ_0, to the vacuum. It was then a truly great achievement of Maxwell to recognize that his equations implied that disturbances of the electric and magnetic fields in vacuum propagate with speed $c = 1/\sqrt{\epsilon_0\mu_0}$ and that these disturbances could be identified with light. However, this relation between ϵ_0, μ_0, and c means that there is redundancy in these constants. Consequently, the appearance of formulas in SI units can be changed in nontrivial-looking ways by using this redundancy. For example, in SI units, one of the Maxwell equations is usually written as $\nabla \cdot \boldsymbol{E} = \rho/\epsilon_0$. However, this equation could equally well be written as $\nabla \cdot \boldsymbol{E} = \mu_0 c^2 \rho$. The latter form may seem rather jarring, as it seems to suggest that the magnetic permeability of the vacuum and the speed of light enter a basic equation of electrostatics. In any case, one must make a choice of which of these constants to use in any formula. The usual convention is to use ϵ_0 in the above Maxwell equation and use μ_0 in the Maxwell equation involving the current density $\boldsymbol{J}$. However, this convention cannot be maintained when one writes Maxwell's equations in special relativistically covariant form, since the 4-current J^μ enters these equations, and it makes no sense to use different conventions for different components of this 4-vector. Indeed, from chapter 8 on, I dispense entirely with ϵ_0 and use μ_0 and c in all formulas. To avoid the unpleasantness associated with this redundancy of ϵ_0, μ_0, and c, I used Gaussian units in the original versions of my lecture notes. However, although Gaussian units were in quite prevalent use decades ago, SI units are used nearly universally now. Thus, the unpleasantness of SI units is outweighed by the unfamiliarity of students with Gaussian units—as well as the possibility that someone using my book may be led to purchase the wrong size of electromagnetic equipment if the formulas were written in Gaussian units. So, I have chosen to use SI units.

Ordinary vectors in 3-dimensional space will be denoted in boldface (e.g., the electric field will be denoted as $\boldsymbol{E}$, as in the previous paragraph). Cartesian components of vectors will be denoted with Latin subscripts and without boldface symbols (e.g., E_i, with $i = 1, 2, 3$, denotes the components of $\boldsymbol{E}$ in a Cartesian basis). Beginning in chapter 8, I introduce the notion of spacetime vectors. For the reasons explained in section 8.1, it then will be essential to explicitly introduce the notion of dual vectors and to distinguish clearly between vectors and dual vectors in our notation. I will then adhere to standard special relativistic index notation, wherein spacetime vectors are denoted with Greek superscripts (e.g., W^μ) and spacetime dual vectors are denoted with Greek subscripts (e.g., U_μ). Some additional special relativistic notational conventions are stated at the end of section 8.1.

I am greatly indebted to numerous colleagues for reading parts (and, in some cases, all) of the manuscript and providing me with valuable feedback. These include Sam Gralla, Abe Harte, Jim Isenberg, Istvan Racz, and Gautam Satishchandran, as well as numerous students who took my course. Among the latter, Tixuan Tan deserves special thanks for reading the manuscript with great care and asking many penetrating questions about the exposition.

CHAPTER 1

Introduction: Electromagnetic Theory without Myths

The full development of the theory of electromagnetism in the nineteenth century stands as one of the greatest achievements in the history of physics. The theory of electromagnetism as formulated by Maxwell is a mathematically consistent theory that provides an excellent description of an extremely wide range of physical phenomena. Of course, Maxwell's electromagnetism is a classical theory that cannot properly describe phenomena in which the quantum properties of the electromagnetic field play an important role, but the quantum field theory of the electromagnetic field is built upon the foundation of the classical theory.

Maxwell's equations relate the electric and magnetic fields, $\boldsymbol{E}$ and $\boldsymbol{B}$, to each other and to the charge density, ρ, and the current density, $\boldsymbol{J}$. That is, $\rho(\boldsymbol{x})$ is the electric charge per unit volume at $\boldsymbol{x}$, and for any unit vector $\hat{\boldsymbol{n}}$ at $\boldsymbol{x}$, $\boldsymbol{J}(\boldsymbol{x}) \cdot \hat{\boldsymbol{n}}$ gives the flux of charge per unit area through an area element perpendicular to $\hat{\boldsymbol{n}}$. Maxwell's equations in SI units[1] are as follows:

$$\nabla \cdot \boldsymbol{E} = \frac{\rho}{\epsilon_0}, \tag{1.1}$$

$$\nabla \times \boldsymbol{B} - \frac{1}{c^2}\frac{\partial \boldsymbol{E}}{\partial t} = \mu_0 \boldsymbol{J}, \tag{1.2}$$

$$\nabla \cdot \boldsymbol{B} = 0, \tag{1.3}$$

$$\nabla \times \boldsymbol{E} + \frac{\partial \boldsymbol{B}}{\partial t} = 0. \tag{1.4}$$

Here ρ and $\boldsymbol{J}$ must satisfy the charge-current conservation equation

$$\frac{\partial \rho}{\partial t} + \nabla \cdot \boldsymbol{J} = 0, \tag{1.5}$$

[1] As discussed in the preface, SI units have the unfortunate feature that the three constants $\epsilon_0 \approx 8.85 \times 10^{-12}$ F/m (the vacuum permittivity), $\mu_0 \approx 1.26 \times 10^{-6}$ H/m (the vacuum permeability), and $c \approx 3.00 \times 10^8$ m/s (the speed of light) appearing in the equations of electromagnetism are not independent but satisfy $\epsilon_0 \mu_0 c^2 = 1$. Consequently, the appearance of formulas in SI units can be changed in nontrivial-looking ways by using this identity.

since otherwise, no solutions to eqs. (1.1) and (1.2) exist. Apart from this restriction, $\rho(t, \boldsymbol{x})$ and $\boldsymbol{J}(t, \boldsymbol{x})$ can be specified arbitrarily.

Maxwell's equations have survived without modification for more than one and a half centuries (i.e., the equations I have written above are equivalent to those given by Maxwell). However, our understanding of electromagnetism at a fundamental level has progressed greatly since the time of Maxwell. Despite this fact, many outdated ways of thinking about electromagnetism remain prevalent. This is strongly reinforced by the quasi-historical way in which electromagnetism is usually taught, even at the graduate level: One normally starts with Coulomb's law in electrostatics, with point charges taken as "fundamental." This motivates the introduction of an electric field $\boldsymbol{E}$ satisfying eq. (1.1) as well as $\nabla \times \boldsymbol{E} = 0$ (i.e., eq. (1.4) with $\partial \boldsymbol{B}/\partial t = 0$). Energy is assigned to the electrostatic interaction via an analysis of the mechanical work done when moving point charges quasi-statically. Similarly, in magnetostatics, one normally starts with the Biot-Savart law for the force between current elements. This motivates the introduction of a magnetic field $\boldsymbol{B}$ satisfying eq. (1.2) with $\partial \boldsymbol{E}/\partial t = 0$ as well as eq. (1.3). The dynamical terms in $\boldsymbol{E}$ and $\boldsymbol{B}$ are then introduced to get the full Maxwell equations in the form given above. A scalar potential, ϕ, and vector potential, $\boldsymbol{A}$, satisfying

$$\boldsymbol{E} = -\nabla \phi - \frac{\partial \boldsymbol{A}}{\partial t}, \tag{1.6}$$

$$\boldsymbol{B} = \nabla \times \boldsymbol{A}, \tag{1.7}$$

also are introduced at some stage as a convenient way of solving the Maxwell equations (1.3) and (1.4).

This manner in which the theory of electromagnetism is presented encourages a number of unhealthy ways of thinking about the theory, which I have referred to as "myths" in the title of this chapter. The most pernicious of these myths are the following: (i) The field strengths, $\boldsymbol{E}$ and $\boldsymbol{B}$, are taken to be fundamental, whereas the potentials, ϕ and $\boldsymbol{A}$, are viewed as quantities that are introduced merely as a convenience. (ii) The energy, momentum, and stress properties of the electromagnetic field are considered to be properties derived or guessed from interactions with charged matter and conservation laws rather than properties of the electromagnetic field having a fundamental status comparable to that of Maxwell's equations themselves. For example, in this regard, it is often stated that the momentum density of the electromagnetic field is ambiguous up to the addition of the curl of a vector field, since it is not uniquely determined by energy conservation. (iii) Electromagnetic fields are considered to be *produced* by charged matter (as opposed to the fact that electromagnetic fields *interact* with charged matter). (iv) Point charges are taken to be a fundamental description of charged matter, despite blatant mathematical inconsistencies associated with them, such as infinite self-energy.

In the following sections, I do my best to debunk these myths. There is, of course, a serious pedagogical problem with my doing this, since to fully follow all of the discussion in this chapter, readers will need to have a considerable knowledge of electromagnetic theory. While it would be reasonable to hope that readers will have a considerable knowledge of electromagnetic theory by the time they have gotten to the end of this book, it is not reasonable to assume such knowledge at the beginning. Indeed, many of the points discussed here will be properly explained in detail only in the last two chapters of this book. It is not necessary that the reader follow all details of the discussion in this chapter—since everything said in this chapter will be elucidated in the remainder of the book—but it is important that the reader gain a sense of the viewpoint on classical electromagnetism that I take. I feel that it is highly preferable to begin this book in this way rather than to get started on the wrong foot by taking the usual quasi-historical path. In the succeeding chapters, I develop the subject in a largely conventional way—starting with electrostatics and magnetostatics before moving to full electrodynamics—but the viewpoint taken will always be fully compatible with the discussion in this chapter.

Before discussing the above myths, I wish to make some comments about the relationship of classical electrodynamics to special relativity. Maxwell's equations are not compatible with the spacetime structure of pre-relativity physics unless one has a "preferred rest frame." This, by itself, was not troubling in the nineteenth century, since it was believed that there was a mechanical medium—the "luminiferous aether"—through which electromagnetic fields propagated. Such an aether would naturally provide a preferred rest frame. However, the lack of evidence in the Michelson-Morley experiment for a preferred rest frame as well as other problems with the theory of the aether resulted in severe difficulties that were ultimately resolved by the theory of special relativity. In the theory of special relativity, the Newtonian time function t (defining an "absolute notion" of simultaneity) and the metric of space are replaced by a single quantity: the metric of spacetime. Classical electrodynamics is fully compatible with the spacetime structure of special relativity, without the need for an aether.

The structure of classical electrodynamics is considerably simpler when formulated within the framework of special relativity. I wait until chapter 8 to give a proper discussion of the formulation of electromagnetism within special relativity, but I wish to make a few remarks here, so that the reader can get some flavor of what this formulation looks like without waiting until near the end of this book. In special relativity, the scalar potential, ϕ, and vector potential, $\boldsymbol{A}$, are seen to be the time and space components of a single "4-(dual-)vector potential"

$$A_\mu = (-\phi/c, \boldsymbol{A}). \tag{1.8}$$

The electric and magnetic fields are seen to arise from a single field strength tensor

$$F_{\mu\nu} = \frac{\partial A_\nu}{\partial x^\mu} - \frac{\partial A_\mu}{\partial x^\nu}, \tag{1.9}$$

with $x^\mu = (x^0 = ct, x^1, x^2, x^3)$. Since $F_{\mu\nu} = -F_{\nu\mu}$, it has 6 independent components. For an observer at rest in these coordinates, the electric field corresponds to the 3 time-space components of $F_{\mu\nu}$

$$E_i = cF_{i0}, \qquad i = 1, 2, 3, \tag{1.10}$$

whereas the magnetic field corresponds to the 3 independent space-space components of $F_{\mu\nu}$,

$$B_i = F_{jk}, \qquad i = 1, 2, 3, \tag{1.11}$$

where (i, j, k) is a cyclic permutation of $(1, 2, 3)$. In particular, since observers who move relative to each other define different "time directions" in spacetime, what would be claimed by one observer to be a "pure electric field" will be seen by another observer to be a combination of electric and magnetic fields. The "invariant description" of the field strengths is given by $F_{\mu\nu}$. Maxwell's equations can then be written in terms of $F_{\mu\nu}$, the spacetime metric, and the charge-current 4-vector:

$$J^{\mu} = (c\rho, \boldsymbol{J}). \tag{1.12}$$

Although the special relativistic formulation of classical electrodynamics has the major advantage of simplicity, it has the major disadvantage of unfamiliarity. Most readers are unlikely to be familiar with the distinction between, for example, vectors and dual vectors, and the role played by the spacetime metric in the equations of physics. While these concepts are not inordinately difficult to explain—and I explain them in chapter 8—it would be too much of a distraction to do so before presenting the theory of electromagnetism. Therefore, I defer the discussion of special relativity until chapter 8 and, with the exception of a few side comments, I do not use special relativistic notation for classical electrodynamics until that point. However, it is important that the reader be aware of the fact that classical electrodynamics is compatible with the spacetime structure of special relativity even if we use a notation that does not make it manifestly so.

1.1 The Fundamental Electromagnetic Variables Are the Potentials, Not the Field Strengths

The electromagnetic field is a fundamental constituent of nature. Its existence does not need to be justified or explained any more (or less) than the existence of, say, electrons needs to be justified or explained. The electromagnetic field is a "gauge field," the same basic type of field that also describes the W and Z bosons and gluons. Indeed, the electromagnetic field together with the W and Z fields comprise a unified "electroweak gauge field" that describes both the electromagnetic and weak interactions. However, for the ("low-energy") phenomena of interest in this book, the electromagnetic field effectively decouples from its electroweak partners and can be considered on its own.

I defer giving a mathematically complete discussion of electromagnetism as a gauge field until chapter 9. What is necessary for the reader to be aware of now is that the fundamental description of the electromagnetic field is given in terms of the *potentials* ϕ and $\boldsymbol{A}$, not the *field strengths* $\boldsymbol{E}$ and $\boldsymbol{B}$. As explained below, there are situations where the potentials contain more information than can be obtained from the field strengths. However, ϕ and $\boldsymbol{A}$ do not uniquely describe the electromagnetic field: the potentials $\phi', \boldsymbol{A}'$ and $\phi, \boldsymbol{A}$ are considered to be physically equivalent (i.e., they represent the same electromagnetic field) if they differ by a *gauge transformation*, that is, if for some function $\chi(t, \boldsymbol{x})$, we have[2]

$$\phi' = \phi - \frac{\partial \chi}{\partial t}, \qquad \boldsymbol{A}' = \boldsymbol{A} + \boldsymbol{\nabla}\chi. \tag{1.13}$$

In other words, an electromagnetic field is an equivalence class of potentials $\phi, \boldsymbol{A}$ under the transformation eq. (1.13).

It is easily verified that the field strengths, $\boldsymbol{E}$ and $\boldsymbol{B}$, defined by eqs. (1.6) and (1.7), are gauge invariant. Furthermore, it is not difficult to show that in any simply connected[3]

[2] In special relativistic notation, a gauge transformation can be expressed more simply as $A_{\mu} \to A_{\mu} + \partial\chi/\partial x^{\mu}$.

[3] A simply connected region is one in which every closed loop can be continuously deformed to a point.

spacetime region, if $\phi_1, \boldsymbol{A}_1$ and $\phi_2, \boldsymbol{A}_2$ give rise to the same field strengths $\boldsymbol{E}$ and $\boldsymbol{B}$, then $\phi_1, \boldsymbol{A}_1$ and $\phi_2, \boldsymbol{A}_2$ can differ at most by a gauge transformation. Thus, in any simply connected region, $\boldsymbol{E}$ and $\boldsymbol{B}$ contain all of the information contained in ϕ and $\boldsymbol{A}$. Since all physically measurable quantities must be gauge invariant, it is very convenient in many circumstances to work with $\boldsymbol{E}$ and $\boldsymbol{B}$ rather than ϕ and $\boldsymbol{A}$. In many contexts, electromagnetic phenomena can be fully described in terms of $\boldsymbol{E}$ and $\boldsymbol{B}$.

However, as we shall see in chapter 9, the coupling of the electromagnetic field to fundamental charged matter (namely, charged fields) can be described only in terms of the potentials, not the field strengths. Furthermore, there are physically relevant situations where $\boldsymbol{E}$ and $\boldsymbol{B}$ do not contain all of the information about the electromagnetic field. As an example, consider the region outside an infinite solenoid. Suppose that inside the solenoid, there is a nonvanishing, uniform magnetic field, but outside the solenoid, we have $\boldsymbol{E} = \boldsymbol{B} = 0$. Since the region outside the solenoid is not simply connected, the fact that $\boldsymbol{E}$ and $\boldsymbol{B}$ vanish in that region does not imply that the potentials are gauge equivalent to zero there. Indeed, eq. (1.7) implies, via Stokes's theorem, that when $\boldsymbol{B} \neq 0$ *inside* the solenoid, we have $\oint \boldsymbol{A} \cdot \boldsymbol{dl} \neq 0$ for any loop *outside* the solenoid that encloses it. (Note that $\oint \boldsymbol{A} \cdot \boldsymbol{dl}$ is gauge invariant, i.e., its value does not change under eq. (1.13).) A quantum mechanical charged particle that stays entirely outside the solenoid will be affected by this vector potential, as it will produce a relative phase shift in the parts of the wave function that go around the solenoid in different directions, producing a physically measurable shift in the resulting interference pattern. This phenomenon, known as the Aharonov-Bohm effect, is sometimes attributed to the weirdness of quantum mechanics. However, the effect has nothing to do with quantum mechanics—the same effect would occur for a classical charged field. And there is nothing weird about the effect, once one recognizes that the electromagnetic field is represented, at a fundamental level, by the potentials $\phi, \boldsymbol{A}$ (modulo gauge), not the field strengths $\boldsymbol{E}, \boldsymbol{B}$.

Thus, while for many purposes, it is convenient to introduce and work with the field strengths $\boldsymbol{E}$ and $\boldsymbol{B}$, it is important to recognize that the fundamental description of the electromagnetic field is given by the potentials ϕ and $\boldsymbol{A}$. The Maxwell equations (1.3) and (1.4) should be viewed as consequences of the definitions of $\boldsymbol{E}$ and $\boldsymbol{B}$ given by eqs. (1.6) and (1.7).

1.2 Electromagnetic Energy, Momentum, and Stress Are an Integral Part of the Theory

The electromagnetic field, like all other forms of matter, has energy, momentum, and stress properties. These properties, like Maxwell's equations (1.1)–(1.4), are an integral part of the theory.

As discussed much more fully in chapter 9, classical electrodynamics can be viewed as arising from the Lagrangian density

$$\mathcal{L} = \frac{1}{2}\left(\epsilon_0 |\boldsymbol{E}|^2 - \frac{1}{\mu_0}|\boldsymbol{B}|^2\right) - \phi\rho + \boldsymbol{A} \cdot \boldsymbol{J}. \tag{1.14}$$

Here, as discussed in section 1.1, the dynamical variables are ϕ and $\boldsymbol{A}$, and the Euler-Lagrange equations are obtained by varying $\mathcal{L}$ with respect to these variables; $\boldsymbol{E}$ and $\boldsymbol{B}$ are viewed as the functions of ϕ and $\boldsymbol{A}$ defined by eqs.(1.6) and (1.7). In eq. (1.14), the charge density ρ and current density $\boldsymbol{J}$ are treated as externally prescribed, nondynamical quantities.[4] The Euler-Lagrange equations arising from the variation of eq. (1.14) with respect to ϕ and $\boldsymbol{A}$ are precisely Maxwell's equations (1.1)–(1.2). The additional Maxwell equations (1.3)–(1.4) follow from the definitions (1.6) and (1.7) of $\boldsymbol{E}$ and $\boldsymbol{B}$, respectively. The fact that the Lagrangian must be viewed as a function of ϕ and $\boldsymbol{A}$—and the terms representing the coupling of the electromagnetic field to charged matter cannot even be written down in terms of $\boldsymbol{E}$ and $\boldsymbol{B}$—is further manifestation of the fact that the fundamental dynamical variables in electomagnetism are ϕ and $\boldsymbol{A}$.

The energy, momentum, and stress properties of the electromagnetic field are determined by its coupling to gravity. The coupling to gravity is obtained by generalizing the Lagrangian (1.14) for the spacetime of special relativity to curved spacetime. This can be done in a very simple and natural way, which is unique if one does not allow derivatives of the metric to appear in the Maxwell Lagrangian. The stress-energy-momentum tensor of the electromagnetic field is then obtained by functional differentiation of the Lagrangian with respect to the spacetime metric, since this is what appears as a source term for gravity in Einstein's equation of general relativity. I briefly indicate how this works in section 9.1. The only point I wish to make here is that, just as the Lagrangian (1.14) gives rise to Maxwell's equations, its natural generalization to curved spacetime gives rise to the following formulas for the energy density $\mathcal{E}$, momentum density $\mathcal{P}$, and stress tensor Θ_{ij} of the electromagnetic field:

$$\mathcal{E} = \frac{1}{2}\left(\epsilon_0 |\boldsymbol{E}|^2 + \frac{1}{\mu_0}|\boldsymbol{B}|^2\right), \tag{1.15}$$

$$\mathcal{P} = \epsilon_0 \boldsymbol{E} \times \boldsymbol{B}, \tag{1.16}$$

$$\Theta_{ij} = \epsilon_0 E_i E_j + \frac{1}{\mu_0} B_i B_j - \frac{1}{2}\delta_{ij}\left(\epsilon_0 |\boldsymbol{E}|^2 + \frac{1}{\mu_0}|\boldsymbol{B}|^2\right). \tag{1.17}$$

These formulas should be viewed as having fundamental status in the theory of electromagnetism, comparable to that of Maxwell's equations.

In principle, the validity of eqs. (1.15)–(1.17) could be tested by observing the gravitational effects of electromagnetic fields. Electromagnetic fields make nontrivial contributions to the mass-energy of ordinary matter—certainly large enough to produce observable gravitational effects for macroscopic bodies. However, there is no way to observe these effects separately from the gravitational effects of the nonelectromagnetic constituents of matter. Thus, it would be necessary to observe the gravitational effects of free electromagnetic fields if one wishes to test eqs. (1.15)–(1.17). The gravitational effects of free electromagnetic fields are far too small to be measured in laboratory experiments. However, in the early universe, the thermally distributed electromagnetic radiation that presently constitutes the cosmic microwave background made a dominant contribution to the energy density and pressure in the universe, both of which affect the expansion of the universe. The expansion history of the universe is observed to be in accord with the electromagnetic energy density and pressure of thermal radiation obtained from the above formulas.

[4] Of course, the charged matter should really have its own dynamical degrees of freedom, and there should be additional terms in the Lagrangian involving the fields representing the charged matter. The coupling terms between the charged matter and electromagnetic field should then be represented in terms of $\phi, \boldsymbol{A}$, and the dynamical fields describing the charged matter. This will be seen explicitly in chapter 9.

There are important conservation laws associated with eqs. (1.15)–(1.17). In special relativity, the "flow of mass" (momentum) and "flow of energy" represent the same quantity apart from a factor of c^2, so

$$\mathcal{S} \equiv c^2 \mathcal{P} = c^2 \epsilon_0 \boldsymbol{E} \times \boldsymbol{B} = \frac{1}{\mu_0} \boldsymbol{E} \times \boldsymbol{B} \tag{1.18}$$

represents the flux of energy per unit volume of the electromagnetic field. A computation using Maxwell's equations yields (see section 5.1 for details)

$$\frac{\partial \mathcal{E}}{\partial t} + \nabla \cdot \mathcal{S} = -\boldsymbol{J} \cdot \boldsymbol{E}, \tag{1.19}$$

$$\frac{\partial \mathcal{P}_i}{\partial t} - \sum_{j=1}^{3} \partial_j \Theta_{ij} = -\,[\rho E_i + (\boldsymbol{J} \times \boldsymbol{B})_i]. \tag{1.20}$$

In the absence of charges and currents (i.e., when $\rho = \boldsymbol{J} = 0$), the right sides of eqs. (1.19) and (1.20) vanish. In this case, eqs. (1.19) and (1.20) have the interpretation of expressing local conservation of energy and momentum of the electromagnetic field. To see this more explicitly, note that in a small volume δV about $\boldsymbol{x}$, the quantity $\delta V \nabla \cdot \mathcal{S}$ represents the net flux of energy out of δV. By eq. (1.19), this is equal to $-\delta V \partial \mathcal{E}/\partial t$ when $\rho = \boldsymbol{J} = 0$, thus expressing local conservation of energy. Global energy conservation for the electromagnetic field is obtained by integrating eq. (1.19) over all of space, assuming that $\boldsymbol{E}$ and $\boldsymbol{B}$ vanish sufficiently rapidly near infinity. In that case, the integral over all of space of $\nabla \cdot \mathcal{S}$ vanishes by Gauss's theorem (see chapter 2), and we obtain

$$\frac{d}{dt} \int \mathcal{E}\, d^3x = 0, \tag{1.21}$$

provided that $\rho = \boldsymbol{J} = 0$. Similarly, when $\rho = \boldsymbol{J} = 0$, eq. (1.20) expresses local conservation of momentum, and integration of eq. (1.20) over all of space yields the global momentum conservation law

$$\frac{d}{dt} \int \mathcal{P}\, d^3x = 0. \tag{1.22}$$

When ρ and $\boldsymbol{J}$ are nonvanishing, the right sides of eqs. (1.19) and (1.20) are, in general, nonvanishing, and electromagnetic energy and momentum are not conserved by themselves. This is because electromagnetic energy and momentum can be exchanged with the energy and momentum of the charged matter. For the total (electromagnetic and matter) energy to be locally conserved, the electromagnetic field must be transferring energy to the matter at the rate

$$\frac{\partial \mathcal{E}_{\text{matter}}}{\partial t} = \boldsymbol{J} \cdot \boldsymbol{E}. \tag{1.23}$$

Similarly, for total momentum to be conserved, the electromagnetic field must be transferring momentum to the matter at the rate given by minus the right side of eq. (1.20); that is, it must be exerting a force per unit volume, $\boldsymbol{f}$, on the matter, given by

$$\boldsymbol{f} = \rho \boldsymbol{E} + \boldsymbol{J} \times \boldsymbol{B}. \tag{1.24}$$

In standard treatments of electromagnetism, the order of the arguments presented above is reversed, giving rise to some serious difficulties. Instead of starting with the eqs. (1.15)–(1.17) for energy density, momentum density, and stress, and then deriving the Lorentz force, eq. (1.24), standard treatments start with the Lorentz force—or rather, the Coulomb's law version of this expression for static point charges in electrostatics. The "work done" in quasi-statically bringing charges together from infinity is then calculated and is associated with the energy contained in the electromagnetic field. This argument eventually leads to the correct formula $\frac{\epsilon_0}{2}\int |\boldsymbol{E}|^2 d^3x$ for the energy of the electromagnetic field in electrostatics. However, this argument works in electrostatics because it is possible to move a charged body in an electric field in such a way that its rest mass (i.e., internal energy) does not change. Although this may seem obvious, the corresponding result does *not* hold in magnetostatics, because energy is required to maintain the currents in a body. This is true for permanent magnets as well as current loops. The rest mass of a magnetic dipole will change as it moves in a nonuniform magnetic field, as we shall see explicitly in section 4.3 and again in section 10.3.2. As I show in section 4.3, the electromagnetic interaction energy of a magnetic dipole $\boldsymbol{\mu}$ in an external magnetic field $\boldsymbol{B}^{\text{ext}}$ can be derived straightforwardly from eq. (1.15) and yields the value $+\boldsymbol{\mu}\cdot\boldsymbol{B}^{\text{ext}}$. However, many references give the incorrect formula $-\boldsymbol{\mu}\cdot\boldsymbol{B}^{\text{ext}}$ based on arguments using "work done," failing to take into account the change in rest mass.

The formulas (1.15) for the electromagnetic energy density and (1.16) for the electromagnetic momentum density $\mathcal{P}$ are justified in many standard treatments by taking eq. (1.23) as a starting point. It is then natural to interpret eq. (1.19) (which is derived directly from Maxwell's equations) as representing local energy conservation. One thereby can identify $\mathcal{E}$ and $\mathcal{S} \equiv c^2\mathcal{P}$ with electromagnetic energy density and energy flux, respectively. However, this argument has the serious drawback that $\mathcal{P}$ appears in eq. (1.19) only in the form $\nabla\cdot\mathcal{P}$. This leads many authors to suggest that $\mathcal{P}$ is undefined up to the addition of the curl of a vector field. This is not correct; formulas for $\mathcal{P}$ that differ by a curl of a vector field will have different gravitational consequences, so if one has two formulas for $\mathcal{P}$ that differ by a curl, at most one of them can be valid.

In summary, rather than attempt to derive eqs. (1.15)–(1.17) from Maxwell's equations by assuming that eq. (1.23) and eq. (1.24) hold, it is much healthier to view formulas (1.15)–(1.17) as an integral part of the specification of the theory, with eq. (1.23) and eq. (1.24) then following as consequences. The conservation laws (1.19) and (1.20) provide important consistency relations between Maxwell's equations and eqs. (1.15)–(1.17), but they do not enable one to derive eqs. (1.15)–(1.17) from Maxwell's equations. Equations (1.15)–(1.17) should be viewed as fundamental aspects of electromagnetic theory, with a status similar to that of Maxwell's equations.

1.3 Electromagnetic Fields Should Not Be Viewed as Being Produced by Charged Matter

Maxwell's equations (1.1)–(1.4) together with eqs. (1.23) and (1.24) describe the *interaction* of the electromagnetic field with matter. The electromagnetic field does not, in any sense, play a subordinate role in this interaction. The electromagnetic field has its own independent dynamical degrees of freedom, and these should be thought of as being on an equal footing with the dynamical degrees of freedom of the charged matter. The electromagnetic field should not be thought of as being *produced* by charges and currents—despite the fact that ρ and $\boldsymbol{J}$ are commonly referred to as "source terms" in Maxwell's equations (and I use this terminology in this book).

The independent dynamical degrees of freedom of the electromagnetic field are characterized by the initial value formulation of Maxwell's equations, which is discussed in section 5.4. The theorem at the end of that section states the following: Specify $\rho(t,\boldsymbol{x})$ and $\boldsymbol{J}(t,\boldsymbol{x})$ on spacetime, subject to the conservation equation (1.5). Let $\boldsymbol{E}_0(\boldsymbol{x})$ and $\boldsymbol{B}_0(\boldsymbol{x})$ be arbitrary vector fields on space such that $\nabla\cdot\boldsymbol{E}_0=\rho(t=0,\boldsymbol{x})/\epsilon_0$, and $\nabla\cdot\boldsymbol{B}_0=0$. Then there exists a unique solution $(\boldsymbol{E}(t,\boldsymbol{x}),\boldsymbol{B}(t,\boldsymbol{x}))$ to Maxwell's equations (1.1)–(1.4) such that $\boldsymbol{E}(t=0,\boldsymbol{x})=\boldsymbol{E}_0(\boldsymbol{x})$, and $\boldsymbol{B}(t=0,\boldsymbol{x})=\boldsymbol{B}_0(\boldsymbol{x})$. Thus, there are as many solutions to Maxwell's equations with a specified ρ and $\boldsymbol{J}$ as there are vector fields $(\boldsymbol{E}_0(\boldsymbol{x}),\boldsymbol{B}_0(\boldsymbol{x}))$ satisfying the above conditions on their divergence. The fact that this initial data for the electromagnetic field can be freely specified shows that the electromagnetic field has its own independent dynamical degrees of freedom. Solutions to Maxwell's equations are *not* determined by ρ and $\boldsymbol{J}$.

The dynamical degrees of freedom of the electromagnetic field are not visible in electrostatics and magnetostatics, since no time-independent solutions of Maxwell's equations with $\rho=\boldsymbol{J}=0$ go to zero at infinity. Thus, if one specifies time-independent sources, $\partial\rho/\partial t=\partial\boldsymbol{J}/\partial t=0$, then solutions to Maxwell's equations for $\boldsymbol{E}$ and $\boldsymbol{B}$ with $\partial\boldsymbol{E}/\partial t=\partial\boldsymbol{B}/\partial t=0$ and with $\boldsymbol{E}$ and $\boldsymbol{B}$ going to zero at infinity are uniquely determined by ρ and $\boldsymbol{J}$. Consequently, one can uniquely associate a stationary electric field $\boldsymbol{E}$ with a stationary charge distribution ρ, and one can uniquely associate a stationary magnetic field $\boldsymbol{B}$ with a stationary current distribution $\boldsymbol{J}$. Therefore, it is possible to view the electric field in electrostatics as being "produced" by charges, and it is possible to view the magnetic field in magnetostatics as being "produced" by currents. In electrostatics, one can even get away with saying—as is frequently done—that charges exert forces on one another. This, of course, is not the case: The electromagnetic force (1.24) on a charged body is exerted by the electromagnetic field that is present at the location of the body, not by other, distant charges.

In electrodynamics, one is frequently interested in considering situations in which there is "no incoming electromagnetic radiation." As discussed in depth in section 5.2, solutions with no incoming radiation are given by the retarded Green's function applied to ρ and $\boldsymbol{J}$, and these solutions are uniquely determined by ρ and $\boldsymbol{J}$. Again, this makes it possible to take the view that, in the absence of incoming radiation, the electromagnetic fields are "produced" by the charges and currents. However, while the "no incoming radiation" condition is a useful idealization applicable to many problems, it should not be taken seriously as an initial condition for our universe. Although we certainly do not know the precise initial conditions at the "big bang," we do know that matter in the very early universe was an extremely hot and dense plasma. In such a hot and dense plasma, the electromagnetic field "produces" charges (e.g., electron-positron pairs) to much the same degree as charges "produce" electromagnetic fields. It certainly does not make any sense to think of the charges as coming first and then producing the electromagnetic fields.

Thus, although there are circumstances where one could take the view that electromagnetic fields are produced by charges, it is far healthier to think of the electromagnetic field and charged matter as independent entities that interact via Maxwell's equations and eqs. (1.23) and (1.24). Indeed, the view that electromagnetic fields are produced by charges is particularly untenable in quantum field theory, since it is essential for the understanding of such phenomena as the vacuum fluctuations of the electromagnetic field that the electromagnetic field have its own dynamical degrees of freedom, independently of the existence of charged matter.

1.4 At a Fundamental Level, Classical Charged Matter Must Be Viewed as Continuous Rather Than Point-Like

Maxwell's equations (1.1)–(1.4) were formulated above using a continuum notion of charge density ρ and current density $\boldsymbol{J}$; that is, ρ and $\boldsymbol{J}$ were taken to be smooth functions of $(t, \boldsymbol{x})$. These equations have a mathematically well-posed initial value formulation, as already mentioned in section 1.3 and as discussed in depth in section 5.4. However, in a complete theory, one must also specify the form of the charged matter and its equations of motion. As discussed further in chapter 9, at a fundamental level, charged matter is believed to consist of charged (quantum) fields. However, one can also consider "phenomenological models" of charged matter, such as a charged fluid. In any case, the equations of motion of the charged matter together with Maxwell's equations comprise a coupled system that must be solved simultaneously—since the motion of the charged matter depends on the electromagnetic field, but the dynamical evolution of the electromagnetic field depends on the charges and currents of the matter. It is essential that the coupled Maxwell–charged-matter system have a well-posed initial value formulation, so that there is no difficulty, in principle, in obtaining solutions to the coupled system for given initial conditions.

However, at least 90% of what is normally treated in electromagnetism courses does not consider the full, coupled Maxwell–matter system but instead considers the following two idealized problems:

- Type I. For a given externally specified ρ and $\boldsymbol{J}$, find the corresponding electromagnetic fields (i.e., the unique stationary solution in electrostatics and magnetostatics and/or the retarded solution in electrodynamics).
- Type II. Find the motion of a charged body for given externally specified fields $\boldsymbol{E}$ and $\boldsymbol{B}$ (i.e., neglecting the self-fields associated with the presence of the charged body).

For these idealized problems, it is very useful to introduce the notion of a point charge.

By a "point charge of charge q" moving on the worldline $\boldsymbol{X}(t)$ (with $|d\boldsymbol{X}/dt| < c$ for all t) is meant the charge-current

$$\rho(t, \boldsymbol{x}) = q\delta(\boldsymbol{x} - \boldsymbol{X}(t)), \qquad \boldsymbol{J}(t, \boldsymbol{x}) = q\frac{d\boldsymbol{X}}{dt}(t)\delta(\boldsymbol{x} - \boldsymbol{X}(t)), \tag{1.25}$$

where δ denotes the 3-dimensional Dirac delta function. This may be thought of as a limit of a charge distribution that at each t becomes more and more concentrated at the point $\boldsymbol{X}(t)$. This limit does not define a function, but it has a well-defined meaning as a distribution.[5] The charge-current eq. (1.25) satisfies eq. (1.5) in a well-defined, distributional sense.

[5] A distribution is a linear map from "test functions" (i.e., smooth functions that are nonvanishing only in a bounded region) into numbers that depends continuously on the test function in an appropriate sense. The Dirac delta function is simply the evaluation map on test functions; that is, $\delta(\boldsymbol{x} - \boldsymbol{X})$ maps the test function f into the number $f(\boldsymbol{X})$.

It can be seen from Gauss's law that for solutions to eq. (1.1) with ρ given by eq. (1.25), the electric field $\boldsymbol{E}$ must diverge near the charge as $1/|\boldsymbol{x}-\boldsymbol{X}|^2$. Consequently, by eq. (1.15), the electromagnetic energy density diverges as $1/|\boldsymbol{x}-\boldsymbol{X}|^4$, which is not integrable. Thus, the total electromagnetic energy of a point charge is infinite, and so point charges cannot be considered to be physical objects in classical electrodynamics. Nevertheless, they can be introduced in the context of problems of type I or type II above.

In problems of type I, since Maxwell's equations are linear in $\boldsymbol{E}$ and $\boldsymbol{B}$, these equations make perfectly good mathematical sense when ρ and $\boldsymbol{J}$ (and hence, $\boldsymbol{E}$ and $\boldsymbol{B}$) are distributions rather than functions. It is extremely useful to consider solutions to Maxwell's equations with a point charge-current, eq. (1.25). Such solutions are of direct interest for describing situations where the charge-current is highly localized, and more general solutions can be obtained by "superposition" (again using the linearity of Maxwell's equations).

For problems of type II, it would be quite complicated to analyze the motion of an extended charged body described by a causal dynamics compatible with special relativity, since the different electromagnetic forces on the different parts of the body would induce internal oscillations. One might therefore be tempted to take a limit of vanishing size of the body, wherein the complications due to internal dynamics should become negligible. However, such a limit at fixed q leads one back to the problem of infinite self-energy and also would require infinite mechanical stresses to keep the body from flying apart. Nevertheless, we will see in chapter 10 that it is possible to take a limit in which the size, charge, and mass of the body all scale to zero in a suitable manner. To leading order, the motion $\boldsymbol{x}(t)$ of the body becomes independent of its internal structure and is given by the Lorentz force equation[6]

$$\frac{d\boldsymbol{p}}{dt} = \boldsymbol{F} = q\left(\boldsymbol{E}^{\text{ext}} + \boldsymbol{v}\times\boldsymbol{B}^{\text{ext}}\right), \tag{1.26}$$

where $\boldsymbol{v} = d\boldsymbol{x}/dt$, and $\boldsymbol{p} = \gamma m\boldsymbol{v}$ with $\gamma = (1 - v^2/c^2)^{-1/2}$. Here $\boldsymbol{E}^{\text{ext}}$ and $\boldsymbol{B}^{\text{ext}}$ are the "externally prescribed" fields, with the influence of the charge-current of eq. (1.25) ignored. Note that the force $\boldsymbol{F}$ appearing in eq. (1.26) corresponds to eq. (1.24), with ρ and $\boldsymbol{J}$ given by eq. (1.25). The equation of motion (1.26) makes good mathematical sense and, for specified external fields, has a unique solution for a given initial position and velocity of the point charge.

One might attempt to go beyond the context of problems of type I or type II to consider the full, coupled Maxwell-matter system with point charges. In other words, one could attempt to solve Maxwell's equations with source eq. (1.25) simultaneously with eq. (1.26), where now $\boldsymbol{E}$ and $\boldsymbol{B}$ represent the full electromagnetic field, including the effects of the point charge. However, this system of equations does not make mathematical sense, since Maxwell's equations imply that $\boldsymbol{E}$ must be singular at the location of the charge, in which case, eq. (1.26) is ill defined. This is a reflection of the fact that the coupled Maxwell-matter system is nonlinear, and distributional solutions make sense for a nonlinear system only in very limited circumstances—which do not apply here.

---

[6]I also show in chapter 10 how to obtain leading-order corrections to Lorentz force motion taking into account the self-field of the body.

This difficulty is resolved by simply recognizing that, at a fundamental level in classical electrodynamics,[7] ρ and $\boldsymbol{J}$ must be taken to be quantities smoothly distributed in spacetime. No difficulties of the sort mentioned in the previous paragraph arise when one considers continuum charged matter. In particular, as we shall see in chapter 9, the self-consistent coupled system of Maxwell's equations and the equation of motion of a charged scalar field is well posed. The notion of a point charge is convenient to introduce in the circumstances described above, but it cannot be viewed as a fundamental description of charged matter.

---

[7]The same is true in quantum electrodynamics in the sense that for any physically acceptable state of charged matter, $\langle\rho\rangle$ and $\langle\boldsymbol{J}\rangle$ must be smoothly distributed in spacetime.

CHAPTER 2

Electrostatics

Before considering full electrodynamics, it is very instructive to give a comprehensive analysis of the case where the charge density, ρ, and current density, $\boldsymbol{J}$, are time independent ($\partial\rho/\partial t = \partial\boldsymbol{J}/\partial t = 0$), and the potentials ϕ and $\boldsymbol{A}$ are also time independent ($\partial\phi/\partial t = \partial\boldsymbol{A}/\partial t = 0$). In that case, equation (1.1) for $\boldsymbol{E} = -\nabla\phi$ completely decouples from equation (1.2) for $\boldsymbol{B} = \nabla \times \boldsymbol{A}$. Thus, it suffices to separately consider the cases where, in addition to stationarity, either we have $\boldsymbol{J} = \boldsymbol{A} = 0$ (electrostatics) or we have $\rho = \phi = 0$ (magnetostatics). This chapter is concerned with electrostatics. We will treat magnetostatics in chapter 4.

Most treatments of electrostatics start with point charges and Coulomb's law and eventually work up to Poisson's equation. We start with Maxwell's equations, which immediately reduces to Poisson's equation. I introduce point charges in section 2.2 and obtain Coulomb's law at the end of section 2.3.

Section 2.1 establishes key properties of solutions to Poisson's equation. The notion of a Green's function will be introduced in section 2.2. Some key results on force and interaction energy will be obtained in section 2.3. The multipole expansion of the Green's function for Poisson's equation is derived in section 2.4. Conducting cavities are treated in section 2.5. This topic is of interest in its own right in electrostatics, but it also serves as a prototype of many other problems that arise in physics.

2.1 Uniqueness of Solutions in Electrostatics

We set $\boldsymbol{J} = \boldsymbol{A} = 0$ and $\partial\rho/\partial t = \partial\phi/\partial t = 0$. The only nontrivial equations of electromagnetism in this case are the first Maxwell equation (1.1),

$$\nabla \cdot \boldsymbol{E} = \rho/\epsilon_0, \tag{2.1}$$

and the relation between $\boldsymbol{E}$ and ϕ, eq. (1.6),

$$\boldsymbol{E} = -\nabla\phi. \tag{2.2}$$

These may be combined into the single equation

$$\nabla^2\phi = -\rho/\epsilon_0, \tag{2.3}$$

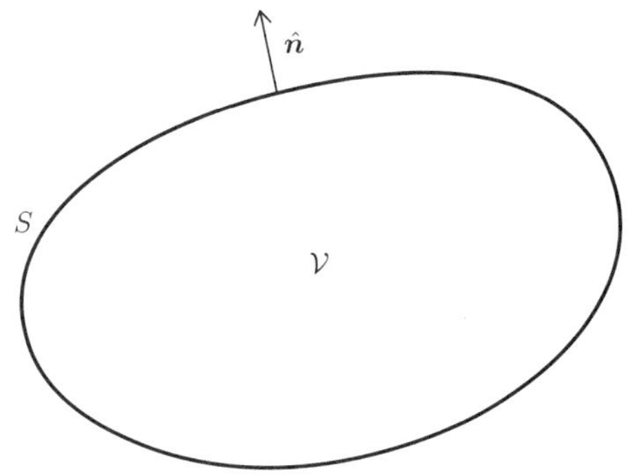

Figure 2.1. A volume $\mathcal{V}$ bounded by a surface S with outward normal $\hat{n}$ as in Gauss's theorem.

where the *Laplacian operator*, ∇^2, is given in Cartesian coordinates by

$$\nabla^2 \equiv \nabla \cdot \nabla = \frac{\partial^2}{\partial x^2} + \frac{\partial^2}{\partial y^2} + \frac{\partial^2}{\partial z^2}. \tag{2.4}$$

Equation (2.3) is known as *Poisson's equation.*

Note that the gauge freedom eq. (1.13) is greatly restricted in electrostatics by the requirement that ϕ be time independent and $\boldsymbol{A} = 0$. Thus, the only allowed gauge transformations are generated by $\chi(t, \boldsymbol{x}) = t \times \text{const.}$, so the only gauge freedom in ϕ is

$$\phi \to \phi' = \phi + \text{const.} \tag{2.5}$$

The following mathematical theorem underlies many results in electrostatics.

Theorem (Gauss's Theorem): *Let $\boldsymbol{v}$ be an arbitrary differentiable vector field on $\mathbb{R}^3$. Let $\mathcal{V} \subset \mathbb{R}^3$ be a bounded region whose boundary, $S = \partial\mathcal{V}$, is a regular 2-dimensional surface (see the comment below this theorem). Then we have*

$$\int_{\mathcal{V}} \nabla \cdot \boldsymbol{v}\, d^3x = \int_S \boldsymbol{v} \cdot \hat{\boldsymbol{n}}\, dS, \tag{2.6}$$

where $\hat{\boldsymbol{n}}$ is the "outward pointing" (i.e., away from $\mathcal{V}$) unit normal to S (see figure 2.1), and dS denotes the area element on S.

Side Comments on Gauss's Theorem: The mathematically precise condition in Gauss's theorem that $\mathcal{V}$ is a bounded region with regular boundary S is that $\mathcal{V}$ and S comprise a compact manifold with boundary. However, it would take me too far afield to attempt to define a "compact manifold with boundary" here. Although we have stated Gauss's theorem here only for 3 dimensions, it holds in arbitrary dimensions, and indeed, we will use a 4-dimensional (spacetime) version of it later in this book. Furthermore, Gauss's theorem in n dimensions is equivalent to the n-dimensional Stokes's theorem, which applies to differential forms of degree $n-1$ on an n-dimensional compact manifold with boundary. (What is usually referred to as "Stokes's theorem" in physics literature is the 2-dimensional version of this theorem—which is what Stokes considered. Again, it would take me too far afield to define the notion of a "differential form" here.) The generalized Stokes's theorem formulation of Gauss's theorem has the advantage that it manifestly does not depend on the metrical structure of space (i.e., it depends only on the topological/differential structure of $\mathcal{V}$ and S). Thus, if one can prove Gauss's theorem for a cube (as is easily done by direct computation), it automatically holds for any region with the topology of a cube and—by breaking an arbitrary (compact) region into a finite union of topological cubes—any (compact) region whatsoever. Similarly, it follows that Gauss's theorem holds in any curved space (with an appropriate definition of $\nabla\cdot$).

Using Gauss's theorem, we can prove the following uniqueness theorem for solutions to Poisson's equation.

Theorem (Uniqueness in Electrostatics): *For a given charge density $\rho(\boldsymbol{x})$, there is at most one solution to Poisson's equation (2.3) such that $\phi \to 0$ as $r \to \infty$ in such a way*[1] *that $r\phi$ remains bounded and $r|\nabla\phi| \to 0$.*

Proof. For the given ρ, let ϕ and ϕ' be solutions to eq. (2.3) that go to zero at infinity, as specified in the statement of the theorem. Let

$$\psi = \phi - \phi'. \tag{2.7}$$

Then ψ satisfies *Laplace's equation*,

$$\nabla^2\psi = 0, \tag{2.8}$$

and we also have that ψ and $|\nabla\psi|$ go to zero at infinity as specified in the statement of the theorem. Now multiply eq. (2.8) by ψ and integrate over a ball of radius R:

$$0 = \int_{r\le R} \psi\nabla^2\psi\, d^3x = \int_{r\le R} \left[\nabla\cdot(\psi\nabla\psi) - |\nabla\psi|^2\right] d^3x. \tag{2.9}$$

By Gauss's theorem, we have

$$\int_{r\le R} \nabla\cdot(\psi\nabla\psi) d^3x = \int_{r=R} \psi\hat{\boldsymbol{n}}\cdot\nabla\psi\, dS. \tag{2.10}$$

Since the area element of a sphere of radius R varies as R^2, the asymptotic conditions on ψ and $\nabla\psi$ imply that the right side of eq. (2.10) goes to zero as $R \to \infty$. Thus, taking the limit of eq. (2.9) as $R \to \infty$, we obtain

$$0 = -\int |\nabla\psi|^2 d^3x, \tag{2.11}$$

[1] This uniqueness result remains valid under the weaker assumption that $\phi \to 0$ as $r \to \infty$ (i.e., without any further requirements on the rate of approach to 0), but other methods would be needed to show this.

where the integral now extends over all of space. This implies $\nabla\psi = 0$, which implies $\psi = \text{const}$. Since $\psi \to 0$ at infinity, this implies that $\psi = 0$, as we desired to show. □

Note that this theorem does not establish the existence of solutions. However, existence of solutions when $\rho \to 0$ sufficiently rapidly as $r \to \infty$ will follow from the results obtained in the next section. In fact, solutions to eq. (2.3) exist for any (smooth) ρ (irrespective of behavior near infinity), but this is not straightforward to prove—and if ρ does not go to zero sufficiently rapidly at infinity, then ϕ cannot go to zero, so uniqueness is lost.

2.2 Point Charges and Green's Functions

As discussed in section 1.4, it is useful for many purposes to consider the point charge-current, eq. (1.25). In electrostatics, this corresponds to considering the stationary charge density

$$\rho(\boldsymbol{x}) = q\delta(\boldsymbol{x} - \boldsymbol{x}_0). \tag{2.12}$$

We wish to solve Poisson's equation (2.3) with this source. For simplicity, we may choose the origin of coordinates to coincide with the location of the charge, so that $\boldsymbol{x}_0 = \boldsymbol{0}$. (We will restore $\boldsymbol{x}_0$ later.) By the uniqueness theorem of section 2.1, there can be at most one solution to Poisson's equation with this source[2] such that ϕ goes to zero at infinity. This solution, if it exists, must be spherically symmetric, since otherwise we could generate new solutions by rotating the solution. In fact, the solution can be found quite easily as follows. Integrating eq. (2.1) over a ball of radius R about the origin, we obtain

$$\int_{r\le R} \nabla \cdot \boldsymbol{E}\, d^3x = \frac{1}{\epsilon_0}\int_{r\le R} q\delta(\boldsymbol{x})d^3x = \frac{q}{\epsilon_0}. \tag{2.13}$$

By Gauss's theorem, the left side of eq. (2.13) can be converted to a surface integral. Since $\boldsymbol{E}$ is rotationally invariant, it must be of the form $\boldsymbol{E} = E(r)\hat{\boldsymbol{r}}$, where $\hat{\boldsymbol{r}}$ is the unit radial vector, which coincides with the outward-pointing normal to spheres. Hence, we obtain

$$\int_{r\le R} \nabla \cdot \boldsymbol{E}\, d^3x = \int_{r=R} \boldsymbol{E}\cdot\hat{\boldsymbol{n}}\, dS = E(R)\int_{r=R} dS = 4\pi R^2 E(R). \tag{2.14}$$

Thus, we obtain

$$\boldsymbol{E}(\boldsymbol{x}) = \frac{1}{4\pi\epsilon_0}\frac{q}{r^2}\hat{\boldsymbol{r}}. \tag{2.15}$$

The corresponding potential can then be obtained by integrating eq. (2.2), subject to the condition that $\phi \to 0$ as $r \to \infty$, which yields

$$\phi(\boldsymbol{x}) = \frac{1}{4\pi\epsilon_0}\frac{q}{r}. \tag{2.16}$$

---

[2] In the proof of the uniqueness theorem, it was assumed that ψ is an ordinary function—since nonlinear operations were performed on ψ, which, in general, are ill defined for distributions. If ρ is distributional, then ϕ will necessarily also be distributional. However, ψ will still satisfy Laplace's equation (2.8), from which it can be shown that ψ must be given by a smooth function. Thus, the proof of the uniqueness theorem of section 2.1 remains valid when ρ is distributional, as in eq. (2.12).

It can be verified that ϕ satisfies eq. (2.3) in the distributional sense, with ρ given by eq. (2.12). Thus, eqs. (2.15) and (2.16) provide the unique solution for the electric field and potential of a point charge located at the origin. The corresponding solutions for a point charge located at $\boldsymbol{x}_0$ are

$$\boldsymbol{E}(\boldsymbol{x}) = \frac{1}{4\pi\epsilon_0} \frac{q(\boldsymbol{x}-\boldsymbol{x}_0)}{|\boldsymbol{x}-\boldsymbol{x}_0|^3} \tag{2.17}$$

and

$$\phi(\boldsymbol{x}) = \frac{1}{4\pi\epsilon_0} \frac{q}{|\boldsymbol{x}-\boldsymbol{x}_0|}. \tag{2.18}$$

If n point charges of charge q_i at locations $\boldsymbol{x}_{0i}$ are present, it follows from the linearity of Poisson's equation (2.3) that the superposition of the individual point charge solutions,

$$\phi(\boldsymbol{x}) = \frac{1}{4\pi\epsilon_0} \sum_{i=1}^{n} \frac{q_i}{|\boldsymbol{x}-\boldsymbol{x}_{0i}|} \tag{2.19}$$

is a solution to Poisson's equation with the n point charges present. By the second theorem of section 2.1, this solution is unique.

The point charge solution eq. (2.18) can be used to obtain solutions to Poisson's equation (2.3) for which $\rho(\boldsymbol{x})$ is continuously distributed. The basic idea is that if we break up space into small volumes $\Delta\mathcal{V}_i$ centered about $\boldsymbol{x}_{0i}$, the charge located in each volume should act much like a point charge of charge $\Delta q_i = \rho(\boldsymbol{x}_{0i})\Delta\mathcal{V}_i$ located at $\boldsymbol{x}_{0i}$. This suggests that the solution to Poisson's equation for a continuously distributed $\rho(\boldsymbol{x})$ should be given by

$$\phi(\boldsymbol{x}) = \frac{1}{4\pi\epsilon_0} \int \frac{\rho(\boldsymbol{x}')}{|\boldsymbol{x}-\boldsymbol{x}'|} d^3x'. \tag{2.20}$$

If $|\rho(\boldsymbol{x})| \leq C/r^{2+\epsilon}$ as $r \to \infty$ for some $C, \epsilon > 0$, then the integral on the right side of eq. (2.20) will converge, and it can be verified that $\phi(\boldsymbol{x})$ yields a solution to eq. (2.3) with charge density $\rho(\boldsymbol{x})$. This proves existence of solutions to Poisson's equation for the case where ρ falls off sufficiently rapidly at infinity. If the fall-off of ρ at infinity is sufficiently slow that the right side of eq. (2.20) does not converge, then solutions to Poisson's equation still exist, but they are not given by eq. (2.20).

When used in this manner to generate solutions with continuous $\rho(\boldsymbol{x})$, the point charge solution with unit charge is called a *Green's function*. Thus, the Green's function for electrostatics is

$$G(\boldsymbol{x}, \boldsymbol{x}') = \frac{1}{4\pi\epsilon_0} \frac{1}{|\boldsymbol{x}-\boldsymbol{x}'|}. \tag{2.21}$$

More generally, if $\mathcal{L}$ is any linear partial differential operator, a Green's function for the linear partial differential equation $\mathcal{L}\psi = f$ is a (distributional) solution to the equation

$$\mathcal{L}G(\boldsymbol{x}, \boldsymbol{x}') = \delta(\boldsymbol{x}-\boldsymbol{x}') \tag{2.22}$$

that satisfies required boundary (or asymptotic) conditions in $\boldsymbol{x}$, where $\mathcal{L}$ in eq. (2.22) acts on the $\boldsymbol{x}$ variable in $G(\boldsymbol{x}, \boldsymbol{x}')$. If $G(\boldsymbol{x}, \boldsymbol{x}')$ is known, then a solution to $\mathcal{L}\psi = f$ can then be obtained as

$$\psi(\boldsymbol{x}) = \int G(\boldsymbol{x}, \boldsymbol{x}') f(\boldsymbol{x}') d^3x', \tag{2.23}$$

provided, of course, that this integral converges. We will discuss Green's functions in electrostatics for cavities in section 2.5, and Green's functions will play a central role in our analysis of electrodynamics in chapter 5.

2.3 Interaction Energy and Force

The energy density of the electric field in electrostatics is given by eq. (1.15) with $\boldsymbol{B} = 0$, namely,

$$\mathcal{E} = \frac{\epsilon_0}{2}|\boldsymbol{E}|^2. \tag{2.24}$$

If the charge density ρ is smoothly distributed over spacetime and the potential ϕ goes to zero at infinity in the manner specified in the uniqueness theorem of section 2.1, then the formula for the total energy $\mathscr{E}$ contained in the electric field can be put in the following useful form:

$$\begin{aligned}
\mathscr{E} &\equiv \int \mathcal{E} d^3x = \frac{\epsilon_0}{2} \lim_{R\to\infty} \int_{r\leq R} |\boldsymbol{E}|^2 d^3x \\
&= \frac{\epsilon_0}{2} \lim_{R\to\infty} \int_{r\leq R} |\nabla\phi|^2 d^3x \\
&= \frac{\epsilon_0}{2} \lim_{R\to\infty} \int_{r\leq R} \left[\nabla\cdot(\phi\nabla\phi) - \phi\nabla^2\phi\right] d^3x \\
&= \frac{\epsilon_0}{2} \lim_{R\to\infty} \left[\int_{r=R} (\phi\nabla\phi)\cdot\hat{\boldsymbol{n}}\, dS\right] - \frac{\epsilon_0}{2}\int \phi\nabla^2\phi\, d^3x \\
&= \frac{1}{2}\int \phi\rho\, d^3x,
\end{aligned} \tag{2.25}$$

where Poisson's equation (2.3) and the asymptotic conditions on ϕ were used in the last line.

Now suppose that the charge density ρ is written as a sum of two charge densities ρ_1 and ρ_2,

$$\rho = \rho_1 + \rho_2 \tag{2.26}$$

as would be natural to do if we had two disjoint charged bodies. Let ϕ_1 be the solution to Poisson's equation with source ρ_1, and let ϕ_2 be the solution to Poisson's equation with source ρ_2. Then the solution with source ρ is

$$\phi = \phi_1 + \phi_2. \tag{2.27}$$

The total energy is thus

$$\begin{aligned}
\mathscr{E} &= \frac{\epsilon_0}{2}\int |\boldsymbol{E}_1 + \boldsymbol{E}_2|^2 d^3x \\
&= \frac{\epsilon_0}{2}\int |\boldsymbol{E}_1|^2 d^3x + \frac{\epsilon_0}{2}\int |\boldsymbol{E}_2|^2 d^3x + \epsilon_0 \int \boldsymbol{E}_1\cdot\boldsymbol{E}_2\, d^3x.
\end{aligned} \tag{2.28}$$

The first two terms are the total energy associated with the fields of the charge distributions ρ_1 and ρ_2, respectively. The last term may be interpreted as the *interaction energy* of the fields of the two charge distributions:

$$\mathscr{E}^{\text{int}} = \epsilon_0 \int \boldsymbol{E}_1 \cdot \boldsymbol{E}_2 \, d^3x. \tag{2.29}$$

By the same manipulations as led to eq. (2.25), we obtain

$$\mathscr{E}^{\text{int}} = \int \rho_1 \phi_2 \, d^3x = \int \rho_2 \phi_1 \, d^3x. \tag{2.30}$$

Now consider a charged body (with smooth charge density $\rho(\boldsymbol{x})$) that is placed in a given external field. By an "external field," we mean a potential, ϕ^{ext}, that satisfies Laplace's equation in a neighborhood of the charged body:

$$\nabla^2 \phi^{\text{ext}} = 0. \tag{2.31}$$

This potential may be thought of as arising from charges that are held in fixed positions at a large (but finite) distance from the charged body. If we assume that ϕ^{ext} does arise from such distant charges, then $\phi^{\text{ext}} \to 0$ as $|\boldsymbol{x}| \to \infty$, and we may use eq. (2.30) to evaluate the interaction energy of the field of the charged body and the external field.[3] We obtain

$$\mathscr{E}^{\text{int}} = \int \rho \phi^{\text{ext}} d^3x. \tag{2.32}$$

We may Taylor expand ϕ^{ext} about the origin as

$$\phi^{\text{ext}} = \phi^{\text{ext}}\Big|_{\boldsymbol{x}=0} + \sum_i \frac{\partial \phi^{\text{ext}}}{\partial x_i}\Big|_{\boldsymbol{x}=0} x_i + \frac{1}{2} \sum_{i,j} \frac{\partial^2 \phi^{\text{ext}}}{\partial x_i \partial x_j}\Big|_{\boldsymbol{x}=0} x_i x_j + \cdots . \tag{2.33}$$

Note that the Taylor coefficients $\partial^n \phi^{\text{ext}} / \partial x_{i_1} \ldots \partial x_{i_n}$ are trace-free in all pairs of indices on account of eq. (2.31). If the charged body is small in extent (relative to the scale of variation of the external field) and is located near the origin, it will be a good approximation to replace ϕ^{ext} by the first few terms of its Taylor expansion. Substituting eq. (2.33) in eq. (2.32), we obtain the following series formula for the interaction energy:

$$\mathscr{E}^{\text{int}} = q\phi^{\text{ext}}\Big|_{\boldsymbol{x}=0} + \boldsymbol{p} \cdot \nabla \phi^{\text{ext}}\Big|_{\boldsymbol{x}=0} + \frac{1}{6} \sum_{i,j} Q_{ij} \frac{\partial^2 \phi^{\text{ext}}}{\partial x_i \partial x_j}\Big|_{\boldsymbol{x}=0} + \cdots . \tag{2.34}$$

Here, the charge q, electric dipole moment $\boldsymbol{p}$, and electric quadrupole moment Q_{ij} of the charged body are defined by

$$q = \int \rho(\boldsymbol{x}) d^3x, \tag{2.35}$$

[3] If we assume that ϕ^{ext} is source free over all of space, then ϕ^{ext} cannot go to zero at infinity, and we must use eq. (2.29) to calculate the interaction energy. This will, in general, give a different answer than eq. (2.30) (see problem 7).

$$\boldsymbol{p} = \int \boldsymbol{x}\rho(\boldsymbol{x})d^3x, \tag{2.36}$$

$$Q_{ij} = \int \left(3x_i x_j - r^2\delta_{ij}\right)\rho(\boldsymbol{x})d^3x. \tag{2.37}$$

Note that Q_{ij} is symmetric (i.e., $Q_{ij} = Q_{ji}$) and has been defined so as to be trace-free, $\sum_i Q_{ii} = 0$. The term proportional to the Kronecker delta, δ_{ij}, in eq. (2.37) does not contribute to eq. (2.34) on account of eq. (2.31). The higher multipole moments—whose contributions to eq. (2.34) are represented by "· · · "—also are defined so as to be totally symmetric in all indices and trace-free over any pair of indices. We will see in section 2.4 that these "Cartesian multipole moments" are directly related to "spherical multipole moments" defined in terms of spherical harmonics.

The force on a charged body of charge density ρ in an external field $\boldsymbol{E}^{\text{ext}}$ is given by eq. (1.24) with $\boldsymbol{B} = 0$, namely,

$$\boldsymbol{F} = \int \boldsymbol{f} d^3x = \int \rho(\boldsymbol{x})\boldsymbol{E}(\boldsymbol{x})d^3x. \tag{2.38}$$

Here $\boldsymbol{E}$ is the *total* electric field, $\boldsymbol{E} = \boldsymbol{E}^{\text{self}} + \boldsymbol{E}^{\text{ext}}$, where $\boldsymbol{E}^{\text{self}}$ is the "self-field" of the body, that is, the electric field associated with the body of charge density ρ:

$$\boldsymbol{E}^{\text{self}}(\boldsymbol{x}) = -\nabla\phi^{\text{self}}(\boldsymbol{x}) = -\frac{1}{4\pi\epsilon_0}\nabla\int \frac{\rho(\boldsymbol{x}')}{|\boldsymbol{x}-\boldsymbol{x}'|}d^3x' = \frac{1}{4\pi\epsilon_0}\int \frac{\rho(\boldsymbol{x}')}{|\boldsymbol{x}-\boldsymbol{x}'|^3}(\boldsymbol{x}-\boldsymbol{x}')d^3x'. \tag{2.39}$$

The contribution to the total force on the body arising from $\boldsymbol{E}^{\text{self}}$ (i.e., the total electromagnetic "self-force" exerted on the body by its own field) is given by

$$\boldsymbol{F}^{\text{self}} = \int \rho(\boldsymbol{x})\boldsymbol{E}^{\text{self}}(\boldsymbol{x})d^3x = \frac{1}{4\pi\epsilon_0}\int \frac{\rho(\boldsymbol{x}')\rho(\boldsymbol{x})}{|\boldsymbol{x}-\boldsymbol{x}'|^3}(\boldsymbol{x}-\boldsymbol{x}')d^3x'd^3x. \tag{2.40}$$

Since the integrand at $(\boldsymbol{x}, \boldsymbol{x}')$ is minus the integrand at $(\boldsymbol{x}', \boldsymbol{x})$, it follows that the self-force vanishes:[4]

$$\boldsymbol{F}^{\text{self}} = 0. \tag{2.41}$$

Thus, the only net contribution to the total force arises from the external field, and we have

$$\boldsymbol{F} = \int \rho(\boldsymbol{x})\boldsymbol{E}^{\text{ext}}(\boldsymbol{x})d^3x. \tag{2.42}$$

Taylor expanding $\boldsymbol{E}^{\text{ext}} = -\nabla\phi^{\text{ext}}$ as in eq. (2.33), we obtain

$$F_i = -q\frac{\partial\phi^{\text{ext}}}{\partial x_i}\bigg|_{\boldsymbol{x}=0} - \sum_j p_j \frac{\partial^2\phi^{\text{ext}}}{\partial x_i \partial x_j}\bigg|_{\boldsymbol{x}=0} - \frac{1}{6}\sum_{j,k} Q_{jk}\frac{\partial^3\phi^{\text{ext}}}{\partial x_i \partial x_j \partial x_k}\bigg|_{\boldsymbol{x}=0} + \cdots. \tag{2.43}$$

[4] As discussed in depth in chapter 10, the self-force will not, in general, vanish in electrodynamics.

Thus, we have

$$F = qE^{\text{ext}} + (p \cdot \nabla)E^{\text{ext}} + \frac{1}{6}\sum_{j,k} Q_{jk}\frac{\partial^2 E^{\text{ext}}}{\partial x_j \partial x_k} + \cdots . \tag{2.44}$$

Similarly, we find that the torque τ exerted on charged body placed in an external field E^{ext} is given by

$$\begin{aligned}\tau &= \int x \times f d^3x = \int x \times [\rho(x)E(x)]d^3x = \int x \times [\rho(x)E^{\text{ext}}(x)]d^3x \\ &= p \times E^{\text{ext}} + \cdots .\end{aligned} \tag{2.45}$$

Finally, we consider the interaction energy and force for the case of point charges. The energy contained in the electric field of a point charge is

$$\mathscr{E} = \frac{\epsilon_0}{2}\int |E|^2 d^3x = \frac{1}{32\pi^2\epsilon_0}\int \frac{q^2}{r^4} d^3x = \infty. \tag{2.46}$$

As already discussed in section 1.4, this is a manifestation of the fact that, apart from some important idealized situations, point charges do not yield a mathematically sensible description of charged matter. Nevertheless, the interaction energy of the fields of two point charges is well defined. For point charges q_1 and q_2 located, respectively, at x_1 and x_2, the field interaction energy is

$$\mathscr{E}^{\text{int}} = \int \rho_1\phi_2 d^3x = \frac{1}{4\pi\epsilon_0}\frac{q_1q_2}{|x_1 - x_2|}. \tag{2.47}$$

The force on a point charge in an external field E^{ext} is given by

$$F = qE^{\text{ext}}. \tag{2.48}$$

In particular, if a point charge q_1 is located at x_1 and the external field corresponds to the field eq. (2.17) of a point charge q_2 located at x_2, then the force on the first charge is given by *Coulomb's law*

$$F_{12} = \frac{1}{4\pi\epsilon_0}\frac{q_1q_2}{|x_1 - x_2|^2}\hat{x}_{12}, \tag{2.49}$$

where $\hat{x}_{12}$ is the unit vector pointing in the direction $x_1 - x_2$.

Equation (2.49) is usually taken as the starting point for the discussion of electrostatics. The notion of electromagnetic energy is then usually introduced by considering the work that must be done to bring charge q_1 in from infinity to x_1, holding q_2 fixed at x_2. This calculation yields a result in agreement with formula (2.47), and it is usually taken as a "derivation" of eq. (2.47). However, it is highly misleading to view this as a derivation, because in order to equate the "work done" to the "change in electromagnetic energy," one must assume that there is no change in the internal energy (i.e., the rest mass) of a charged body under quasi-static motion in an external electrostatic field. Although this is indeed true in electrostatics, as already mentioned in section 1.2, the corresponding result does *not* hold in magnetostatics. As discussed in section 1.2, eq. (2.24) should be viewed as a fundamental property of the electromagnetic field, on par with Maxwell's equations, and it is fundamentally wrong to attempt to derive it from the expression for force.

2.4 Multipole Expansion of the Green's Function

As discussed in section 2.2, if the charge density ρ is such that $\rho \to 0$ sufficiently rapidly at infinity, the unique solution to Poisson's equation (2.3) with $\phi \to 0$ at infinity is given by

$$\phi(\boldsymbol{x}) = \frac{1}{4\pi\epsilon_0} \int \frac{\rho(\boldsymbol{x}')}{|\boldsymbol{x}-\boldsymbol{x}'|} d^3x'. \tag{2.50}$$

If ρ is nonzero only for $|\boldsymbol{x}'| < R$ and we are interested in the solution at $|\boldsymbol{x}| > R$, then $|\boldsymbol{x}| > |\boldsymbol{x}'|$ in eq. (2.50). It is useful to Taylor expand $1/|\boldsymbol{x}-\boldsymbol{x}'|$ as

$$\begin{aligned} \frac{1}{|\boldsymbol{x}-\boldsymbol{x}'|} &= \frac{1}{\left[|\boldsymbol{x}|^2 - 2\boldsymbol{x}'\cdot\boldsymbol{x} + |\boldsymbol{x}'|^2\right]^{1/2}} \\ &= \frac{1}{|\boldsymbol{x}|} \frac{1}{\left[1 - 2\boldsymbol{x}'\cdot\boldsymbol{x}/|\boldsymbol{x}|^2 + |\boldsymbol{x}'|^2/|\boldsymbol{x}|^2\right]^{1/2}} \\ &= \frac{1}{|\boldsymbol{x}|} + \frac{\boldsymbol{x}'\cdot\hat{\boldsymbol{x}}}{|\boldsymbol{x}|^2} + \frac{1}{2}\frac{3(\boldsymbol{x}'\cdot\hat{\boldsymbol{x}})^2 - |\boldsymbol{x}'|^2}{|\boldsymbol{x}|^3} + \ldots, \end{aligned} \tag{2.51}$$

where $\hat{\boldsymbol{x}} = \boldsymbol{x}/|\boldsymbol{x}|$. Plugging this into eq. (2.50), we obtain the following multipole expansion for ϕ for $|\boldsymbol{x}| > R$:

$$\phi(\boldsymbol{x}) = \frac{1}{4\pi\epsilon_0}\left[\frac{q}{|\boldsymbol{x}|} + \frac{\boldsymbol{p}\cdot\hat{\boldsymbol{x}}}{|\boldsymbol{x}|^2} + \frac{1}{2}\sum_{ij} Q_{ij}\frac{\hat{x}_i\hat{x}_j}{|\boldsymbol{x}|^3} + \cdots\right], \tag{2.52}$$

where q, $\boldsymbol{p}$, and Q_{ij} are defined by eqs. (2.35)–(2.37). Note that the lowest nonvanishing multipole moment of the charge distribution will dominate the potential as $|\boldsymbol{x}| \to \infty$.

It is extremely useful to perform this expansion—and the similar expansion for $|\boldsymbol{x}| < |\boldsymbol{x}'|$—in spherical coordinates (r, θ, φ), where spherical coordinates are defined in terms of Cartesian coordinates (x, y, z) by

$$r = \sqrt{x^2 + y^2 + z^2}, \tag{2.53}$$

$$\cos\theta = \frac{z}{\sqrt{x^2+y^2+z^2}}, \tag{2.54}$$

$$\tan\varphi = \frac{y}{x}. \tag{2.55}$$

One might attempt to directly rewrite eq. (2.51) in terms of spherical coordinates. However, it is easier and more instructive to derive an expansion for the Green's function, $G(\boldsymbol{x}, \boldsymbol{x}')$, in spherical coordinates starting from scratch. So, let us ignore the fact that we already know that $G(\boldsymbol{x},\boldsymbol{x}') = 1/(4\pi\epsilon_0|\boldsymbol{x}-\boldsymbol{x}'|)$ and seek to determine $G(\boldsymbol{x},\boldsymbol{x}')$, working in spherical coordinates.

The Laplacian operator, eq. (2.4), acting on a function f is given by the following expression in spherical coordinates:

$$\nabla^2 f = \frac{1}{r^2}\frac{\partial}{\partial r}\left(r^2\frac{\partial f}{\partial r}\right) + \frac{1}{r^2}\left[\frac{1}{\sin\theta}\frac{\partial}{\partial\theta}\left(\sin\theta\frac{\partial f}{\partial\theta}\right) + \frac{1}{\sin^2\theta}\frac{\partial^2 f}{\partial\varphi^2}\right]. \tag{2.56}$$

The term in square brackets defines the Laplacian operator, $\mathcal{D}^2$, on the unit sphere:

$$\mathcal{D}^2 f \equiv \frac{1}{\sin\theta}\frac{\partial}{\partial\theta}\left(\sin\theta\frac{\partial f}{\partial\theta}\right) + \frac{1}{\sin^2\theta}\frac{\partial^2 f}{\partial\varphi^2}. \tag{2.57}$$

The operator $\mathcal{D}^2$ is a self-adjoint operator on the Hilbert space, $\mathcal{H}$, of complex-valued square integrable functions[5] on the unit sphere, with inner product defined by

$$\langle f_1, f_2\rangle = \int_0^{2\pi}\int_0^{\pi} f_1^*(\theta,\varphi) f_2(\theta,\varphi)\sin\theta\, d\theta\, d\varphi, \tag{2.58}$$

where $*$ denotes complex conjugation. It follows that there exists an orthonormal basis of $\mathcal{H}$ composed of eigenvectors of $\mathcal{D}^2$ (see the boxed side comment below). The eigenvalues of $\mathcal{D}^2$ are $-\ell(\ell+1)$, with $\ell = 0, 1, 2, \ldots$, and the degeneracy of the ℓth eigensubspace is $2\ell+1$. Since there is degeneracy in the eigensubspaces of $\mathcal{D}^2$, the choice of an orthonormal basis of eigenvectors is not unique. However, a very convenient (and extremely standard) choice of basis is given by *spherical harmonics*, defined by

$$Y_{\ell m}(\theta,\varphi) = \sqrt{\frac{2\ell+1}{4\pi}}\sqrt{\frac{(\ell-m)!}{(\ell+m)!}} P_\ell^m(\cos\theta) e^{im\varphi}, \tag{2.59}$$

with

$$P_\ell^m(x) \equiv \frac{(-1)^m}{2^\ell \ell!}(1-x^2)^{m/2}\frac{d^{\ell+m}}{dx^{\ell+m}}(x^2-1)^\ell. \tag{2.60}$$

Here, ℓ ranges over the nonnegative integers, $\ell = 0, 1, 2, \ldots$, and m is an integer with $-\ell \le m \le \ell$. The functions P_ℓ^m are called *associated Legendre functions*; for $m = 0$, they are called *Legendre polynomials* and denoted by P_ℓ. The first few Legendre polynomials are

$$P_0(x) = 1, \qquad P_1(x) = x, \qquad P_2(x) = \frac{1}{2}(3x^2-1). \tag{2.61}$$

The spherical harmonics satisfy $Y_{\ell,-m}(\theta,\varphi) = Y_{\ell m}^*(\theta,\varphi)$. Explicitly, the spherical harmonics for $\ell = 0, 1, 2$ and $m \ge 0$ are

$$Y_{00} = \frac{1}{\sqrt{4\pi}}; \qquad Y_{10} = \sqrt{\frac{3}{4\pi}}\cos\theta, \quad Y_{11} = -\sqrt{\frac{3}{8\pi}}\sin\theta e^{i\varphi}, \tag{2.62}$$

[5] Note that the "completeness" condition in the definition of a Hilbert space requires that $\mathcal{H}$ include *all* square integrable functions (including, e.g., discontinuous functions).

$$Y_{20} = \sqrt{\frac{5}{16\pi}}(3\cos^2\theta - 1), \quad Y_{21} = -\sqrt{\frac{15}{8\pi}}\sin\theta\cos\theta e^{i\varphi}, \quad Y_{22} = \sqrt{\frac{15}{32\pi}}\sin^2\theta e^{2i\varphi}. \tag{2.63}$$

The key properties of spherical harmonics are that (i) they are a basis of $\mathcal{H}$, (ii) they are orthonormal with respect to the inner product (eq. (2.58)) of $\mathcal{H}$, and (iii) they satisfy

$$\mathcal{D}^2 Y_{\ell m} = -\ell(\ell+1) Y_{\ell m}. \tag{2.64}$$

Since the $Y_{\ell m}$ comprise an orthonormal basis of $\mathcal{H}$, any square integrable function $f(\theta, \varphi)$ on a sphere can be expanded as[6]

$$f(\theta, \varphi) = \sum_{\ell, m} c_{\ell m} Y_{\ell m}(\theta, \varphi), \tag{2.65}$$

where

$$c_{\ell m} = \int_0^{2\pi} \int_0^{\pi} Y^*_{\ell m}(\theta, \varphi) f(\theta, \varphi) \sin\theta \, d\theta \, d\varphi. \tag{2.66}$$

Side Comment on Self-Adjointness and Bases of Eigenvectors: Operators such as $\mathcal{D}^2$ are unbounded and cannot be defined on all vectors in $\mathcal{H}$. The precise definition of "self-adjointness" places a nontrivial condition on the domain of definition of the operator (i.e., the vectors on which the operator is defined). The domain of $\mathcal{D}^2$ can be uniquely extended beyond its natural domain of twice differentiable functions so as to satisfy this condition, so $\mathcal{D}^2$ is indeed truly a self-adjoint operator. Despite what is commonly claimed, self-adjoint operators (even if bounded) on an infinite-dimensional Hilbert space need not have a basis of eigenvectors. For example, the position operator in quantum mechanics is self-adjoint but does not admit any eigenvectors at all (i.e., there is no square integrable function that is an eigenvector of the position operator). Nevertheless, there is still a "spectral theorem" that applies to all self-adjoint operators, which can be used to justify the arguments that are usually made by assuming existence of an orthonormal basis of eigenvectors. However, for a self-adjoint, elliptic, partial differential operator (such as $\mathcal{D}^2$) on a compact manifold (such as a sphere), one can prove that there always does exist an orthonormal basis of eigenvectors.

By definition, the Green's function for the Poisson equation satisfies

$$\nabla^2 G(\boldsymbol{x}, \boldsymbol{x}') = -\frac{1}{\epsilon_0}\delta(\boldsymbol{x} - \boldsymbol{x}'), \tag{2.67}$$

---

[6]Convergence of the sum in eq. (2.65) is in the sense that the Hilbert space norm of the difference between the left side and the sum on the right side for $\ell \leq L$ goes to zero as $L \to \infty$.

with the asymptotic condition $G(\boldsymbol{x},\boldsymbol{x}')\to 0$ as $|\boldsymbol{x}|\to\infty$. The delta function may be expressed in spherical coordinates as

$$\delta(\boldsymbol{x}-\boldsymbol{x}')=\frac{1}{r^2\sin\theta}\delta(r-r')\delta(\theta-\theta')\delta(\varphi-\varphi'). \tag{2.68}$$

Here, the inverse of the volume element, $r^2\sin\theta$, in spherical coordinates appears in this formula, because $\delta(\boldsymbol{x}-\boldsymbol{x}')$ is defined so that $\int\delta(\boldsymbol{x}-\boldsymbol{x}')d^3x=\int\delta(\boldsymbol{x}-\boldsymbol{x}')r^2\sin\theta drd\theta d\varphi=1$, whereas the delta functions on the right side are defined so that $\int\delta(r-r')dr=1$, $\int\delta(\theta-\theta')d\theta=1$, and $\int\delta(\varphi-\varphi')d\varphi=1$. Now expand $G(\boldsymbol{x},\boldsymbol{x}')$ in spherical harmonics in the $\boldsymbol{x}$ angular variables:

$$G(\boldsymbol{x},\boldsymbol{x}')=\sum_{\ell,m}G_{\ell m}(r,\boldsymbol{x}')Y_{\ell m}(\theta,\varphi), \tag{2.69}$$

where, by eq. (2.66), we have

$$G_{\ell m}(r,\boldsymbol{x}')=\int_0^{2\pi}\int_0^{\pi}Y^*_{\ell m}(\theta,\varphi)G(r,\theta,\varphi,\boldsymbol{x}')\sin\theta d\theta d\varphi. \tag{2.70}$$

Substituting eq. (2.69) into eq. (2.67), we obtain[7]

$$\sum_{\ell,m}\left[\frac{1}{r^2}\frac{d}{dr}\left(r^2\frac{dG_{\ell m}(r,\boldsymbol{x}')}{dr}\right)-\frac{\ell(\ell+1)}{r^2}G_{\ell m}(r,\boldsymbol{x}')\right]Y_{\ell m}(\theta,\varphi)$$
$$=-\frac{1}{\epsilon_0 r^2\sin\theta}\delta(r-r')\delta(\theta-\theta')\delta(\varphi-\varphi'). \tag{2.71}$$

Multiplying eq. (2.71) by $Y^*_{\ell m}(\theta,\varphi)$, integrating over the sphere (with volume element $\sin\theta d\theta d\varphi$), and using the orthonormality of the $Y_{\ell m}$, we obtain

$$\frac{1}{r^2}\frac{d}{dr}\left(r^2\frac{dG_{\ell m}(r,\boldsymbol{x}')}{dr}\right)-\frac{\ell(\ell+1)}{r^2}G_{\ell m}(r,\boldsymbol{x}')=-\frac{1}{\epsilon_0 r^2}\delta(r-r')Y^*_{\ell m}(\theta',\varphi'). \tag{2.72}$$

Thus, we have

$$G_{\ell m}(r,\boldsymbol{x}')=g_{\ell m}(r,r')Y^*_{\ell m}(\theta',\varphi'), \tag{2.73}$$

where $g_{\ell m}(r,r')$ satisfies

$$\frac{d}{dr}\left(r^2\frac{dg_{\ell m}(r,r')}{dr}\right)-\ell(\ell+1)g_{\ell m}(r,r')=-\frac{1}{\epsilon_0}\delta(r-r'). \tag{2.74}$$

---

[7] It should be noted that the Hilbert space convergence of eq. (2.65) is not adequate to justify interchange of summation and differentiation. However, for smooth functions on the sphere, the $c_{\ell m}$ coefficients go to zero rapidly as $\ell,m\to\infty$, and the convergence occurs in a strong enough sense to justify such an interchange. Since $G(r,\theta,\varphi,\boldsymbol{x}')$ is a smooth function on the sphere when $r\neq r'$, the interchange can be justified.

When the right side of this equation vanishes (as occurs for $r > r'$ and for $r < r'$), the general solution is

$$g_{\ell m}(r, r') = \frac{a_{\ell m}(r')}{r^{\ell+1}} + b_{\ell m}(r') r^{\ell}, \tag{2.75}$$

where $a_{\ell m}(r')$ and $b_{\ell m}(r')$ are arbitrary functions of r'. The solution $a_{\ell m}(r')/r^{\ell+1}$ is unacceptable at $r=0$, whereas the solution $b_{\ell m}(r')r^{\ell}$ is unacceptable at $r \to \infty$. Therefore, $g_{\ell m}$ must be of the form

$$g_{\ell m}(r, r') = \begin{cases} b_{\ell m}(r') r^{\ell}, & \text{if } r < r' \\ a_{\ell m}(r')/r^{\ell+1}, & \text{if } r > r'. \end{cases} \tag{2.76}$$

Integrating eq. (2.74) from $r = r' - \epsilon$ to $r = r' + \epsilon$ and letting $\epsilon \to 0$, we find that the r derivative of $g_{\ell m}(r, r')$ must have a jump discontinuity at $r = r'$:

$$r'^2 \left[\left. \frac{dg_{\ell m}}{dr} \right|_{r \downarrow r'} - \left. \frac{dg_{\ell m}}{dr} \right|_{r \uparrow r'} \right] = -\frac{1}{\epsilon_0}. \tag{2.77}$$

However, $g_{\ell m}(r, r')$ itself must be continuous at $r = r'$, since otherwise the left side of eq. (2.74) would get a contribution proportional to the derivative of a delta function, whereas the right side has only a delta function. The continuity of $g_{\ell m}$ at $r = r'$ yields

$$b_{\ell m}(r') r'^{\ell} = \frac{a_{\ell m}(r')}{r'^{\ell+1}} \tag{2.78}$$

whereas eq. (2.77) yields

$$-(\ell+1) \frac{a_{\ell m}(r')}{r'^{\ell}} - \ell b_{\ell m}(r') r'^{\ell+1} = -\frac{1}{\epsilon_0}. \tag{2.79}$$

The unique solution to these equations is

$$a_{\ell m}(r') = \frac{1}{\epsilon_0} \frac{1}{2\ell+1} r'^{\ell}, \tag{2.80}$$

$$b_{\ell m}(r') = \frac{1}{\epsilon_0} \frac{1}{2\ell+1} \frac{1}{r'^{\ell+1}}. \tag{2.81}$$

Putting everything together, we have obtained the following series expansion for the Green's function, $G(\boldsymbol{x}, \boldsymbol{x}')$:

$$G(\boldsymbol{x}, \boldsymbol{x}') = \frac{1}{4\pi\epsilon_0} \frac{1}{|\boldsymbol{x} - \boldsymbol{x}'|} = \begin{cases} \dfrac{1}{\epsilon_0} \displaystyle\sum_{\ell,m} \frac{1}{2\ell+1} \frac{r^{\ell}}{r'^{\ell+1}} Y^*_{\ell m}(\theta', \varphi') Y_{\ell m}(\theta, \varphi), & \text{if } r < r' \\ \dfrac{1}{\epsilon_0} \displaystyle\sum_{\ell,m} \frac{1}{2\ell+1} \frac{r'^{\ell}}{r^{\ell+1}} Y^*_{\ell m}(\theta', \varphi') Y_{\ell m}(\theta, \varphi), & \text{if } r > r'. \end{cases} \tag{2.82}$$

This formula for the case $r > r'$ is, of course, equivalent to eq. (2.51), but we now have written all of the terms explicitly (instead of resorting to writing "..."). Using this

Green's function, we can immediately write down the solution to Poisson's equation with an arbitrary charge density, $\rho(\boldsymbol{x})$, that falls off sufficiently rapidly at infinity:

$$\begin{aligned}\phi(\boldsymbol{x}) &= \int G(\boldsymbol{x},\boldsymbol{x}')\rho(\boldsymbol{x}')d^3x' \\ &= \frac{1}{\epsilon_0}\sum_{\ell,m}\frac{1}{2\ell+1}\left[\alpha_{\ell m}(r)r^\ell + \frac{\beta_{\ell m}(r)}{r^{\ell+1}}\right]Y_{\ell m}(\theta,\varphi), \end{aligned} \tag{2.83}$$

where

$$\alpha_{\ell m}(r) = \int_0^{2\pi}\int_0^{\pi}\int_r^{\infty}\frac{\rho(r',\theta',\varphi')}{r'^{\ell+1}}Y^*_{\ell m}(\theta',\varphi')r'^2\sin\theta' dr' d\theta' d\varphi', \tag{2.84}$$

$$\beta_{\ell m}(r) = \int_0^{2\pi}\int_0^{\pi}\int_0^{r}\rho(r',\theta',\varphi')r'^\ell Y^*_{\ell m}(\theta',\varphi')r'^2\sin\theta' dr' d\theta' d\varphi'. \tag{2.85}$$

In particular, if $\rho(\boldsymbol{x}')$ is nonzero only when $r' < R$, then for $r > R$, we obtain

$$\phi(\boldsymbol{x}) = \frac{1}{\epsilon_0}\sum_{\ell,m}\frac{1}{2\ell+1}\frac{q_{\ell m}}{r^{\ell+1}}Y_{\ell m}(\theta,\varphi), \tag{2.86}$$

where

$$q_{\ell m} \equiv \int \rho(\boldsymbol{x}')r'^\ell Y^*_{\ell m}(\theta',\varphi')d^3x'. \tag{2.87}$$

Equations (2.86) and (2.87) are very useful, since they allow one to explicitly determine the behavior of ϕ as $r \to \infty$ to whatever accuracy in $1/r^n$ that one wishes. Equation (2.86) is equivalent to eq. (2.52), and there is a direct correspondence between the Cartesian multipole moments $q, \boldsymbol{p}, Q_{ij}, \ldots$ and the spherical multipole moments $q_{\ell m}$. In particular, we have

$$q_{00} = \frac{1}{\sqrt{4\pi}}q, \tag{2.88}$$

$$q_{10} = \sqrt{\frac{3}{4\pi}}p_z, \qquad q_{11} = -q^*_{1,-1} = -\sqrt{\frac{3}{8\pi}}(p_x - ip_y), \tag{2.89}$$

whereas Q_{ij} is directly related to q_{2m}. However, eq. (2.86) has the advantage over eq. (2.52) in that we have now explicitly written down all the terms in the series.

2.5 Conducting Cavities; Dirichlet and Neumann Green's Functions

A *conductor* is a material with charges that are free to move within the material in such a way to make $\boldsymbol{E} = -\nabla\phi = 0$ inside the conductor.[8] Thus, by definition, ϕ is constant inside any conductor. It follows immediately that $\nabla^2\phi = 0$ inside a conductor,

[8]The condition $\boldsymbol{E} = 0$ inside conductors holds in electrostatics, where the conductor is assumed to have reached equilibrium. As we shall see in section 6.3.2, a time-varying electric field can penetrate inside a conductor of finite conductivity.

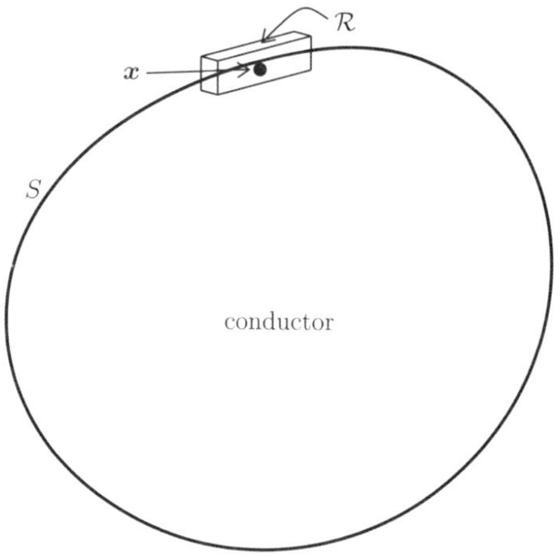

Figure 2.2. A "Gaussian pillbox," $\mathcal{R}$, enclosing a point, x, on the surface, S, of a conductor.

which implies by Poisson's equation that $\rho = 0$ inside a conductor. Thus, any charge placed on a conductor must lie entirely on its surface. The charge density is assumed to take the form[9]

$$\rho(\boldsymbol{x}) = \sigma(\boldsymbol{x})\delta(s), \tag{2.90}$$

where s denotes the normal distance of $\boldsymbol{x}$ to the surface, S, of the conductor. We refer to σ as the *surface charge density*.

Poisson's equation,

$$\nabla^2\phi = -\frac{1}{\epsilon_0}\sigma\delta(s), \tag{2.91}$$

implies that ϕ must be continuous across S, since a discontinuity in ϕ across S would give rise to a derivative of a delta function in $\nabla^2\phi$. Similarly, the derivative of ϕ in a direction tangential to S must be continuous across S, so $\boldsymbol{E}_\parallel$ is continuous across S, where $\boldsymbol{E}_\parallel$ denotes the components of $\boldsymbol{E}$ tangential to S. Since $\boldsymbol{E} = 0$ inside the conductor, it follows that just outside the conductor, we have

$$\boldsymbol{E}_\parallel = 0. \tag{2.92}$$

Now, consider an infinitesimally small rectangular volume, $\mathcal{R}$, (i.e., a "Gaussian pillbox") centered at $\boldsymbol{x} \in S$ and oriented with top and bottom faces parallel to S, and with the top face outside the conductor, as illustrated in figure 2.2. Applying Gauss's theorem to the integral of $\nabla \cdot \boldsymbol{E} = \sigma\delta(s)/\epsilon_0$ over $\mathcal{R}$, we obtain

$$\int_{\mathcal{R}} \nabla \cdot \boldsymbol{E} = \frac{1}{\epsilon_0}\sigma A = \int \boldsymbol{E} \cdot \hat{\boldsymbol{n}} dS = \boldsymbol{E}(\boldsymbol{x}) \cdot \hat{\boldsymbol{n}} A, \tag{2.93}$$

where A denotes the area of the top face, and $\hat{\boldsymbol{n}}$ is the unit normal pointing away from the conductor. Thus, just outside a conductor, we have

$$\boldsymbol{E} \cdot \hat{\boldsymbol{n}} = \sigma/\epsilon_0. \tag{2.94}$$

[9]The assumption here that the right side of eq. (2.90) has only a delta function and has no derivatives of a delta function. (A derivative of a delta function would correspond to a dipole layer of charge.)

For most of the remainder of this section, we will be concerned with solutions inside a conducting cavity. In other words, we consider a conductor that encloses a volume $\mathcal{V}$, and we consider electrostatics within $\mathcal{V}$. This is of some direct physical interest in its own right, but it also is a prototype for problems that arise in many areas of physics when one is solving equations similar to Poisson's equation (i.e., elliptic equations) in a bounded region, with specified boundary conditions. We first state the following key existence and uniqueness theorem

Theorem (Existence and Uniqueness in a Bounded Region): *Let $\mathcal{V} \subset \mathbb{R}^3$ be a bounded region whose boundary, $S = \partial\mathcal{V}$, is a regular 2-dimensional surface as in Gauss's theorem of section 2.1. Let $\rho(\boldsymbol{x})$ be an arbitrary continuous function on $\mathcal{V}$, and consider Poisson's equation*

$$\nabla^2\phi = -\rho/\epsilon_0 \tag{2.95}$$

in $\mathcal{V}$. The following results hold:

(i) Dirichlet conditions: *Let ψ be an arbitrary continuous function on S. Then there exists a unique solution, ϕ, to eq. (2.95) in $\mathcal{V}$ such that $\phi|_S = \psi$.*

(ii) Neumann conditions: *Let χ be an arbitrary continuous function on S such that*

$$\int_S \chi = -\frac{1}{\epsilon_0}\int_{\mathcal{V}} \rho. \tag{2.96}$$

Then there exists a solution, ϕ, to eq. (2.95) in $\mathcal{V}$ such that $[\hat{\boldsymbol{n}} \cdot \nabla\phi]|_S = \chi$. Furthermore, this solution is unique up to the addition of a constant.

The proof of the uniqueness part of this theorem is the same as the proof of the uniqueness theorem of figure 2.1, except that the integral in eq. (2.9) is now taken over $\mathcal{V}$. The proof of existence is beyond the scope of this book.

Now consider a conducting cavity (i.e., a bounded region $\mathcal{V}$ whose boundary, S, is the inner surface of a conducting body that surrounds $\mathcal{V}$). Thus, the setup is similar to figure 2.2, except that the S is now the *inner* boundary of a conductor, and the interior region labeled as "conductor" in that figure is now hollow. (The conductor may take the form of a thin shell at S, i.e., it need not extend beyond S.) It follows immediately from the defining property of a conductor that the electrostatic potential, ϕ, is constant on S (i.e., $\phi|_S = C$, where C is a constant). By the above theorem, if $\rho = 0$ in $\mathcal{V}$, then there is a unique solution of $\nabla^2\phi = 0$ in $\mathcal{V}$ with this boundary value. It can immediately be seen that this unique solution is $\phi = C$ throughout $\mathcal{V}$. Thus, $\boldsymbol{E} = 0$ inside any conducting cavity that contains no charges inside the cavity. Thus, if one wishes to have a region that is free of electrostatic fields, one can simply enclose the region in a conducting shell. Such a shell is usually referred to as a *Faraday cage.*

We now consider the solutions ϕ inside a conducting cavity $\mathcal{V}$ with $\rho \neq 0$ in $\mathcal{V}$. As noted above, ϕ is constant on the boundary S. We can use our gauge freedom to shift ϕ by a constant, so without loss of generality, we may assume $\phi = 0$ on S. Thus, we wish to solve eq. (2.95) subject to the Dirichlet boundary condition $\phi|_S = 0$. We will be able to find the solution for a general charge distribution $\rho(\boldsymbol{x})$ if we obtain the *Dirichlet Green's function* $G_D(\boldsymbol{x}, \boldsymbol{x}')$, defined as the solution to

$$\nabla^2 G_D(\boldsymbol{x}, \boldsymbol{x}') = -\frac{1}{\epsilon_0}\delta(\boldsymbol{x} - \boldsymbol{x}'), \tag{2.97}$$

subject to the boundary condition that $G_D(\boldsymbol{x}, \boldsymbol{x}') = 0$ for all $\boldsymbol{x} \in S$. We may interpret $G_D(\boldsymbol{x}, \boldsymbol{x}')$ as the potential of a unit point charge placed at a point $\boldsymbol{x}'$ inside the conducting cavity. Given $G_D(\boldsymbol{x}, \boldsymbol{x}')$, the solution for a general charge distribution $\rho(\boldsymbol{x})$ inside the conducting cavity is

$$\phi(\boldsymbol{x}) = \int_{\mathcal{V}} G_D(\boldsymbol{x}, \boldsymbol{x}')\rho(\boldsymbol{x}')d^3x'. \tag{2.98}$$

For the case of a spherical cavity of radius R, it turns out that $G_D(\boldsymbol{x}, \boldsymbol{x}')$ is given by the sum of the Green's function $G(\boldsymbol{x}, \boldsymbol{x}') = 1/(4\pi\epsilon_0|\boldsymbol{x} - \boldsymbol{x}'|)$ for unbounded space plus the potential corresponding to that of an "image charge" placed outside the sphere. Explicitly, for a spherical cavity, we have

$$G_D(\boldsymbol{x}, \boldsymbol{x}') = \frac{1}{4\pi\epsilon_0}\frac{1}{|\boldsymbol{x} - \boldsymbol{x}'|} + \frac{1}{4\pi\epsilon_0}\frac{\alpha}{|\boldsymbol{x} - \boldsymbol{x}''|}, \tag{2.99}$$

where $\boldsymbol{x}'' = \boldsymbol{x}'(R/|\boldsymbol{x}'|)^2$, and $\alpha = -R/|\boldsymbol{x}'|$. To prove that this formula for $G_D(\boldsymbol{x}, \boldsymbol{x}')$ is correct, we need only show that eq. (2.97) holds for $|\boldsymbol{x}| < R$ and that $G_D(\boldsymbol{x}, \boldsymbol{x}') = 0$ for $|\boldsymbol{x}| = R$. The first is obvious, since the image charge lies outside the sphere, and the second can be straightforwardly verified. For a few other cases where the cavity has high symmetry (such as a cylindrical cavity or a rectangular cavity), useful eigenfunction expansion expressions for $G_D(\boldsymbol{x}, \boldsymbol{x}')$ can be obtained in a manner similar to our spherical harmonic expansion of section 2.4. However, for general cavities, numerical methods would be needed to evaluate $G_D(\boldsymbol{x}, \boldsymbol{x}')$.

Even though one cannot obtain an explicit formula for $G_D(\boldsymbol{x}, \boldsymbol{x}')$ for general cavities, the basic form of $G_D(\boldsymbol{x}, \boldsymbol{x}')$ can be seen as follows: As we previously found, $1/(4\pi\epsilon_0|\boldsymbol{x} - \boldsymbol{x}'|)$ satisfies eq. (2.97). Of course, $1/(4\pi\epsilon_0|\boldsymbol{x} - \boldsymbol{x}'|)$ does not satisfy the required boundary condition on S. However, we can correct for this by writing

$$G_D(\boldsymbol{x}, \boldsymbol{x}') = \frac{1}{4\pi\epsilon_0}\frac{1}{|\boldsymbol{x} - \boldsymbol{x}'|} + F_D(\boldsymbol{x}, \boldsymbol{x}'), \tag{2.100}$$

where $F_D(\boldsymbol{x}, \boldsymbol{x}')$ satisfies Laplace's equation,

$$\nabla_x^2 F_D(\boldsymbol{x}, \boldsymbol{x}') = 0, \tag{2.101}$$

subject to the boundary condition $F_D(\boldsymbol{x}, \boldsymbol{x}') = -1/(4\pi\epsilon_0|\boldsymbol{x} - \boldsymbol{x}'|)$ for all $\boldsymbol{x} \in S$. Here the subscript "x" on ∇_x^2 in eq. (2.101) is meant to convey the fact that the derivatives are to be taken with respect to the $\boldsymbol{x}$ variable. By the existence and uniqueness theorem stated earlier in this section, a unique solution for $F_D(\boldsymbol{x}, \boldsymbol{x}')$ exists. It further can be shown that $F_D(\boldsymbol{x}, \boldsymbol{x}')$ must be smooth in $(\boldsymbol{x}, \boldsymbol{x}')$. Thus, for any conducting cavity, $G_D(\boldsymbol{x}, \boldsymbol{x}')$ differs from $1/(4\pi\epsilon_0|\boldsymbol{x} - \boldsymbol{x}'|)$ by a smooth function.

Remarkably, we can use the Dirichlet Green's function $G_D(\boldsymbol{x}, \boldsymbol{x}')$ to solve a different mathematical problem, known as the "Dirichlet problem": Suppose, as above, we are given a cavity $\mathcal{V}$ bounded by a surface S. Suppose $\rho(\boldsymbol{x})=0$ inside $\mathcal{V}$, so the potential, ϕ, satisfies Laplace's equation, $\nabla^2\phi=0$, in $\mathcal{V}$. However, instead of S being the inner surface of a conductor (i.e., instead of the requirement that $\phi|_S=\text{const}$), suppose that $\phi|_S$ is specified to be some arbitrary function ψ on S. By the above existence and uniqueness theorem, a unique solution ϕ subject to this boundary condition will exist. As we now shall see, this solution can be found explicitly in terms of $G_D(\boldsymbol{x},\boldsymbol{x}')$. The method we will use to obtain this solution is applied in section 5.4 to obtain the initial value formulation and is used in section 7.3.2 to obtain results in the theory of diffraction.

Our analysis relies on the following simple but powerful result, known as *Green's theorem*.

Theorem (Green's Theorem): *Let $\mathcal{V}\subset\mathbb{R}^3$ be a bounded region whose boundary, $S=\partial\mathcal{V}$, is a regular 2-dimensional surface. Let ρ_1 and ρ_2 be arbitrary continuous functions in $\mathcal{V}$, and suppose that ϕ_1 and ϕ_2 satisfy*

$$\nabla^2\phi_1=-\rho_1/\epsilon_0, \tag{2.102}$$

$$\nabla^2\phi_2=-\rho_2/\epsilon_0 \tag{2.103}$$

in $\mathcal{V}$. (Note that no boundary conditions on S are imposed on ϕ_1 or ϕ_2.) Then we have

$$\int_S \hat{\boldsymbol{n}}\cdot(\phi_1\nabla\phi_2-\phi_2\nabla\phi_1)dS=-\frac{1}{\epsilon_0}\int_{\mathcal{V}}(\phi_1\rho_2-\phi_2\rho_1)d^3x. \tag{2.104}$$

Proof. Using Gauss's theorem, we have

$$\begin{aligned}\int_S \hat{\boldsymbol{n}}\cdot(\phi_1\nabla\phi_2-\phi_2\nabla\phi_1)dS&=\int_{\mathcal{V}}\nabla\cdot(\phi_1\nabla\phi_2-\phi_2\nabla\phi_1)d^3x\\&=\int_{\mathcal{V}}(\phi_1\nabla^2\phi_2-\phi_2\nabla^2\phi_1)d^3x\\&=-\frac{1}{\epsilon_0}\int_{\mathcal{V}}(\phi_1\rho_2-\phi_2\rho_1)d^3x.\end{aligned} \tag{2.105}$$

□

Now, let ϕ be a solution to $\nabla^2\phi=0$ in $\mathcal{V}$ with $\phi|_S=\psi$, as in the Dirichlet problem above. Let $\phi_1(\boldsymbol{x})=\phi(\boldsymbol{x})$, and let[10] $\phi_2(\boldsymbol{x})=G_D(\boldsymbol{x},\boldsymbol{x}')$. Equation (2.104) yields

$$\int_S\phi(\boldsymbol{x})\hat{\boldsymbol{n}}\cdot\nabla_x G_D(\boldsymbol{x},\boldsymbol{x}')dS(\boldsymbol{x})=-\frac{1}{\epsilon_0}\int_{\mathcal{V}}\phi\delta(\boldsymbol{x}-\boldsymbol{x}')d^3x=-\frac{1}{\epsilon_0}\phi(\boldsymbol{x}'), \tag{2.106}$$

where we have used $G_D(\boldsymbol{x},\boldsymbol{x}')|_{\boldsymbol{x}\in S}=0$ and $\rho_1=0$. Thus, reversing the roles of $\boldsymbol{x}$ and $\boldsymbol{x}'$ and using $\phi|_S=\psi$, we obtain

$$\phi(\boldsymbol{x})=-\epsilon_0\int_S\psi(\boldsymbol{x}')\hat{\boldsymbol{n}}'\cdot\nabla_{x'}G_D(\boldsymbol{x}',\boldsymbol{x})dS(\boldsymbol{x}'), \tag{2.107}$$

[10] I am being a bit sloppy here, since $\rho_2(\boldsymbol{x})$ is not continuous at $\boldsymbol{x}'$. However, we can deal with this by removing a ball of radius ϵ about $\boldsymbol{x}'$ from $\mathcal{V}$. In this modified volume, we then have $\rho_2=0$ on the right side of eq. (2.104), but we pick up an extra surface term on the left side from the surface of the ball of radius ϵ. Using eq. (2.100), we can evaluate this surface integral in the limit as $\epsilon\to 0$ and recover eq. (2.106).

which solves the Dirichlet problem. If we wish to solve $\nabla^2\phi = -\rho/\epsilon_0$ subject to $\phi|_S = \psi$, we simply add eq. (2.98) to eq. (2.107); that is, if $\rho \neq 0$ in $\mathcal{V}$, the solution is

$$\phi(\boldsymbol{x}) = \int_{\mathcal{V}} G_D(\boldsymbol{x}, \boldsymbol{x}')\rho(\boldsymbol{x}')d^3x' - \epsilon_0 \int_S \psi(\boldsymbol{x}')\hat{\boldsymbol{n}}' \cdot \nabla_{x'} G_D(\boldsymbol{x}', \boldsymbol{x})dS(\boldsymbol{x}'). \tag{2.108}$$

Green's theorem can also be used to prove that the Dirichlet Green's function is symmetric under interchange of $\boldsymbol{x}$ and $\boldsymbol{x}'$:

$$G_D(\boldsymbol{x}, \boldsymbol{x}') = G_D(\boldsymbol{x}', \boldsymbol{x}), \tag{2.109}$$

that is, the potential at $\boldsymbol{x}$ of a point charge placed at $\boldsymbol{x}'$ is the same as the potential at $\boldsymbol{x}'$ for a point charge placed at $\boldsymbol{x}$. To show this, we use Green's theorem with $\phi_1(\boldsymbol{x}) = G_D(\boldsymbol{x}, \boldsymbol{x}_1)$ and $\phi_2(\boldsymbol{x}) = G_D(\boldsymbol{x}, \boldsymbol{x}_2)$. Since $\phi_1|_S = \phi_2|_S = 0$, the left side of eq. (2.106) vanishes. Since $\rho_1 = \delta(\boldsymbol{x} - \boldsymbol{x}_1)$ and $\rho_2 = \delta(\boldsymbol{x} - \boldsymbol{x}_2)$, the vanishing of the right side immediately yields $\phi_1(\boldsymbol{x}_2) = \phi_2(\boldsymbol{x}_1)$, or

$$G_D(\boldsymbol{x}_2, \boldsymbol{x}_1) = G_D(\boldsymbol{x}_1, \boldsymbol{x}_2), \tag{2.110}$$

as we desired to show.

Side Comment on the Symmetry of the Green's Function: There is an important underlying reason that the equality $G_D(\boldsymbol{x}, \boldsymbol{x}') = G_D(\boldsymbol{x}', \boldsymbol{x})$ holds. Consider the Hilbert space $\mathcal{H}_{\mathcal{V}}$, of square integrable functions on $\mathcal{V}$, that is, the vector space of functions $F : \mathcal{V} \to \mathbb{C}$ with inner product

$$\langle F_1, F_2 \rangle = \int_{\mathcal{V}} F_1^*(\boldsymbol{x})F_2(\boldsymbol{x})d^3x. \tag{2.111}$$

(This is the same sort of inner product as considered when defining the Hilbert space $\mathcal{H}$ in section 2.4—see eq. (2.58)—except that now the integral is taken over $\mathcal{V}$ instead of the unit sphere.) We can view ∇^2 as a linear map from $\mathcal{H}_{\mathcal{V}}$ to $\mathcal{H}_{\mathcal{V}}$, initially defined on all smooth functions in $\mathcal{H}_{\mathcal{V}}$. However, ∇^2 fails to be self-adjoint, because by Green's theorem, eq. (2.106), we have

$$\langle F_1, \nabla^2 F_2 \rangle - \langle \nabla^2 F_1, F_2 \rangle = \int_S \hat{\boldsymbol{n}} \cdot (F_1^* \nabla F_2 - F_2 \nabla F_1^*)dS. \tag{2.112}$$

We can cure this problem by *restricting* the domain of definition of ∇^2. One way of doing this is to restrict its domain so that it acts only on smooth functions that vanish on S, in which case the right side of eq. (2.112) vanishes. The domain of definition can then be extended to suitably regular functions that vanish on S in such a way that the resulting "Dirichlet Laplacian operator" ∇_D^2 is self-adjoint (see the side comment of section 2.4). The subscript "D" is needed here to distinguish ∇_D^2 from other notions of the Laplacian operator on $\mathcal{H}$ that would result from using different boundary conditions to restrict the domain of definition of ∇^2.

Now, let $\mathcal{F}(\boldsymbol{x},\boldsymbol{x}')$ be any integrable function on $\mathcal{V}\times\mathcal{V}$. Then $\mathcal{F}$ defines a linear map $\hat{\mathcal{F}}:\mathcal{H}_{\mathcal{V}}\to\mathcal{H}_{\mathcal{V}}$ via

$$(\hat{\mathcal{F}}F)(\boldsymbol{x})=\int_{\mathcal{V}}\mathcal{F}(\boldsymbol{x},\boldsymbol{x}')F(\boldsymbol{x}')d^3x'. \tag{2.113}$$

Self-adjointness of such an "integral operator" $\hat{\mathcal{F}}$ corresponds to the symmetry property $\mathcal{F}(\boldsymbol{x}',\boldsymbol{x})=\mathcal{F}^*(\boldsymbol{x},\boldsymbol{x}')$ of $\mathcal{F}$. Let $\hat{G}_D$ be the operator on $\mathcal{H}_{\mathcal{V}}$ corresponding to $G_D(\boldsymbol{x},\boldsymbol{x}')$. The delta function $\delta(\boldsymbol{x}-\boldsymbol{x}')$ also corresponds to an operator on $\mathcal{H}_{\mathcal{V}}$ via eq. (2.113), and it is easily seen that $\hat{\delta}=I$, where I denotes the identity operator on $\mathcal{H}_{\mathcal{V}}$.

Putting the above remarks together, we see that eq. (2.97) corresponds to the following operator relation on $\mathcal{H}_{\mathcal{V}}$:

$$\nabla_D^2\circ\hat{G}_D=-I/\epsilon_0, \tag{2.114}$$

where "$\circ$" denotes the composition of operators. Thus, apart from the factor of $-1/\epsilon_0$, the Dirichlet Green's function is just the inverse of the Dirichlet Laplacian operator. Since the Dirichlet Laplacian operator is self-adjoint, so is $\hat{G}_D$. Since ∇_D^2 is real (i.e., it commutes with complex conjugation), so is $\hat{G}_D$. Thus, the symmetry property $G_D(\boldsymbol{x},\boldsymbol{x}')=G_D(\boldsymbol{x}',\boldsymbol{x})$ can be viewed as a direct consequence of the self-adjointness (and reality) of ∇_D^2.

Similar results hold for the case of a cavity where Neumann conditions are imposed on the boundary as in case (ii) of the existence and uniqueness theorem given earlier in this section. Although this case does not arise as naturally in electrostatics as the Dirichlet case, it arises in many other contexts in physics (such as fluid flow problems), so it is certainly worthy of mention here. The Neumann Green's function $G_N(\boldsymbol{x},\boldsymbol{x}')$ is defined as the solution to

$$\nabla^2G_N(\boldsymbol{x},\boldsymbol{x}')=-\frac{1}{\epsilon_0}\delta(\boldsymbol{x}-\boldsymbol{x}') \tag{2.115}$$

in $\mathcal{V}$, subject to the boundary condition

$$(\hat{\boldsymbol{n}}\cdot\nabla_x)G_N(\boldsymbol{x},\boldsymbol{x}')|_{x\in S}=-\frac{1}{\epsilon_0 A}, \tag{2.116}$$

where A denotes the area of S, and with the arbitrary "constant" (i.e., function of $\boldsymbol{x}'$) in $G_N(\boldsymbol{x},\boldsymbol{x}')$ fixed by the requirement that

$$\int_S G_N(\boldsymbol{x},\boldsymbol{x}')dS=0. \tag{2.117}$$

By the same arguments as in the Dirichlet case, if we are given $G_N(\boldsymbol{x},\boldsymbol{x}')$, then the solution to eq. (2.95) in $\mathcal{V}$ with the boundary conditions of case (ii) of the above existence and uniqueness theorem is

$$\phi(\boldsymbol{x}) = \int_{\mathcal{V}} G_N(\boldsymbol{x}, \boldsymbol{x}')\rho(\boldsymbol{x}')d^3x' + \epsilon_0 \int_S \chi(\boldsymbol{x}')G_N(\boldsymbol{x}', \boldsymbol{x})dS(\boldsymbol{x}'), \tag{2.118}$$

which is unique up to the addition of a constant.[11] As in the Dirichlet case, we have $G_N(\boldsymbol{x}, \boldsymbol{x}') = G_N(\boldsymbol{x}', \boldsymbol{x})$.

Finally, let us discuss the corresponding "exterior problems." Instead of considering a cavity $\mathcal{V}$, suppose we consider all of space with $\mathcal{V}$ *removed*. In other words, let S again be the surface of a bounded region $\mathcal{V}$, but we now consider the region exterior to S. We now may allow $\mathcal{V}$ to consist of a finite number of disconnected components (e.g., we may consider the region exterior to a finite number of conducting bodies). (There was no point allowing $\mathcal{V}$ to be disconnected for the interior problem, since we could then consider each connected component separately.) The above theorem on existence and uniqueness in $\mathcal{V}$ then holds for existence and uniqueness *outside* $\mathcal{V}$ with only the following modifications: (a) We require $\rho(\boldsymbol{x}) \to 0$ sufficiently rapidly as $|\boldsymbol{x}| \to \infty$. (b) The asymptotic condition $\phi(\boldsymbol{x}) \to 0$ as $|\boldsymbol{x}| \to \infty$ must be added. (c) For the Neumann condition, we need not impose any restriction analogous to eq. (2.96) on χ (because there is no longer a potential inconsistency with Gauss's theorem, since there can be a boundary contribution from infinity) and the solution is unique (because the constant is fixed by the condition that $\phi(\boldsymbol{x}) \to 0$ as $|\boldsymbol{x}| \to \infty$).

For the exterior problem, the Dirichlet Green's function $G_D(\boldsymbol{x}, \boldsymbol{x}')$ is defined as the solution to eq. (2.97) with $G_D(\boldsymbol{x}, \boldsymbol{x}')|_{\boldsymbol{x}\in S} = 0$ and $G_D(\boldsymbol{x}, \boldsymbol{x}') \to 0$ as $|\boldsymbol{x}| \to \infty$. The solution ϕ to the exterior problem with charge density ρ and $\phi|_S = \psi$ is again given by eq. (2.108) except that (a) the volume integral is now, of course, over $\mathbb{R}^3 \setminus \mathcal{V}$ rather than $\mathcal{V}$, and (b) the unit normal given by Gauss's theorem now points into $\mathcal{V}$ rather than away from $\mathcal{V}$. Thus, if one wishes to use the outward-pointing normal, $\hat{\boldsymbol{n}}$, to S, as is conventional, one must reverse the sign of the surface term in eq. (2.108).

For the exterior problem, the Neumann Green's function $G_N(\boldsymbol{x}, \boldsymbol{x}')$ is defined as the solution to eq. (2.115) with $[\hat{\boldsymbol{n}} \cdot \nabla_x G_N(\boldsymbol{x}, \boldsymbol{x}')]|_{\boldsymbol{x}\in S} = 0$ (rather than eq. (2.116)) and $G_N(\boldsymbol{x}, \boldsymbol{x}') \to 0$ as $|\boldsymbol{x}| \to \infty$ (rather than eq. (2.117)). The solution, ϕ, to the exterior problem with charge density ρ and $[\hat{\boldsymbol{n}} \cdot \nabla\phi]|_S = \chi$ is again given by eq. (2.118) with the volume integral now being over $\mathbb{R}^3 \setminus \mathcal{V}$ and with a sign reversal of the surface term if $\hat{\boldsymbol{n}}$ is taken to be outward pointing.

Problems

1. Let $\mathcal{V}$ be a bounded region of space, and let ϕ be an electrostatic potential that is source free in this region, so that $\nabla^2\phi = 0$ throughout $\mathcal{V}$. Suppose that for all $\boldsymbol{x}$ lying on the boundary $S = \partial\mathcal{V}$, we have $\phi(\boldsymbol{x}) = -f(\boldsymbol{x})\hat{\boldsymbol{n}} \cdot \nabla\phi(\boldsymbol{x})$, where f is a nonnegative function ($f(\boldsymbol{x}) \geq 0$), and $\hat{\boldsymbol{n}}$ is the outward-pointing normal. Show that $\phi = 0$ throughout $\mathcal{V}$.

2. The potential of an electrostatic dipole of dipole moment $\boldsymbol{p}$ located at $\boldsymbol{x}'$ is given by

$$\phi(\boldsymbol{x}) = \frac{1}{4\pi\epsilon_0} \frac{\boldsymbol{p} \cdot (\boldsymbol{x} - \boldsymbol{x}')}{|\boldsymbol{x} - \boldsymbol{x}'|^3}.$$

Suppose a dipole $\boldsymbol{p}_1$ is located at $\boldsymbol{x}_1$ and another dipole $\boldsymbol{p}_2$ is located at $\boldsymbol{x}_2$.

[11] The constant has been fixed in eq. (2.118) by the requirement that $\int_S \phi \, dS = 0$.

(a) What is the electrostatic force on the second dipole?
(b) What is the electrostatic interaction energy of the two dipoles?

3. (a) Consider the following classical model of a proton: The proton is taken to be a uniformly charged ball of radius $R = 10^{-15}$ m and total charge $e = 1.6 \times 10^{-19}$ Coulombs. What is the electromagnetic (self-)energy of the proton? Compare this with the mass of the proton; that is in this model, what fraction of the mass of the proton consists of electromagnetic energy?
(b) Consider the following classical model of a hydrogen atom: The electron is taken to have charge density $\rho = -e|\psi|^2$, where ψ is the ground state wave function, $\psi(r) = \exp(-r/a)/\sqrt{\pi a^3}$, with $a = \hbar^2/me^2 = 5.3 \times 10^{-11}$m. The proton is as in part (a), but since $R \ll a$, you may treat it as a point charge located at the origin. What is the electromagnetic interaction energy of the proton and electron? Compare this with the ground state energy of hydrogen. (Note: The electromagnetic interaction energy should not be equal to the ground state energy, because the electron also has "kinetic energy." However, there is a very simple "virial theorem" relationship between them.)

4. Consider the charge density $\rho(\boldsymbol{x})$ given by

$$\rho(\boldsymbol{x}) = \alpha(R - r)[1 - \cos(\theta)]^2 \quad \text{for } |\boldsymbol{x}| \leq R,$$

and $\rho(\boldsymbol{x}) = 0$ for $|\boldsymbol{x}| \geq R$, where α is a constant. Find the electrostatic potential $\phi(\boldsymbol{x})$ of this charge distribution at all $\boldsymbol{x}$.

5. Let $\mathcal{V}$ be an arbitrary bounded region of space, and suppose that a total charge Q is to be distributed in $\mathcal{V}$ an arbitrary way, with $\rho = 0$ outside $\mathcal{V}$. Show that the total energy is minimized if the charge is distributed the way that it would be if $\mathcal{V}$ were a conductor, so that $\phi = \text{const}$ within $\mathcal{V}$ (and thus, in particular, all of the charge lies on the boundary of $\mathcal{V}$). [Hint: Let $\phi_0(\boldsymbol{x})$ be the potential one would obtain if $\mathcal{V}$ were filled by a conducting body. Consider the energy of $\phi_0 + \phi'$, where the source ρ' of ϕ' vanishes outside $\mathcal{V}$ and has no net charge within $\mathcal{V}$.]

6. Charge is distributed on a (nonconducting) sphere of radius R, that is, the charge density throughout space is of the form $\rho(\boldsymbol{x}) = \sigma(\theta, \varphi)\delta(r - R)$. The surface charge distribution σ on the sphere is chosen in such a way that the electrostatic potential on this sphere is $\phi(r = R, \theta, \varphi) = \alpha \cos\theta$, where α is a constant.

(a) Find the electrostatic potential $\phi(\boldsymbol{x})$ at all $r \leq R$.
(b) Find the electrostatic potential $\phi(\boldsymbol{x})$ at all $r \geq R$.
(c) Find the surface charge density $\sigma(\theta, \varphi)$ that is required to produce this potential ϕ.
(d) Find the total electrostatic energy.

7. An uncharged conducting sphere of radius R is placed in an otherwise uniform electric field $\boldsymbol{E}_0 = C\hat{\boldsymbol{z}}$.

(a) Find the surface charge density, $\sigma(\theta, \varphi)$, on the conductor and the electric field, $\boldsymbol{E}_1$, due to this surface charge density.
(b) If $\boldsymbol{E}_0$ were produced by a finite charge distribution (e.g., by capacitor plates of finite size) so that it is uniform only in the vicinity of the conductor and goes to zero at infinity, then the interaction energy between $\boldsymbol{E}_0$ and $\boldsymbol{E}_1$ would be given by eq. (2.30). Calculate this interaction energy.
(c) If $\boldsymbol{E}_0$ extends to infinity as a uniform electric field, then the dropping of the boundary term in the derivation of eq. (2.25) would not be justified, and one would have to use eq. (2.29) for the interaction energy. Calculate the interaction energy in this case.

8. A point charge of charge q is placed at point $\boldsymbol{x}'$ inside a conducting spherical shell of radius R. (Treat the shell as a perfect conductor of very small but finite thickness.) The shell is held at vanishing potential, so the potential inside the shell is given by $qG_D(\boldsymbol{x}, \boldsymbol{x}')$, where the explicit formula for $G_D(\boldsymbol{x}, \boldsymbol{x}')$ is given by eq. (2.99).

   (a) Find the surface charge density $\sigma(\theta, \varphi)$ on the inner surface of the conducting shell.
   (b) Find the force $\boldsymbol{F}$ that must be exerted on the point charge to hold it in place.

9. (a) Prove the *mean value theorem*: For any solution ϕ to $\nabla^2\phi = 0$, the value of ϕ at $\boldsymbol{x}$ is equal to the average value of ϕ on a sphere of radius R (for any R) centered at $\boldsymbol{x}$. [Hint: Apply Green's theorem to ϕ and $1/(4\pi\epsilon_0|\boldsymbol{x} - \boldsymbol{x}'|)$ for a suitable choice of region and a suitable choice of $\boldsymbol{x}'$.]
   (b) Use the result in part (a) to show that a point charge can never be in stable equilibrium if placed in an electric field $\boldsymbol{E}$ that is source free in a neighborhood of the charge—and, indeed, it can be in neutral equilibrium only if $\boldsymbol{E} = 0$ in this neighborhood.

10. Use the mean value theorem (see problem 9) to show that if ϕ satisfies $\nabla^2\phi = 0$ in a region $\mathcal{V}$, then ϕ must achieve its maximum and minimum values on the boundary S of $\mathcal{V}$. (This result is referred to as the *maximum principle.*) Use this result to show that the Dirichlet Green's function satisfies

$$0 \le G_D(\boldsymbol{x}, \boldsymbol{x}') \le \frac{1}{4\pi\epsilon_0} \frac{1}{|\boldsymbol{x} - \boldsymbol{x}'|}.$$

[Hint: For the upper bound, apply the maximum principle to F_D, eq. (2.100). For the lower bound, apply the maximum principle to G_D in the region outside a ball of radius ϵ about $\boldsymbol{x}'$.]

11. Consider a collection of N arbitrary, disjoint conductors. Suppose that the jth conductor is held at unit potential, and the other conductors are held at vanishing potential. Let ϕ_j denote the unique solution to Laplace's equation in the region exterior to the conductors with these boundary conditions and with

$\phi_j \to 0$ as $|\boldsymbol{x}| \to \infty$. Let C_{ij} denote the total charge on the ith conductor in the solution ϕ_j.

(a) Show that if all the conductors are now held at arbitrary potentials V_j, the charge on the ith conductor will be given by

$$Q_i = \sum_{j=1}^{N} C_{ij} V_j \,.$$

We refer to C_{ij} as the *capacitance matrix* of the system of conductors.

(b) Show that if the conductors are held at potentials V_j, the total electromagnetic energy is

$$\mathscr{E} = \frac{1}{2} \sum_{i,j} C_{ij} V_i V_j \,.$$

(c) Prove that the capacitance matrix is symmetric: $C_{ij} = C_{ji}$. [Hint: Apply Green's theorem to ϕ_i and ϕ_j.]

(d) Show that C_{ij} satisfies (i) $C_{jj} \geq 0$, (ii) $C_{ij} \leq 0$ if $i \neq j$, and (iii) $\sum_j C_{ij} \geq 0$. [Hint: For (i) and (ii), use the maximum principle (see problem 10) applied to ϕ_j to argue that the surface charge density of the jth conductor cannot be negative and the surface charge density of the ith conductor cannot be positive for $i \neq j$. Part (iii) can be shown by a similar argument.]

CHAPTER 3

Dielectrics

A *dielectric* is a material composed of microscopic constituents that have—or are capable of having—a nonvanishing electric dipole moment. For linguistic simplicity, I refer to these microscopic constituents as "atoms," but they need not literally be atoms (e.g., they may be molecules). The electric dipole moment of each atom may arise because its ground state charge distribution has a permanent electric dipole moment (as occurs, e.g., for a water molecule) or because an electric field external to the atom is present, which polarizes the charge distribution of the atom.

All information about the electromagnetic fields of dielectrics in electrostatics can be obtained by solving Poisson's equation, eq. (2.3), with ρ given by the exact charge distribution of all the atoms composing the dielectric. However, if one is interested in the macroscopic behavior of a dielectric, this detailed, microscopic description would not be any more useful than attempting to determine the macroscopic properties of a gas by following the motion of each of the individual particles. For a gas, we can give a very useful macroscopic description by averaging over the behavior of many particles. Similarly, for a dielectric, we seek to obtain a macroscopic description of the averaged properties of the electromagnetic field by performing spatial averages over many atoms. To do so, it is very important to keep in mind the following 3 distinct scales:

$$d = \text{size of the atoms},$$

$$L = \text{scale over which the averaging is done},$$

$$R = \text{size of the dielectric body}.$$

To give a macroscopic description of the averaged field properties of a dielectric, it is essential that

$$d \ll L \ll R. \tag{3.1}$$

3.1 Macroscopic Description of Dielectrics

To give a macroscopic description of dielectrics, we will perform averages of microscopically defined quantities, Ψ—such as ρ and ϕ—over the scale L. One way to define the average of Ψ at $\boldsymbol{x}$ would be to average it over a ball, $B_L(\boldsymbol{x})$, of radius L centered at $\boldsymbol{x}$; that is, we could attempt to define the average of Ψ at $\boldsymbol{x}$ to be $\int_{B_L} \Psi(\boldsymbol{x}')d^3x'/(\frac{4}{3}\pi L^3)$. However, the sharp boundaries of $B_L(\boldsymbol{x})$ would make this average vary rapidly with $\boldsymbol{x}$ when individual atoms cross the boundary of $B_L(\boldsymbol{x})$. It is much better to perform the

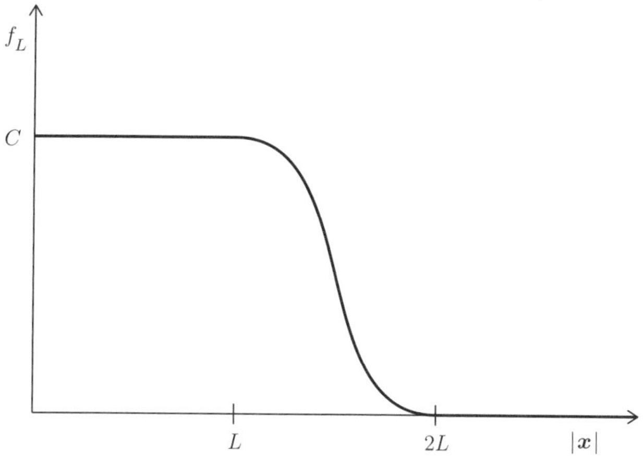

Figure 3.1. A sketch of the function f_L used to define averaging.

average in a smoother way, so that the effects of an atom are fully counted within distance L of $\boldsymbol{x}$, partially counted between L and $2L$, and not counted at all at distances greater than $2L$. To make this precise, let $f_L(\boldsymbol{x})$ be a smooth, positive, spherically symmetric function with the following properties: (i) We have $f_L(\boldsymbol{x}) = C = \text{const} > 0$ for $|\boldsymbol{x}| \leq L$. (ii) We have $f_L(\boldsymbol{x}) = 0$ for $|\boldsymbol{x}| \geq 2L$. (iii) For $L < |\boldsymbol{x}| < 2L$, $f_L(\boldsymbol{x})$ monotonically interpolates between C and 0 in a sufficiently smooth way that nth derivatives of f_L never get much bigger than $2^n C/L^n$ anywhere. (This condition is imposed to ensure that f_L does not have any unnecessary "wiggles" or "sharp turns," so that all of its derivatives are near the minimum size needed to interpolate between C and 0.) (iv) We have $\int f_L(\boldsymbol{x}) d^3x = 1$. A sketch of a function f_L satisfying these conditions is shown in figure 3.1.

Let $\Psi(\boldsymbol{x})$ be any microscopically defined quantity, such as the exact charge density, $\rho(\boldsymbol{x})$, or the exact potential, $\phi(\boldsymbol{x})$. We define the average, $\langle\Psi\rangle$, of Ψ over the scale L by

$$\langle\Psi\rangle(\boldsymbol{x}) \equiv \int \Psi(\boldsymbol{x}') f_L(\boldsymbol{x} - \boldsymbol{x}') d^3x', \tag{3.2}$$

where the integral is taken over all space (but, of course, $f_L(\boldsymbol{x} - \boldsymbol{x}') = 0$ when $|\boldsymbol{x} - \boldsymbol{x}'| > 2L$, so the integrand is nonvanishing only in a bounded region). An important property of this averaging procedure is that it commutes with differentiation, as can be seen from the following computation:

$$\begin{aligned}
\frac{\partial\langle\Psi\rangle}{\partial x_i}(\boldsymbol{x}) &= \int \Psi(\boldsymbol{x}') \frac{\partial}{\partial x_i} f_L(\boldsymbol{x} - \boldsymbol{x}') d^3x' \\
&= -\int \Psi(\boldsymbol{x}') \frac{\partial}{\partial x_i'} f_L(\boldsymbol{x} - \boldsymbol{x}') d^3x' \\
&= \int \left[\frac{\partial\Psi}{\partial x_i'}(\boldsymbol{x}')\right] f_L(\boldsymbol{x} - \boldsymbol{x}') d^3x' \\
&= \left\langle \frac{\partial\Psi}{\partial x_i} \right\rangle (\boldsymbol{x}).
\end{aligned} \tag{3.3}$$

Here, in the second line, we changed the x_i-derivative of f_L to an x'_i-derivative (at the price of a minus sign), because f_L is a function of $\boldsymbol{x}-\boldsymbol{x}'$. In the third line, we integrated by parts; no boundary term appeared, because f_L is nonvanishing only in a bounded region.

Taking the average of both sides of the exact Poisson's equation (2.3) and commuting $\langle\rangle$ and ∇^2, we obtain

$$\nabla^2\langle\phi\rangle(\boldsymbol{x}) = -\frac{1}{\epsilon_0}\langle\rho\rangle(\boldsymbol{x}), \tag{3.4}$$

so Poisson's equation continues to hold for the macroscopically averaged quantities over scale L. To proceed further, we must make more precise our assumption about the atomic nature of the exact charge density ρ. We assume that the exact charge density takes the form

$$\rho(\boldsymbol{x}) = \sum_j \rho_j(\boldsymbol{x}), \tag{3.5}$$

where each ρ_j is localized on a scale $d \ll L$ about the point $\boldsymbol{x}_j$. It is not necessary to assume that each ρ_j takes the same functional form (i.e., the material may be composed of different types of "atoms"). Although we are primarily interested in the case where the total charge, $q_j = \int \rho_j$, of each atom vanishes, there is no harm in allowing some of the atoms to have a nonvanishing q_j as would occur if excess charge were placed in the dielectric. Such excess charge in a dielectric is normally referred to as "free charge," although this terminology[1] is misleading, because this excess charge is *not* assumed to be "free" to move about the dielectric, as it would be in a conductor.

Let us now consider the contribution of an individual ρ_j to $\langle\rho\rangle$,

$$\langle\rho_j\rangle(\boldsymbol{x}) = \int \rho_j(\boldsymbol{x}')f_L(\boldsymbol{x}-\boldsymbol{x}')d^3x'. \tag{3.6}$$

We Taylor expand $f_L(\boldsymbol{x}-\boldsymbol{x}')$ about $\boldsymbol{x}'=\boldsymbol{x}_j$:

$$f_L(\boldsymbol{x}-\boldsymbol{x}') = f_L(\boldsymbol{x}-\boldsymbol{x}_j) + (\boldsymbol{x}'-\boldsymbol{x}_j)\cdot\left[\nabla_{x'}f_L(\boldsymbol{x}-\boldsymbol{x}')\right]\Big|_{\boldsymbol{x}'=\boldsymbol{x}_j} + \cdots. \tag{3.7}$$

However, we have

$$\left[\nabla_{x'}f_L(\boldsymbol{x}-\boldsymbol{x}')\right]\Big|_{\boldsymbol{x}'=\boldsymbol{x}_j} = -\left[\nabla_x f_L(\boldsymbol{x}-\boldsymbol{x}')\right]\Big|_{\boldsymbol{x}'=\boldsymbol{x}_j} = -\nabla_x f_L(\boldsymbol{x}-\boldsymbol{x}_j). \tag{3.8}$$

Substituting eq. (3.7) and eq. (3.8) into eq. (3.6), we obtain

$$\langle\rho_j\rangle(\boldsymbol{x}) = q_j f_L(\boldsymbol{x}-\boldsymbol{x}_j) - \boldsymbol{p}_j\cdot\nabla_x f_L(\boldsymbol{x}-\boldsymbol{x}_j) + \cdots, \tag{3.9}$$

where q_j is the charge of the jth atom, and $\boldsymbol{p}_j$ is its electric dipole moment about $\boldsymbol{x}_j$. The terms in eq. (3.9) represented by "$\cdots$" arise from higher order terms in the Taylor expansion of f_L. These terms will be successively smaller by factors of d/L as a

[1] See the paragraph below eq. (6.7) in chapter 6 for a discussion of how the terminology of "free charges" and "free currents" is used in electrodynamics.

consequence of our assumption that ρ_j is nonvanishing only over the atomic scale d, whereas differentiation of f_L yields factors of order $1/L$. Since we assume that $d \ll L$, we shall neglect these terms. Thus, we obtain

$$\langle\rho\rangle(\boldsymbol{x}) = \sum_j \langle\rho_j\rangle(\boldsymbol{x}) = \sum_j \left[q_j f_L(\boldsymbol{x}-\boldsymbol{x}_j) - \boldsymbol{p}_j \cdot \nabla_x f_L(\boldsymbol{x}-\boldsymbol{x}_j)\right]. \tag{3.10}$$

We define the (microscopic) "free charge density," ρ_f, of the dielectric by

$$\rho_f(\boldsymbol{x}) = \sum_j q_j \delta(\boldsymbol{x}-\boldsymbol{x}_j) \tag{3.11}$$

and the (microscopic) "dipole moment density," or *polarization* $\boldsymbol{P}$, by

$$\boldsymbol{P}(\boldsymbol{x}) = \sum_j \boldsymbol{p}_j \delta(\boldsymbol{x}-\boldsymbol{x}_j), \tag{3.12}$$

that is, we assign the charge q_j and dipole moment $\boldsymbol{p}_j$ of the jth atom to the site of each atom. Equation (3.10) then can be written as

$$\begin{aligned}
\langle\rho\rangle(\boldsymbol{x}) &= \int \left[\rho_f(\boldsymbol{x}')f_L(\boldsymbol{x}-\boldsymbol{x}') - \boldsymbol{P}(\boldsymbol{x}') \cdot \nabla_x f_L(\boldsymbol{x}-\boldsymbol{x}')\right] d^3x' \\
&= \int \rho_f(\boldsymbol{x}')f_L(\boldsymbol{x}-\boldsymbol{x}')d^3x' - \nabla \cdot \int \boldsymbol{P}(\boldsymbol{x}')f_L(\boldsymbol{x}-\boldsymbol{x}')d^3x' \\
&= \langle\rho_f\rangle - \nabla \cdot \langle\boldsymbol{P}\rangle,
\end{aligned} \tag{3.13}$$

where $\langle\rho_f\rangle$ and $\langle\boldsymbol{P}\rangle$ are the macroscopic averages of eq. (3.11) and eq. (3.12), respectively. Thus, the macroscopic average of the charge density ρ gets an expected contribution from the average of the "free charge density," but it also can get a contribution from the average of the dipole moment density if that average is spatially varying. The quantity $-\nabla \cdot \langle\boldsymbol{P}\rangle$ is called the *polarization charge density*.

Thus, Poisson's equation (3.4) for the macroscopically averaged potential in terms of the averages of the free charge density and dipole moment density is

$$\nabla^2\langle\phi\rangle = -\frac{1}{\epsilon_0}\langle\rho_f\rangle + \frac{1}{\epsilon_0}\nabla \cdot \langle\boldsymbol{P}\rangle. \tag{3.14}$$

This equation is the entire content of Maxwell's equations for dielectrics in electrostatics. We can rewrite this equation in a form that should look very familiar to most readers as follows. We have $\langle\boldsymbol{E}\rangle = -\nabla\langle\phi\rangle$, since, as shown above, averaging commutes with differentiation. We can work with $\langle\boldsymbol{E}\rangle$ instead of $\langle\phi\rangle$ and replace $\langle\boldsymbol{E}\rangle = -\nabla\langle\phi\rangle$ with

$$\nabla \times \langle\boldsymbol{E}\rangle = 0. \tag{3.15}$$

It is then conventional to define[2]

$$\langle\boldsymbol{D}\rangle = \epsilon_0\langle\boldsymbol{E}\rangle + \langle\boldsymbol{P}\rangle. \tag{3.16}$$

[2] We could define the "microscopic quantity" $\boldsymbol{D} = \epsilon_0\boldsymbol{E} + \boldsymbol{P}$—although this is never done! Then $\langle\boldsymbol{D}\rangle$ would be the macroscopic average of $\boldsymbol{D}$.

Equation (3.14) then becomes

$$\nabla \cdot \langle \boldsymbol{D} \rangle = \langle \rho_f \rangle. \tag{3.17}$$

Equation (3.15)–(3.17) are the usual form of Maxwell's equations for dielectrics in electrostatics.

If we wish to solve eq. (3.14) (or equivalently, eqs. (3.15)–(3.17)), we must know more about $\langle \rho_f \rangle$ and $\langle \boldsymbol{P} \rangle$. The free charge $\langle \rho_f \rangle$ would normally be taken to be independently specified—it is normally taken to be charge externally inserted into the dielectric. However, $\langle \boldsymbol{P} \rangle$ is the average of the dipole moment of the microscopic constituents and is not independently specified. In most dielectric materials, $\langle \boldsymbol{P} \rangle$ would be uniquely determined by $\langle \boldsymbol{E} \rangle$. In the case where the microscopic constituents have permanent dipole moments, $\langle \boldsymbol{E} \rangle$ would govern the amount and direction of their coherent orientation; if the microscopic constituents have dipole moments arising from polarization, $\langle \boldsymbol{E} \rangle$ would govern the amount and direction of this polarization. A simple situation—which is a good approximation in many materials if $\langle \boldsymbol{E} \rangle$ is not too large—is that of a "linear medium" in which $\langle \boldsymbol{P} \rangle$ is linearly related to $\langle \boldsymbol{E} \rangle$; that is, there is a symmetric matrix $\chi_{ik}(\boldsymbol{x}) = \chi_{ki}(\boldsymbol{x})$, called the *electric susceptibility tensor*, such that

$$\langle P_i \rangle (\boldsymbol{x}) = \epsilon_0 \sum_k \chi_{ik}(\boldsymbol{x}) \langle E_k \rangle (\boldsymbol{x}). \tag{3.18}$$

An even simpler situation is one where the material is homogeneous (so that $\chi_{ik}(\boldsymbol{x})$ is independent of $\boldsymbol{x}$) and isotropic (so that $\chi_{ik} = \chi \delta_{ik}$), in which case

$$\langle \boldsymbol{P} \rangle = \epsilon_0 \chi \langle \boldsymbol{E} \rangle. \tag{3.19}$$

It follows immediately that

$$\langle \boldsymbol{D} \rangle = \epsilon \langle \boldsymbol{E} \rangle, \tag{3.20}$$

where the *electric permittivity*, ϵ, is given by

$$\epsilon = \epsilon_0 (1 + \chi). \tag{3.21}$$

The *dielectric constant* (also referred to as the *relative electric permittivity*) of the medium is defined to be

$$\kappa \equiv \frac{\epsilon}{\epsilon_0} = 1 + \chi. \tag{3.22}$$

In this case, inside a dielectric medium, Poisson's equation (3.14) becomes

$$\nabla^2 \langle \phi \rangle = -\frac{1}{\epsilon} \langle \rho_f \rangle. \tag{3.23}$$

In particular, in an infinite, linear, homogeneous, isotropic dielectric medium, the macroscopically averaged Maxwell equations take exactly the same form as in vacuum, except that the ("free") charge density is "renormalized" by the factor ϵ_0/ϵ. All of the results of chapter 2 apply with this renormalization.

In the more realistic case of a dielectric medium of finite size—assumed to be linear, homogeneous, and isotropic—we must solve eq. (3.23) inside the dielectric and eq. (2.3) outside the dielectric.[3] We will obtain a solution to Maxwell's equations with a dielectric of finite size present if we have interior and exterior solutions that match properly at the boundary of the dielectric. The matching conditions can be determined as follows. First, we assume that there is no "surface layer" of "free charge" (i.e., there is no surface contribution[4] to $\langle \rho_f \rangle$). However, $\langle \boldsymbol{P} \rangle$ may now be discontinuous[5] across the boundary of the dielectric; that is, $\langle \boldsymbol{P} \rangle$ may have a finite, nonzero value inside the dielectric and jump discontinuously to 0 outside the dielectric. Thus, derivatives of $\langle \boldsymbol{P} \rangle$ can give rise to a delta function and consequently, to an effective surface charge density of the polarization charge density $-\nabla \cdot \langle \boldsymbol{P} \rangle$. Nevertheless, as in the case of conductors, $\langle \phi \rangle$ must be continuous across the boundary—and hence $\langle \boldsymbol{E} \rangle_{\parallel}$ must be continuous across the boundary. Furthermore, since we assume that there is no free surface charge density, by eq. (3.17), $\hat{\boldsymbol{n}} \cdot \langle \boldsymbol{D} \rangle$ must be continuous across the boundary (where, of course, $\langle \boldsymbol{D} \rangle = \epsilon_0 \boldsymbol{E}$ outside the dielectric). Thus, the appropriate matching conditions at the boundary of a dielectric are continuity of $\langle \phi \rangle$ (and hence, $\langle \boldsymbol{E} \rangle_{\parallel}$) and $\hat{\boldsymbol{n}} \cdot \langle \boldsymbol{D} \rangle$.

3.2 Force and Interaction Energy

Consider a dielectric that is placed in an external field ϕ^{ext}, as in section 2.3. Thus, ϕ^{ext} is a solution to Poisson's equation (2.3) whose source ρ^{ext} consists of "distant charges" (i.e., $\rho^{\text{ext}} = 0$ near the dielectric). We will treat ρ^{ext} and ϕ^{ext} as given, fixed quantities throughout this section (i.e., we will not consider varying them). We assume that within the dielectric, ϕ^{ext} is slowly varying over the averaging scale L used to define macroscopic quantities in the dielectric.

By eq. (2.42), the total force on the dielectric is

$$\boldsymbol{F} = \int \rho(\boldsymbol{x}) \boldsymbol{E}^{\text{ext}}(\boldsymbol{x}) d^3x. \tag{3.24}$$

Since ϕ^{ext} is slowly varying over the scale L, we will make a negligible error if we replace ρ by $\langle \rho \rangle$ in eq. (3.24). We thereby obtain

$$\begin{aligned} \boldsymbol{F} &= \int \langle \rho \rangle (\boldsymbol{x}) \boldsymbol{E}^{\text{ext}}(\boldsymbol{x}) d^3x \\ &= \int \left[\langle \rho_f \rangle - \nabla \cdot \langle \boldsymbol{P} \rangle \right] \boldsymbol{E}^{\text{ext}} \, d^3x, \end{aligned} \tag{3.25}$$

where eq. (3.13) was used. However, we have

$$(\nabla \cdot \langle \boldsymbol{P} \rangle)(E^{\text{ext}})_i = \sum_k (\partial_k \langle P_k \rangle)(E^{\text{ext}})_i = \sum_k \partial_k \left[\langle P_k \rangle (E^{\text{ext}})_i \right] - \sum_k \langle P_k \rangle \partial_k (E^{\text{ext}})_i. \tag{3.26}$$

[3] Outside the dielectric, there is no longer any rapid variation of electromagnetic quantities on the scale d, so we need not distinguish between $\langle \boldsymbol{E} \rangle$ and $\boldsymbol{E}$ in this case.

[4] If there were a surface contribution, this could be handled in the same way as for the surface of a conductor and would lead to a jump in $\hat{\boldsymbol{n}} \cdot \langle \boldsymbol{D} \rangle$.

[5] The averaged quantities defined by eq. (3.2) will always be smooth. What is meant here by "discontinuous" is a large change over the averaging scale L.

The integral over all space of the first term vanishes by Gauss's theorem, so we obtain our final expression

$$F = \int \left[\langle \rho_f \rangle E^{\rm ext} + (\langle P \rangle \cdot \nabla) E^{\rm ext}\right] d^3x. \tag{3.27}$$

In a sense, this formula is obvious: We have modeled the dielectric as being composed of "atoms" of charge q_j and electric dipole moment p_j, with the higher multipole moments being negligible. The total force on a dielectric in an external field therefore must be the sum of the external field forces on the charge and dipole of each atom, as given by eq. (2.44). This agrees with eq. (3.27). Note that no assumptions, such as $\langle P \rangle = \epsilon_0 \chi \langle E \rangle$, have been made in this derivation (i.e., eq. (3.27) holds for an arbitrary dielectric medium).

We also can rewrite this expression for F as follows. Let $\langle E^{\rm self} \rangle$ be the averaged self-field of the dielectric, that is, the field associated with its own averaged charge density $\langle \rho \rangle$:

$$\langle E^{\rm self} \rangle(x) = -\nabla \langle \phi^{\rm self} \rangle(x) = -\frac{1}{4\pi\epsilon_0} \nabla_x \int \frac{\langle \rho_f \rangle(x') - \nabla_{x'} \cdot \langle P \rangle(x')}{|x - x'|} d^3x'. \tag{3.28}$$

By the same argument that led to eq. (2.41) (replacing ρ by $\langle \rho \rangle$ and $E^{\rm self}$ by $\langle E^{\rm self} \rangle$), we have[6]

$$\int \langle \rho \rangle \langle E^{\rm self} \rangle d^3x = 0. \tag{3.29}$$

If we add the left side of eq. (3.29) to the right side of the first line of eq. (3.25), we obtain

$$F = \int \langle \rho \rangle \left[E^{\rm ext} + \langle E^{\rm self} \rangle\right] d^3x = \int \langle \rho \rangle \langle E \rangle d^3x. \tag{3.30}$$

Carrying through the same manipulations as led to eq. (3.27), we obtain the following equivalent expression for F:

$$F = \int \left[\langle \rho_f \rangle \langle E \rangle + (\langle P \rangle \cdot \nabla) \langle E \rangle\right] d^3x, \tag{3.31}$$

that is, we can replace $E^{\rm ext}$ by $\langle E \rangle$ in eq. (3.27). Again, this formula holds for an arbitrary dielectric medium (i.e., linearity has not been assumed).

Before considering the interaction energy of the dielectric, we should clarify what is meant by "energy" and what entities contribute to it. The energy density, momentum density, and stress properties of any form of matter (including electromagnetic fields) are defined by the coupling of the matter to gravity. As emphasized in section 1.2, these quantities are part of the specification of the theory. In the case at hand, in the presence of a dielectric, the total energy density is

$$\mathcal{E}^{\rm tot} = \mathcal{E}_{\rm M}^{\rm self} + \mathcal{E}_{\rm EM}, \tag{3.32}$$

[6]The true self-force, of course, is $\int \rho E^{\rm self} d^3x = 0$. Since $\langle \rho E^{\rm self} \rangle \neq \langle \rho \rangle \langle E^{\rm self} \rangle$, eq. (3.29) holds not because of the vanishing of the true self-force but because the same arguments used to prove the vanishing of true self-force yield eq. (3.29).

where $\mathcal{E}_{\mathrm{M}}^{\mathrm{self}}$ is the nonelectromagnetic energy density of the matter composing the dielectric, and $\mathcal{E}_{\mathrm{EM}}$ is the total electromagnetic energy density. If the dielectric is placed in an external field, then, as in eq. (2.28), the electromagnetic field energy decomposes as

$$\mathcal{E}_{\mathrm{EM}} = \mathcal{E}_{\mathrm{EM}}^{\mathrm{self}} + \mathcal{E}^{\mathrm{ext}} + \mathcal{E}^{\mathrm{int}}, \tag{3.33}$$

where $\mathcal{E}_{\mathrm{EM}}^{\mathrm{self}}$ is the electromagnetic self-energy density of the dielectric (i.e., the electromagnetic self-energy density of the atoms and the electromagnetic interaction energy density between the atoms), $\mathcal{E}^{\mathrm{ext}}$ is the electromagnetic energy density of the external field, and $\mathcal{E}^{\mathrm{int}}$ is the electromagnetic interaction energy density of the external field with the electromagnetic field of the dielectric.

The quantity $\mathcal{E}^{\mathrm{ext}}$ is fixed, and it has nothing to do with the dielectric, so it is not of any interest. The quantities $\mathcal{E}_{\mathrm{M}}^{\mathrm{self}}$ and $\mathcal{E}_{\mathrm{EM}}^{\mathrm{self}}$ and their macroscopic averages are not calculable without a detailed microscopic model of the dielectric. With regard to the calculation of $\mathcal{E}_{\mathrm{EM}}^{\mathrm{self}}$, it should be noted that for any microscopic quantity Ψ that varies significantly on scales $\ll L$, the quantity $\langle|\Psi|^2\rangle$ will be very different from the quantity $|\langle\Psi\rangle|^2$. Since $\mathcal{E}_{\mathrm{EM}}^{\mathrm{self}} \propto |\boldsymbol{E}^{\mathrm{self}}|^2$, it is clear that a knowledge of the average of linear quantities, such as $\boldsymbol{E}^{\mathrm{self}}$, tells us virtually nothing about the average of quadratic quantities, such as $\mathcal{E}_{\mathrm{EM}}^{\mathrm{self}}$.

However, the electromagnetic interaction energy of the external field and the dielectric is both of interest and calculable. By eq. (2.30), we have

$$\mathscr{E}^{\mathrm{int}} = \int \mathcal{E}^{\mathrm{int}} d^3x = \int \rho(\boldsymbol{x})\phi^{\mathrm{ext}}(\boldsymbol{x}) d^3x. \tag{3.34}$$

Again, since ϕ^{ext} is slowly varying over the scale L, we will make a negligible error if we replace ρ by $\langle\rho\rangle$. We thereby obtain

$$\begin{aligned}
\mathscr{E}^{\mathrm{int}} &= \int \langle\rho\rangle(\boldsymbol{x})\phi^{\mathrm{ext}}(\boldsymbol{x}) d^3x \\
&= \int \left[\langle\rho_f\rangle - \nabla\cdot\langle\boldsymbol{P}\rangle\right]\phi^{\mathrm{ext}}\, d^3x \\
&= \int \left[\langle\rho_f\rangle\phi^{\mathrm{ext}} - \nabla\cdot(\langle\boldsymbol{P}\rangle\phi^{\mathrm{ext}}) + \langle\boldsymbol{P}\rangle\cdot\nabla\phi^{\mathrm{ext}}\right] d^3x \\
&= \int \left[\langle\rho_f\rangle\phi^{\mathrm{ext}} - \langle\boldsymbol{P}\rangle\cdot\boldsymbol{E}^{\mathrm{ext}}\right] d^3x.
\end{aligned} \tag{3.35}$$

Again, this formula is obvious in view of our model of the dielectric as being composed of "atoms" of charge q_j and electric dipole moment $\boldsymbol{p}_j$. The total electromagnetic interaction energy must be the sum of the interaction energies of the individual "atoms," each of which is given by eq. (2.34). Again, this formula holds for an arbitrary dielectric medium.

Remarkably, we will now show that, although $\mathcal{E}_{\mathrm{M}}^{\mathrm{self}}$ and $\mathcal{E}_{\mathrm{EM}}^{\mathrm{self}}$ are not calculable without a microscopic model, for the case of a *linear* dielectric medium (i.e., a medium satisfying eq. (3.18)) it is possible to derive an expression for the *change* in $\mathcal{E}_{\mathrm{M}}^{\mathrm{self}} + \mathcal{E}_{\mathrm{EM}}^{\mathrm{self}}$ when the dielectric is moved to different locations in the external field ϕ^{ext}. For simplicity, in the following, let us assume that $\langle\rho_f\rangle = 0$ (i.e., that there is no "free charge") and that the linear medium is homogeneous and isotropic (so $\langle\boldsymbol{P}\rangle = \epsilon_0\chi\langle\boldsymbol{E}\rangle$ with χ constant in the body). By eq. (3.27), the force on the dielectric is

$$\boldsymbol{F} = \int (\langle\boldsymbol{P}\rangle\cdot\nabla)\boldsymbol{E}^{\text{ext}} d^3x. \tag{3.36}$$

We may rewrite the integrand in component notation as

$$\begin{aligned}[(\langle\boldsymbol{P}\rangle\cdot\nabla)\boldsymbol{E}^{\text{ext}}]_i &= -\sum_k \langle P_k\rangle \frac{\partial^2\phi^{\text{ext}}}{\partial x_k \partial x_i} \\ &= -\sum_k \langle P_k\rangle \frac{\partial^2\phi^{\text{ext}}}{\partial x_i \partial x_k} \\ &= \sum_k \langle P_k\rangle \frac{\partial E_k^{\text{ext}}}{\partial x_i}. \end{aligned} \tag{3.37}$$

Thus, if an external agent were to infinitesimally displace the dielectric body by $\delta\boldsymbol{x}$, the work done by this agent in overcoming the electromagnetic force is

$$\delta W = -\boldsymbol{F}\cdot\delta\boldsymbol{x} = -\int \langle\boldsymbol{P}\rangle\cdot\delta\boldsymbol{E}^{\text{ext}}\, d^3x, \tag{3.38}$$

where $\delta\boldsymbol{E}^{\text{ext}} \equiv (\delta\boldsymbol{x}\cdot\nabla)\boldsymbol{E}^{\text{ext}}$ is the change in the external field as felt by the body as a consequence of the displacement. In other words, we may evaluate the right side of eq. (3.38) by viewing the dielectric body as being in a fixed position, with the external field changing by $\delta\boldsymbol{E}^{\text{ext}}$. It is convenient to take this view in the following calculation.

We now make use of the (not at all obvious!) fact that for a linear, homogeneous, isotropic dielectric body, if the external field is changed by $\delta\boldsymbol{E}^{\text{ext}}$, then we have

$$\int \langle\boldsymbol{P}\rangle\cdot\delta\boldsymbol{E}^{\text{ext}}\, d^3x = \int \delta\langle\boldsymbol{P}\rangle\cdot\boldsymbol{E}^{\text{ext}}\, d^3x, \tag{3.39}$$

where $\delta\langle\boldsymbol{P}\rangle$ denotes the change in $\langle\boldsymbol{P}\rangle$ resulting from the change $\delta\boldsymbol{E}^{\text{ext}}$ of the external field. (Here $\delta\boldsymbol{E}^{\text{ext}}$ is allowed to be an arbitrary infinitesimal change in $\boldsymbol{E}^{\text{ext}}$, not necessarily one corresponding to a displacement of the dielectric body.) To prove eq. (3.39), we use

$$\langle\boldsymbol{P}\rangle = \epsilon_0\chi\langle\boldsymbol{E}\rangle = \epsilon_0\chi(\boldsymbol{E}^{\text{ext}} + \langle\boldsymbol{E}^{\text{self}}\rangle), \tag{3.40}$$

where $\langle\boldsymbol{E}^{\text{self}}\rangle$ is given by eq. (3.28). Since we have assumed $\langle\rho_f\rangle = 0$, we may rewrite $\langle\boldsymbol{E}^{\text{self}}\rangle$ as

$$\langle\boldsymbol{E}^{\text{self}}\rangle(\boldsymbol{x}) = \frac{1}{4\pi\epsilon_0}\nabla_x \int \frac{\nabla_{x'}\cdot\langle\boldsymbol{P}\rangle(\boldsymbol{x}')}{|\boldsymbol{x}-\boldsymbol{x}'|} d^3x' = -\frac{1}{4\pi\epsilon_0}\nabla_x \int \langle\boldsymbol{P}\rangle(\boldsymbol{x}')\cdot\nabla_{x'}\frac{1}{|\boldsymbol{x}-\boldsymbol{x}'|} d^3x', \tag{3.41}$$

where we integrated by parts to obtain the second equality. From eq. (3.40), we thus obtain the following formula for $\epsilon_0\chi\boldsymbol{E}^{\text{ext}}$:

$$\epsilon_0 \chi \boldsymbol{E}^{\text{ext}}(\boldsymbol{x}) = \langle \boldsymbol{P} \rangle(\boldsymbol{x}) + \frac{\chi}{4\pi} \nabla_x \int \langle \boldsymbol{P} \rangle(\boldsymbol{x}') \cdot \nabla_{x'} \frac{1}{|\boldsymbol{x} - \boldsymbol{x}'|} d^3x'. \tag{3.42}$$

Taking the inner product of this equation with $\delta \langle \boldsymbol{P} \rangle$ and integrating over $\boldsymbol{x}$, we obtain

$$\epsilon_0 \chi \int \delta \langle \boldsymbol{P} \rangle \cdot \boldsymbol{E}^{\text{ext}}(\boldsymbol{x}) d^3x = \int \delta \langle \boldsymbol{P} \rangle \cdot \langle \boldsymbol{P} \rangle d^3x$$
$$+ \frac{\chi}{4\pi} \int [\delta \langle \boldsymbol{P} \rangle(\boldsymbol{x}) \cdot \nabla_x] \left[\langle \boldsymbol{P} \rangle(\boldsymbol{x}') \cdot \nabla_{x'} \right] \frac{1}{|\boldsymbol{x} - \boldsymbol{x}'|} d^3x d^3x'. \tag{3.43}$$

But $\epsilon_0 \chi \int \langle \boldsymbol{P} \rangle \cdot \delta \boldsymbol{E}^{\text{ext}}(\boldsymbol{x}) d^3x$ is given by the same expression with $\delta \langle \boldsymbol{P} \rangle$ and $\langle \boldsymbol{P} \rangle$ interchanged. However, the right side of eq. (3.43) is manifestly symmetric in $\delta \langle \boldsymbol{P} \rangle$ and $\langle \boldsymbol{P} \rangle$. This proves eq. (3.39).

Returning to our calculation of δW, we see that eq. (3.38) can be written as

$$\begin{aligned} \delta W &= -\int \langle \boldsymbol{P} \rangle \cdot \delta \boldsymbol{E}^{\text{ext}} \, d^3x \\ &= -\frac{1}{2} \int \langle \boldsymbol{P} \rangle \cdot \delta \boldsymbol{E}^{\text{ext}} \, d^3x - \frac{1}{2} \int \delta \langle \boldsymbol{P} \rangle \cdot \boldsymbol{E}^{\text{ext}} \, d^3x \\ &= -\frac{1}{2} \delta \int \langle \boldsymbol{P} \rangle \cdot \boldsymbol{E}^{\text{ext}} \, d^3x. \end{aligned} \tag{3.44}$$

The key point is that we have now written δW as a "total variation," so we can easily integrate this expression to obtain the total work done for a finite displacement $\Delta \boldsymbol{x}$ of the body:

$$\Delta W = -\frac{1}{2} \Delta \int \langle \boldsymbol{P} \rangle \cdot \boldsymbol{E}^{\text{ext}}. \tag{3.45}$$

By conservation of energy, this work done by the external agent must equal the change in the total energy of the remainder of the system,

$$\Delta W = \Delta \left(\mathcal{E}^{\text{self}}_{\text{M}} + \mathcal{E}^{\text{self}}_{\text{EM}} + \mathcal{E}^{\text{int}} \right) \tag{3.46}$$

(where we have excluded $\mathcal{E}^{\text{ext}}$ from the right side, since its value doesn't change). Thus, we obtain

$$\Delta \left(\mathcal{E}^{\text{self}}_{\text{M}} + \mathcal{E}^{\text{self}}_{\text{EM}} + \mathcal{E}^{\text{int}} \right) = -\frac{1}{2} \Delta \int \langle \boldsymbol{P} \rangle \cdot \boldsymbol{E}^{\text{ext}} \, d^3x. \tag{3.47}$$

However, by eq. (3.35), we have

$$\Delta \mathcal{E}^{\text{int}} = -\Delta \int \langle \boldsymbol{P} \rangle \cdot \boldsymbol{E}^{\text{ext}} \, d^3x. \tag{3.48}$$

Thus, we deduce that

$$\Delta \left(\mathcal{E}^{\text{self}}_{\text{M}} + \mathcal{E}^{\text{self}}_{\text{EM}} \right) = +\frac{1}{2} \Delta \int \langle \boldsymbol{P} \rangle \cdot \boldsymbol{E}^{\text{ext}} \, d^3x. \tag{3.49}$$

Although, in the absence of a microscopic theory, we cannot calculate $\mathcal{E}_{\rm M}^{\rm self}$ or $\mathcal{E}_{\rm EM}^{\rm self}$—nor can we calculate their individual changes—we see that for a linear dielectric, the change of their sum is always $-\frac{1}{2}$ times the change of the electromagnetic interaction energy.

The energy formula eq. (3.47) is usually written without the Δ and without it being made clear exactly what is supposed to be included in this energy. We can rewrite this formula as follows. We have

$$\begin{aligned}\langle \boldsymbol{P}\rangle \cdot \boldsymbol{E}^{\rm ext} &= (\langle \boldsymbol{D}\rangle - \epsilon_0\langle \boldsymbol{E}\rangle)\cdot \boldsymbol{E}^{\rm ext} \\ &= -\langle \boldsymbol{D}\rangle\cdot\langle \boldsymbol{E}\rangle + \epsilon_0|\boldsymbol{E}^{\rm ext}|^2 + \left(\langle \boldsymbol{D}\rangle - \epsilon_0\boldsymbol{E}^{\rm ext}\right)\cdot\left(\langle \boldsymbol{E}\rangle + \boldsymbol{E}^{\rm ext}\right), \end{aligned} \tag{3.50}$$

where the second equality is merely an algebraic identity. Since we have assumed that $\langle\rho_f\rangle = 0$ and since $\langle \boldsymbol{D}\rangle = \epsilon_0\langle \boldsymbol{E}\rangle = \epsilon_0 \boldsymbol{E}$ outside the dielectric, we have $\boldsymbol{\nabla}\cdot(\langle \boldsymbol{D}\rangle - \epsilon_0\boldsymbol{E}^{\rm ext}) = 0$. Since $\langle \boldsymbol{E}\rangle + \boldsymbol{E}^{\rm ext} = -\boldsymbol{\nabla}(\langle\phi\rangle + \phi^{\rm ext})$, it can be seen that the integral of the last term vanishes. Therefore, we obtain

$$\Delta\left(\mathcal{E}_{\rm M}^{\rm self} + \mathcal{E}_{\rm EM}^{\rm self} + \mathcal{E}^{\rm int}\right) = \frac{1}{2}\Delta\int \langle \boldsymbol{D}\rangle\cdot\langle \boldsymbol{E}\rangle d^3x. \tag{3.51}$$

This formula (without the Δ) is the usual formula for the "energy of a dielectric" that can be found in essentially all texts.

In summary, eq. (3.27), eq. (3.31), and eq. (3.35) hold for an arbitrary dielectric, whereas eq. (3.47), eq. (3.49), and eq. (3.51) hold for linear dielectrics.

Problems

1. A dielectric ball of radius R and dielectric constant ϵ/ϵ_0 is placed in a uniform external electric field $\boldsymbol{E}^{\rm ext} = C_0\hat{\boldsymbol{z}}$, where C_0 is a constant. Find the resulting electric field $\langle \boldsymbol{E}\rangle$ everywhere, that is, both inside $(r < R)$ and outside $(r > R)$ the dielectric. Show that the induced dipole moment of the dielectric ball is

$$\boldsymbol{p} = 4\pi\epsilon_0\frac{\epsilon - \epsilon_0}{\epsilon + 2\epsilon_0}R^3\boldsymbol{E}^{\rm ext}.$$

2. Consider a dielectric ball of radius R with dielectric constant ϵ/ϵ_0. Obtain a multipole expansion for the field, $\phi(\boldsymbol{x})$, of a point charge q placed at a point $\boldsymbol{x}'$ with $|\boldsymbol{x}'| = d > R$ (so the charge is outside the dielectric ball). [Hint: Follow the same logic as led to eq. (2.82), but now consider the 3 regions: $r \le R$, $R \le r \le d$, and $d \le r$. Obtain the form of the solution in these regions and match suitably.]

3. A dielectric ball of radius R and dielectric constant ϵ/ϵ_0 is placed in the external electrostatic potential $\phi_0 = \alpha(2z^2 - x^2 - y^2)$ (where α is a constant), with the center of the ball at $\boldsymbol{x} = 0$.

   (a) Find the total electrostatic potential ϕ everywhere. [Hint: Note that the external potential is proportional to $r^2 Y_{20}(\theta,\varphi)$.]

(b) Calculate the interaction energy between the field produced by the dielectric and the external field. (Assume that the external potential arises from "distant charges" and goes to zero at infinity, so that eq. (2.30) holds.)
(c) Show that the total force on the dielectric ball vanishes.

4. The half-space $z < 0$ is filled by a homogeneous, isotropic dielectric material of dielectric constant ϵ/ϵ_0, whereas the half-space $z > 0$ is vacuum. A point charge of charge q is placed in the vacuum region at point $\boldsymbol{x}_0 = (0, 0, D)$. Show that the resulting electromagnetic field can be obtained by the following "image charge" ansatz. For $z > 0$, seek a solution of the form

$$\phi(\boldsymbol{x}) = \frac{1}{4\pi\epsilon_0}\frac{q}{|\boldsymbol{x}-\boldsymbol{x}_0|} + \frac{1}{4\pi\epsilon_0}\frac{q'}{|\boldsymbol{x}+\boldsymbol{x}_0|},$$

whereas for $z < 0$, seek a solution of the form

$$\langle\phi\rangle(\boldsymbol{x}) = \frac{1}{4\pi\epsilon}\frac{q''}{|\boldsymbol{x}-\boldsymbol{x}_0|}.$$

Show that q' and q'' can be chosen so that the matching conditions hold at $z = 0$, thereby obtaining the desired solution.

5. Two concentric spherical conducting shells have radii a and b. A dielectric material of dielectric constant ϵ/ϵ_0 fills a cone of angle θ between the shells, as shown in the figure. The inner conducting shell at radius a is held at potential V, and the outer conducting shell at radius b is held at potential 0.

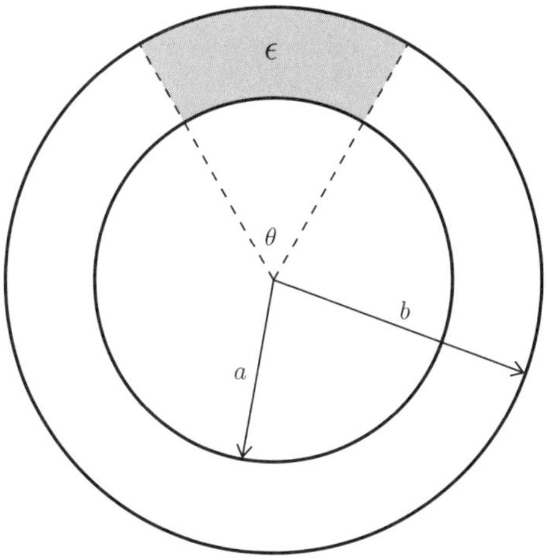

(a) Find $\langle\phi\rangle$ everywhere for $a < r < b$ (i.e., both inside and outside the dielectric). [Hint: A spherically symmetric solution can be found that satisfies the boundary conditions and the matching conditions.]
(b) Find the surface charge density everywhere on both conductors.
(c) Find the polarization charge density everywhere on the boundary of the dielectric (i.e., at the boundaries $r = a, b$ and at the boundary of the cone of angle θ).

CHAPTER 4

Magnetostatics

In this chapter, we analyze magnetostatics, that is, stationary solutions to Maxwell's equations with $\rho = \phi = 0$, so we have only a stationary current $\boldsymbol{J}$ and a stationary vector potential $\boldsymbol{A}$. As we shall see, many results in magnetostatics are direct analogs of results in electrostatics, but there also are some significant differences.

In section 4.1, we reduce the equations of magnetostatics to solving a vector version of Poisson's equation for $\boldsymbol{A}$ in the Coulomb gauge. We will give a multipole expansion of $\boldsymbol{A}$ in section 4.2. The Cartesian multipole expansion will be carried out only to dipole order, but the spherical multipole expansion is given to all orders. In the course of giving the spherical expansion, we will introduce the notion of vector spherical harmonics. Magnetic interaction energy and force are analyzed in section 4.3, with a special emphasis on the consequences of the sign difference between electrostatics and magnetostatics with regard to the electromagnetic interaction energy of a dipole placed in an external field. Magnetic materials are discussed in section 4.4.

4.1 The Equations of Magnetostatics

If we set $\rho = \phi = 0$ and also set the time derivatives of $\boldsymbol{A}$ and $\boldsymbol{J}$ to zero, Maxwell's equations reduce to

$$\nabla \times \boldsymbol{B} = \mu_0 \boldsymbol{J}, \tag{4.1}$$

where

$$\boldsymbol{B} \equiv \nabla \times \boldsymbol{A}. \tag{4.2}$$

Charge-current conservation, eq. (1.5), reduces to

$$\nabla \cdot \boldsymbol{J} = 0, \tag{4.3}$$

that is, only stationary $\boldsymbol{J}$ that satisfy eq. (4.3) are allowed to appear on the right side of eq. (4.1). In terms of $\boldsymbol{A}$, the full content of Maxwell's equations in magnetostatics is given by the single equation

$$\nabla \times (\nabla \times \boldsymbol{A}) = \mu_0 \boldsymbol{J}. \tag{4.4}$$

In magnetostatics, the gauge freedom eq. (1.13) reduces to

$$\boldsymbol{A} \to \boldsymbol{A}' = \boldsymbol{A} + \nabla \chi, \tag{4.5}$$

where χ is independent of t.

The triple cross product appearing in eq. (4.4) is the first of many expressions involving cross-products that we will encounter that can be rewritten in a much more useful form. Therefore, before analyzing the above equations of magnetostatices, we make some comments on cross-product identities.

Side Comment on Cross-Product Identities: There are many identities involving cross-products. In particular, any expression with two cross-products always can be rewritten as an expression involving no cross products. One could compile a long list of cross-product identities—as can be found on the inside cover of various electromagnetism texts—but unless one is prepared to carry around (or memorize) this list, it is probably more useful to explain the basic identity that underlies all cross-product identities.

Define the quantity ϵ_{ijk}—with i, j, k each ranging over the Cartesian spatial directions $1, 2, 3$—by

$$\epsilon_{ijk} = \begin{cases} 1, & \text{if } i,j,k \text{ is an even permutation of } 1,2,3 \\ -1, & \text{if } i,j,k \text{ is an odd permutation of } 1,2,3 \\ 0, & \text{otherwise.} \end{cases} \tag{4.6}$$

This is equivalent to defining $\epsilon_{123} = 1$ and requiring that ϵ_{ijk} be antisymmetric in every pair of indices. The cross-product of vectors $\boldsymbol{a}$ and $\boldsymbol{b}$ is defined by

$$(\boldsymbol{a} \times \boldsymbol{b})_i = \sum_{j,k} \epsilon_{ijk} a_j b_k \,. \tag{4.7}$$

Since ϵ_{ijk} is invariant under even (i.e., cyclic) permutations of its indices, it follows immediately from eq. (4.7) that for any vector $\boldsymbol{c}$, we have

$$\boldsymbol{c} \cdot (\boldsymbol{a} \times \boldsymbol{b}) = \sum_{i,j,k} \epsilon_{ijk} c_i a_j b_k = \boldsymbol{a} \cdot (\boldsymbol{b} \times \boldsymbol{c}) = \boldsymbol{b} \cdot (\boldsymbol{c} \times \boldsymbol{a}) \,. \tag{4.8}$$

The fundamental identity that governs all formulas involving two cross-products is

$$\begin{aligned} \epsilon_{ijk}\epsilon_{lmn} &= \sum_{\text{perm } P \text{ of } i,j,k} (-1)^{\text{sgn}(P)} \delta_{il}\delta_{jm}\delta_{kn} \\ &= \delta_{il}\delta_{jm}\delta_{kn} + \delta_{jl}\delta_{km}\delta_{in} + \delta_{kl}\delta_{im}\delta_{jn} - \delta_{jl}\delta_{im}\delta_{kn} \\ &\quad - \delta_{kl}\delta_{jm}\delta_{in} - \delta_{il}\delta_{km}\delta_{jn}, \end{aligned} \tag{4.9}$$

where δ_{ij} denotes the Kronecker delta. (This identity can be proven, by "brute force" verification, to hold for all possible values of the indices.) If we set $l = i$ and sum over i, we get the auxiliary identity

$$\sum_i \epsilon_{ijk}\epsilon_{imn} = \delta_{jm}\delta_{kn} - \delta_{km}\delta_{jn} \,. \tag{4.10}$$

Most identities involving two cross-product follow directly from eq. (4.10). For example, we have

$$\begin{aligned}(\boldsymbol{a}\times\boldsymbol{b})\cdot(\boldsymbol{c}\times\boldsymbol{d}) &= \sum_{i,j,k,m,n}\epsilon_{ijk}\epsilon_{imn}a_j b_k c_m d_n\\ &= \sum_{j,k,m,n}\left(\delta_{jm}\delta_{kn}-\delta_{km}\delta_{jn}\right)a_j b_k c_m d_n\\ &= (\boldsymbol{a}\cdot\boldsymbol{c})(\boldsymbol{b}\cdot\boldsymbol{d})-(\boldsymbol{b}\cdot\boldsymbol{c})(\boldsymbol{a}\cdot\boldsymbol{d})\,. \end{aligned}\tag{4.11}$$

We also obtain the following formula for the triple cross-product:

$$\begin{aligned}[\boldsymbol{a}\times(\boldsymbol{b}\times\boldsymbol{c})]_i &= \sum_{j,k,m,n}\epsilon_{ijk}a_j\epsilon_{kmn}b_m c_n\\ &= \sum_{j,k,m,n}\epsilon_{kij}\epsilon_{kmn}a_j b_m c_n\\ &= \sum_{j,k,m,n}\left(\delta_{im}\delta_{jn}-\delta_{jm}\delta_{in}\right)a_j b_m c_n, \end{aligned}\tag{4.12}$$

and hence

$$\boldsymbol{a}\times(\boldsymbol{b}\times\boldsymbol{c}) = (\boldsymbol{a}\cdot\boldsymbol{c})\boldsymbol{b}-(\boldsymbol{a}\cdot\boldsymbol{b})\boldsymbol{c}\,. \tag{4.13}$$

Similar identities hold if any vector is replaced by $\boldsymbol{\nabla}$, but we have to be careful to make sure that the derivatives are acting on precisely the quantities that they are supposed to act on. For example, by the same calculation as in eq. (4.12), we have

$$\begin{aligned}[\boldsymbol{\nabla}\times(\boldsymbol{b}\times\boldsymbol{c})]_i &= \sum_{j,k,m,n}\left(\delta_{im}\delta_{jn}-\delta_{jm}\delta_{in}\right)\partial_j(b_m c_n)\\ &= \sum_{j,k,m,n}\left(\delta_{im}\delta_{jn}-\delta_{jm}\delta_{in}\right)\left[(\partial_j b_m)c_n + b_m\partial_j c_n\right], \end{aligned}\tag{4.14}$$

so

$$\boldsymbol{\nabla}\times(\boldsymbol{b}\times\boldsymbol{c}) = (\boldsymbol{c}\cdot\boldsymbol{\nabla})\boldsymbol{b}+\boldsymbol{b}(\boldsymbol{\nabla}\cdot\boldsymbol{c})-(\boldsymbol{b}\cdot\boldsymbol{\nabla})\boldsymbol{c}-\boldsymbol{c}(\boldsymbol{\nabla}\cdot\boldsymbol{b})\,. \tag{4.15}$$

Similarly, we obtain

$$\boldsymbol{\nabla}\times(\boldsymbol{\nabla}\times\boldsymbol{c}) = \boldsymbol{\nabla}(\boldsymbol{\nabla}\cdot\boldsymbol{c})-\nabla^2\boldsymbol{c}\,. \tag{4.16}$$

We now return to eq. (4.4) and use the identity eq. (4.16) to obtain

$$\boldsymbol{\nabla}(\boldsymbol{\nabla}\cdot\boldsymbol{A})-\nabla^2\boldsymbol{A} = \mu_0\boldsymbol{J}\,. \tag{4.17}$$

This equation is greatly simplified by imposing the *Coulomb gauge condition* $\boldsymbol{\nabla}\cdot\boldsymbol{A}=0$; that is, we replace $\boldsymbol{A}$ by a gauge-equivalent vector potential $\boldsymbol{A}'$ satisfying $\boldsymbol{\nabla}\cdot\boldsymbol{A}'=0$. To do this, we must find a function $\chi(\boldsymbol{x})$ so that

$$\nabla \cdot (\boldsymbol{A} + \nabla \chi) = 0, \tag{4.18}$$

that is, we must solve

$$\nabla^2 \chi = -\nabla \cdot \boldsymbol{A} . \tag{4.19}$$

But this is just Poisson's equation, which always can be solved.[1] Thus, a gauge transformation $\boldsymbol{A} \to \boldsymbol{A}'$ with $\nabla \cdot \boldsymbol{A}' = 0$ can always be achieved. In the following, we drop the "prime" on $\boldsymbol{A}'$; that is, we denote by $\boldsymbol{A}$ the vector potential in which our Coulomb gauge choice $\nabla \cdot \boldsymbol{A} = 0$ has been made. Note that if we require $\boldsymbol{A} \to 0$ at infinity, then the Coulomb gauge choice is unique, since a further gauge transformation $\boldsymbol{A} \to \boldsymbol{A} + \nabla \psi$ that preserves the Coulomb gauge condition would have to satisfy $\nabla^2 \psi = 0$ with $\nabla \psi \to 0$ at infinity, which implies $\nabla \psi = 0$.

In the Coulomb gauge, $\boldsymbol{A}$ satisfies

$$\nabla^2 \boldsymbol{A} = -\mu_0 \boldsymbol{J} \tag{4.20}$$

together with

$$\nabla \cdot \boldsymbol{A} = 0 . \tag{4.21}$$

Thus, each Cartesian component A_i of $\boldsymbol{A}$ satisfies Poisson's equation. Consequently, we can directly apply many of the results obtained for electrostatics in chapter 2. In particular, we see that if $\boldsymbol{J} \to 0$ sufficiently rapidly at infinity, the unique solution to eq. (4.20) is

$$\boldsymbol{A}(\boldsymbol{x}) = \frac{\mu_0}{4\pi} \int \frac{\boldsymbol{J}(\boldsymbol{x}')}{|\boldsymbol{x} - \boldsymbol{x}'|} d^3x' . \tag{4.22}$$

This solution automatically satisfies eq. (4.21), because

$$\begin{aligned}
\nabla \cdot \boldsymbol{A} &= \frac{\mu_0}{4\pi} \int \boldsymbol{J}(\boldsymbol{x}') \cdot \nabla_x \frac{1}{|\boldsymbol{x} - \boldsymbol{x}'|} d^3x' \\
&= -\frac{\mu_0}{4\pi} \int \boldsymbol{J}(\boldsymbol{x}') \cdot \nabla_{x'} \frac{1}{|\boldsymbol{x} - \boldsymbol{x}'|} d^3x' \\
&= \frac{\mu_0}{4\pi} \int \frac{\nabla_{x'} \cdot \boldsymbol{J}(\boldsymbol{x}')}{|\boldsymbol{x} - \boldsymbol{x}'|} d^3x' \\
&= 0 .
\end{aligned} \tag{4.23}$$

Thus, if $\boldsymbol{J} \to 0$ sufficiently rapidly at infinity, eq. (4.22) is the unique solution to the equations of magnetostatics.

4.2 Multipole Expansion

We can now perform a multipole expansion of the solution eq. (4.22) in parallel with what was done in electrostatics in section 2.4. To put the results in the desired form, it

[1] As noted at the end of section 2.1, solutions exist even if the the source term does not decay to zero at infinity.

is important to note that since $\nabla \cdot \boldsymbol{J} = 0$, it follows that if $\boldsymbol{J}$ vanishes sufficiently rapidly at infinity, then

$$\int \boldsymbol{J}(\boldsymbol{x}) d^3x = 0 \,. \tag{4.24}$$

To prove this result, we note that if $r^3|\boldsymbol{J}| \to 0$ as $r \to \infty$, then by Gauss's theorem, we have

$$\int \nabla \cdot (x_j \boldsymbol{J}) d^3x = \lim_{R\to\infty} \int_{r=R} x_j \boldsymbol{J} \cdot \hat{\boldsymbol{r}} dA = 0 \,. \tag{4.25}$$

Equation (4.24) then follows from the following computation:

$$\begin{aligned} 0 = \int \nabla \cdot (x_j \boldsymbol{J}) d^3x &= \int \left[x_j (\nabla \cdot \boldsymbol{J}) + \boldsymbol{J} \cdot \nabla x_j \right] d^3x \\ &= \sum_i \int J_i \partial_i x_j d^3x = \sum_i \int J_i \delta_{ij} d^3x = \int J_j d^3x \,. \end{aligned} \tag{4.26}$$

By a similar computation assuming $r^4|\boldsymbol{J}| \to 0$ as $r \to \infty$ and using $0 = \int \nabla \cdot (x_j x_k \boldsymbol{J}) d^3x$, we obtain

$$\int x_i J_j d^3x = -\int x_j J_i d^3x \,. \tag{4.27}$$

To perform the Cartesian multipole expansion, we substitute eq. (2.51) into eq. (4.22). Each Cartesian component A_i of $\boldsymbol{A}$ then will be given by a formula corresponding to eq. (2.52). Keeping only the terms up to dipole order, we obtain

$$A_i = \frac{\alpha_i}{|\boldsymbol{x}|} + \frac{\sum_j \beta_{ij} \hat{x}_j}{|\boldsymbol{x}|^2} + \cdots \,, \tag{4.28}$$

where $\hat{\boldsymbol{x}} = \boldsymbol{x}/|\boldsymbol{x}|$ and

$$\alpha_i = \frac{\mu_0}{4\pi} \int J_i(\boldsymbol{x}') d^3x' \,, \tag{4.29}$$

$$\beta_{ij} = \frac{\mu_0}{4\pi} \int J_i(\boldsymbol{x}') x'_j d^3x' \,. \tag{4.30}$$

By eq. (4.24), the monopole term α_i, vanishes, so the leading-order contribution to $\boldsymbol{A}$ at large $|\boldsymbol{x}|$ comes from the dipole term β_{ij}. Using eq. (4.27), we may rewrite β_{ij} as

$$\beta_{ij} = \frac{\mu_0}{8\pi} \int [J_i(\boldsymbol{x}') x'_j - J_j(\boldsymbol{x}') x'_i] d^3x' \,. \tag{4.31}$$

Using the vector identity eq. (4.13), we obtain

$$\begin{aligned} \sum_j \hat{x}_j \beta_{ij} &= \frac{\mu_0}{8\pi} \sum_j \hat{x}_j \int [J_i(\boldsymbol{x}') x'_j - J_j(\boldsymbol{x}') x'_i] d^3x' \\ &= \frac{\mu_0}{8\pi} \left[\hat{\boldsymbol{x}} \times \int \boldsymbol{J}(\boldsymbol{x}') \times (\boldsymbol{x}') d^3x' \right]_i \,. \end{aligned} \tag{4.32}$$

Thus, using the antisymmetry of the cross-product, we obtain

$$A = \frac{\mu_0}{4\pi} \frac{\boldsymbol{\mu} \times \hat{\boldsymbol{x}}}{|\boldsymbol{x}|^2} + \cdots, \tag{4.33}$$

where

$$\boldsymbol{\mu} \equiv \frac{1}{2} \int \boldsymbol{x}' \times \boldsymbol{J}(\boldsymbol{x}') d^3x' \tag{4.34}$$

is called the *magnetic dipole moment* of the current distribution $\boldsymbol{J}$. For $|\boldsymbol{x}| > 0$, the magnetic field associated with eq. (4.33) is

$$\begin{aligned} \boldsymbol{B} = \nabla \times \boldsymbol{A} &= \frac{\mu_0}{4\pi} \nabla \times \left[\frac{\boldsymbol{\mu} \times \hat{\boldsymbol{x}}}{|\boldsymbol{x}|^2} \right] \\ &= -\frac{\mu_0}{4\pi} \nabla \left[\frac{\boldsymbol{\mu} \cdot \hat{\boldsymbol{x}}}{|\boldsymbol{x}|^2} \right], \end{aligned} \tag{4.35}$$

where $\nabla \cdot (\hat{\boldsymbol{x}}/|\boldsymbol{x}|^2) = 0$ (for $|\boldsymbol{x}| > 0$) and $\nabla \times (\hat{\boldsymbol{x}}/|\boldsymbol{x}|^2) = 0$ were used.[2] Thus—apart from a factor of $\epsilon_0\mu_0 = 1/c^2$ having to do with units and conventions—the magnetostatic field $\boldsymbol{B}$ of a magnetic dipole, $\boldsymbol{\mu}$, takes exactly the same form for $|\boldsymbol{x}| > 0$ as the electrostatic field $\boldsymbol{E} = -\nabla\phi$ of an electric dipole, $\boldsymbol{p}$, under the substitution $\boldsymbol{p} \to \boldsymbol{\mu}$.

We can also substitute eq. (2.82) into eq. (4.22) to obtain an expansion of each Cartesian component of $\boldsymbol{A}$ in spherical harmonics, in parallel with eq. (2.86). However, although the resulting spherical multipole expansion of the Cartesian components of $\boldsymbol{A}$ in terms of spherical multipole moments of the Cartesian components of $\boldsymbol{J}$ will yield correct formulas, they are much less useful than the corresponding expansion that we obtained in electrostatics. In eq. (2.86), the spherical harmonics of different ℓs and ms completely decouple from each other: If we wish to know the ℓm contribution to ϕ, we only have to know the ℓm multipole moment of ρ. The reason underlying this decoupling is the rotational invariance of Poisson's equation. The equations of magnetostatics are also rotationally invariant. However, the Cartesian basis is not invariant under rotations, and the Cartesian basis components of $\boldsymbol{J}$ are not independent but must satisfy $\nabla \cdot \boldsymbol{J} = 0$. This equation couples the coefficients of spherical harmonics of different ℓs and ms appearing in the expansions of the Cartesian components of $\boldsymbol{J}$. Thus, one cannot restrict consideration to a single ℓ and m in the expansion of Cartesian components when considering properties of the solutions, since, in general, other ℓs and ms must necessarily be present.

Therefore, when considering a vector field satisfying rotationally invariant equations, it is more useful to use a spherical harmonic expansion that does not introduce any non-rotationally-invariant elements, such as a Cartesian basis. Such an expansion for an arbitrary vector field $\boldsymbol{V}$ on space can be obtained as follows. First, we decompose $\boldsymbol{V}$ into its tangential and normal (i.e., radial) parts relative to spheres. We can expand the radial part, $\hat{\boldsymbol{r}} \cdot \boldsymbol{V}$ in spherical harmonics:

$$\hat{\boldsymbol{r}} \cdot \boldsymbol{V} = \sum_{\ell,m} f_{\ell m}(r) Y_{\ell m}(\theta, \varphi). \tag{4.36}$$

[2] Since $\nabla \cdot (\hat{\boldsymbol{x}}/|\boldsymbol{x}|^2) = 4\pi\delta(\boldsymbol{x}) \neq 0$, eq. (4.35) is *not* a correct distributional expression for $\boldsymbol{B}$ (i.e., it holds only for $|\boldsymbol{x}| > 0$).

For the tangential part, we use the fact that any vector field $\boldsymbol{W}$ tangential to a sphere can be written as[3]

$$\boldsymbol{W} = \boldsymbol{\nabla}\psi_1 + \hat{\boldsymbol{r}} \times \boldsymbol{\nabla}\psi_2 \tag{4.37}$$

for some functions $\psi_1(\theta,\varphi)$ and $\psi_2(\theta,\varphi)$. We can then expand ψ_1 and ψ_2 in spherical harmonics. We thereby obtain the following spherical harmonic expansion for an arbitrary vector field $\boldsymbol{V}$:

$$\boldsymbol{V} = \sum_{\ell,m} \left[f_{\ell m}(r) Y_{\ell m}(\theta,\varphi)\hat{\boldsymbol{r}} + g_{\ell m}(r) r \boldsymbol{\nabla} Y_{\ell m}(\theta,\varphi) + h_{\ell m}(r) \boldsymbol{r} \times \boldsymbol{\nabla} Y_{\ell m}(\theta,\varphi) \right] . \tag{4.38}$$

The quantities $Y_{\ell m}\hat{\boldsymbol{r}}$, $r\boldsymbol{\nabla} Y_{\ell m}/\sqrt{\ell(\ell+1)}$, and $\boldsymbol{r} \times \boldsymbol{\nabla} Y_{\ell m}/\sqrt{\ell(\ell+1)}$ are referred to as *vector spherical harmonics*.[4] They comprise an orthonormal basis of the Hilbert space of square integrable vector fields on a unit sphere in the same manner as the $Y_{\ell m}$ comprise an orthonormal basis of the Hilbert space of square integrable scalar functions on a unit sphere. The term $h_{\ell m}(r)\boldsymbol{r} \times \boldsymbol{\nabla} Y_{\ell m}$ has opposite parity from the other terms in eq. (4.38) because of the cross-product. It is said to have *magnetic parity*, whereas the terms $f_{\ell m}(r) Y_{\ell m}\hat{\boldsymbol{r}}$ and $g_{\ell m}(r) r\boldsymbol{\nabla} Y_{\ell m}$ are said to have *electric parity*.

We now expand $\boldsymbol{A}$ and $\boldsymbol{J}$ as in eq. (4.38). Rotationally invariant operators, such as ∇^2 and $\boldsymbol{\nabla}\cdot$, now will not couple the coefficients of different ℓs and ms in this expansion, so solving eq. (4.20) and eq. (4.21) reduces to separate, decoupled problems for each ℓ and m. Furthermore, since ∇^2 and $\boldsymbol{\nabla}\cdot$ are also parity invariant, the terms in the expansion of $\boldsymbol{A}$ and $\boldsymbol{J}$ with magnetic parity decouple from the terms with electric parity. Thus, we may separately consider the magnetic and electric parity contributions for each ℓ and m.

To see more explicitly how this works, consider a magnetic parity current distribution of the form

$$\boldsymbol{J} = j^{\mathrm{M}}_{\ell m}(r)\, \boldsymbol{r} \times \boldsymbol{\nabla} Y_{\ell m}, \tag{4.39}$$

where the superscript M indicates that this is the radial function appearing in the magnetic parity part of $\boldsymbol{J}$. It follows from what we have just said that the corresponding solution $\boldsymbol{A}$ to eq. (4.20) must be of magnetic parity and of the same ℓ and m, and thus must be of the form

$$\boldsymbol{A} = a^{\mathrm{M}}_{\ell m}(r)\, \boldsymbol{r} \times \boldsymbol{\nabla} Y_{\ell m} . \tag{4.40}$$

Note that the right sides of eq. (4.39) and eq. (4.40) automatically satisfy $\boldsymbol{\nabla}\cdot\boldsymbol{J} = \boldsymbol{\nabla}\cdot\boldsymbol{A} = 0$. Thus, current conservation and the Coulomb gauge condition hold automatically in the case of magnetic parity, so we need only solve eq. (4.20). Plugging eq. (4.39) and eq. (4.40) into eq. (4.20), we obtain, after some algebra,

$$\frac{d^2 a^{\mathrm{M}}_{\ell m}}{dr^2} + \frac{2}{r}\frac{d a^{\mathrm{M}}_{\ell m}}{dr} - \frac{\ell(\ell+1)}{r^2} a^{\mathrm{M}}_{\ell m} = -\mu_0 j^{\mathrm{M}}_{\ell m} . \tag{4.41}$$

The operator appearing on the left side of this equation is the same radial Laplace operator as previously encountered in electrostatics in eq. (2.74). From the results of

---

[3]This follows from the fact that, by the Hodge decomposition theorem, any vector field $\boldsymbol{W}$ on a sphere can be written as a gradient plus a divergence-free (with respect to the intrinsic sphere derivative) vector field $\boldsymbol{w}$. It then follows that $\hat{\boldsymbol{r}} \times \boldsymbol{w}$ is curl-free on the sphere and hence can be written as a gradient.

[4]The factor of r has been inserted in $r\boldsymbol{\nabla} Y_{\ell m}/\sqrt{\ell(\ell+1)}$, and we have used $\boldsymbol{r}$ rather than $\hat{\boldsymbol{r}}$ in $\boldsymbol{r} \times \boldsymbol{\nabla} Y_{\ell m}/\sqrt{\ell(\ell+1)}$ to give these vector spherical harmonics the same scaling with r as for scalar spherical harmonics.

section 2.4, it follows that the solution that is regular at the origin and goes to zero at infinity is

$$a^{\mathrm{M}}_{\ell m}(r) = \frac{\mu_0}{2\ell+1}\left[\frac{1}{r^{\ell+1}}\int_0^r j^{\mathrm{M}}_{\ell m}(r')r'^{\ell}r'^2dr' + r^{\ell}\int_r^{\infty} j^{\mathrm{M}}_{\ell m}(r')\frac{1}{r'^{\ell+1}}r'^2dr'\right]. \tag{4.42}$$

In particular, if $j^{\mathrm{M}}_{\ell m}(r) = 0$ for $r > R$, then for $r > R$, we obtain

$$a^{\mathrm{M}}_{\ell m}(r) = -\frac{\mu_0}{\ell(2\ell+1)}\frac{m_{\ell m}}{r^{\ell+1}}, \tag{4.43}$$

where the spherical magnetic multipole moments $m_{\ell m}$ are defined by

$$\begin{aligned} m_{\ell m} &\equiv -\ell\int_0^{\infty} j^{\mathrm{M}}_{\ell m}(r)r^{\ell}r^2dr = -\frac{1}{\ell+1}\int r^{\ell}\left[\boldsymbol{r}\times\nabla Y^*_{\ell m}\right]\cdot\boldsymbol{J}(\boldsymbol{x})d^3x \\ &= \frac{1}{\ell+1}\int r^{\ell}\,Y^*_{\ell m}(\theta,\varphi)\,\boldsymbol{r}\cdot(\nabla\times\boldsymbol{J})\,d^3x. \end{aligned} \tag{4.44}$$

Here we chose the normalization of $m_{\ell m}$ for later convenience (see eq. (4.51) below). We used the orthonormality of the vector spherical harmonics in the second equality, and a factor of $1/\ell(\ell+1)$ arose from their normalization. In the last line, we integrated by parts and used a cross-product identity of the type discussed in section 4.1.

For an electric parity current distribution, we have

$$\boldsymbol{J} = j^{\mathrm{E1}}_{\ell m}(r)Y_{\ell m}\hat{\boldsymbol{r}} + j^{\mathrm{E2}}_{\ell m}(r)r\nabla Y_{\ell m}. \tag{4.45}$$

Here $j^{\mathrm{E1}}_{\ell m}(r)$ and $j^{\mathrm{E2}}_{\ell m}(r)$ are related by current conservation, $\nabla\cdot\boldsymbol{J} = 0$, which yields

$$\frac{dj^{\mathrm{E1}}_{\ell m}(r)}{dr} + 2\frac{j^{\mathrm{E1}}_{\ell m}(r)}{r} = \frac{\ell(\ell+1)}{r}j^{\mathrm{E2}}_{\ell m}(r). \tag{4.46}$$

Thus, we may choose $j^{\mathrm{E1}}_{\ell m}(r)$ to be an arbitrary function of r, but $j^{\mathrm{E2}}_{\ell m}(r)$ is then uniquely determined by eq. (4.46). The vector potential is given by a similar expression. However, in this case, it is much easier to work directly with the magnetic field $\boldsymbol{B} = \nabla\times\boldsymbol{A}$, which takes the magnetic parity form

$$\boldsymbol{B} = b^{\mathrm{E}}_{\ell m}(r)\,\boldsymbol{r}\times\nabla Y_{\ell m}. \tag{4.47}$$

Plugging eq. (4.47) and eq. (4.45) into eq. (4.1), we obtain, after some algebra, the simple result

$$b^{\mathrm{E}}_{\ell m}(r) = -\frac{\mu_0 r}{\ell(\ell+1)}j^{\mathrm{E1}}_{\ell m}(r). \tag{4.48}$$

In particular, for an electric parity current distribution, $\boldsymbol{B}$ is nonvanishing only at radii at which $\boldsymbol{J}$ is nonvanishing.

A completely general current distribution $\boldsymbol{J}$ will be a sum over ℓ and m of currents of the form eq. (4.39) and eq. (4.45). The solution for $\boldsymbol{A}$ will be the sum of the solutions we found above. In particular, if $\boldsymbol{J}(\boldsymbol{x}) = 0$ for $r > R$, then for $r > R$, the electric parity part of $\boldsymbol{J}$ will not contribute to $\boldsymbol{B}$. Its contribution to $\boldsymbol{A}$ will therefore be pure gauge, so we can ignore it. Thus, the vector potential for $r > R$ is given by

$$\boldsymbol{A}(\boldsymbol{x}) = -\sum_{\ell,m} \frac{\mu_0}{\ell(2\ell+1)} \frac{m_{\ell m}}{r^{\ell+1}} \boldsymbol{r} \times \nabla Y_{\ell m}, \tag{4.49}$$

where $m_{\ell m}$ is given by eq. (4.44). After some algebra, it can be seen that the corresponding magnetic field can be written as

$$\boldsymbol{B} = \nabla \times \boldsymbol{A} = -\nabla \phi_{\mathrm{M}}, \tag{4.50}$$

where

$$\phi_{\mathrm{M}} = \sum_{\ell,m} \frac{\mu_0}{2\ell+1} \frac{m_{\ell m}}{r^{\ell+1}} Y_{\ell m}(\theta, \varphi). \tag{4.51}$$

Thus, at radii outside a current distribution, the magnetic field $\boldsymbol{B}$ can be derived from a *magnetic scalar potential* ϕ_{M}. Apart from a factor of $\epsilon_0\mu_0 = 1/c^2$, $\boldsymbol{B}$ takes exactly the same form as the electric field obtained from the potential ϕ, eq. (2.86), outside a charge distribution with the substitution $q_{\ell m} \to m_{\ell m}$.

More generally, one can introduce a magnetic scalar potential, ϕ_{M}, in any simply connected region in which $\boldsymbol{J} = 0$. In some simple problems where $\boldsymbol{J} \neq 0$ only on the boundary between simply connected regions (see problems 9 and 10 at the end of this chapter), it is convenient to introduce separate magnetic scalar potentials for these regions and then do appropriate matching of the field strengths across the boundary (where the magnetic scalar potentials are not defined). However, aside from such problems and from the general multipole expansion eq. (4.51) outside a current distribution, the magnetic scalar potential is of limited applicability. By contrast, the vector spherical harmonic expansion presented above is of completely general applicability. In particular, if one wishes to determine $\boldsymbol{B}$ in a region where $\boldsymbol{J} \neq 0$, one can use the general solution, eq. (4.42), to get the magnetic parity contribution and eq. (4.48) to get the electric parity contribution.

4.3 Interaction Energy and Force

In parallel with the discussion of section 2.3 in electrostatics, we now consider the magnetic field energy in magnetostatics and the magnetic force on a body. The energy density of the magnetic field in magnetostatics is given by eq. (1.15) with $\boldsymbol{E} = 0$, namely,

$$\mathcal{E} = \frac{1}{2\mu_0} |\boldsymbol{B}|^2 . \tag{4.52}$$

If the current density $\boldsymbol{J}$ and vector potential $\boldsymbol{A}$ go to zero sufficiently rapidly at infinity, we obtain the following formula for total energy:

$$\begin{aligned} \mathscr{E} \equiv \int \mathcal{E} d^3x &= \frac{1}{2\mu_0} \int |\boldsymbol{B}|^2 d^3x \\ &= \frac{1}{2\mu_0} \int \boldsymbol{B} \cdot (\nabla \times \boldsymbol{A}) d^3x \end{aligned}$$

$$= \frac{1}{2\mu_0} \int [-\nabla \cdot (\boldsymbol{B} \times \boldsymbol{A}) + \boldsymbol{A} \cdot (\nabla \times \boldsymbol{B})]\, d^3x$$
$$= \frac{1}{2} \int \boldsymbol{A} \cdot \boldsymbol{J} d^3x, \tag{4.53}$$

where Gauss's theorem and eq. (4.1) were used in the last equality.

If the current is the sum of two current distributions, $\boldsymbol{J} = \boldsymbol{J}_1 + \boldsymbol{J}_2$, and if $\boldsymbol{A}_1$ and $\boldsymbol{A}_2$ are the corresponding vector potentials, then the magnetic interaction energy is

$$\mathscr{E}^{\text{int}} = \frac{1}{\mu_0} \int \boldsymbol{B}_1 \cdot \boldsymbol{B}_2\, d^3x\,. \tag{4.54}$$

By the same manipulations as led to eq. (4.53), we obtain

$$\mathscr{E}^{\text{int}} = \int \boldsymbol{A}_1 \cdot \boldsymbol{J}_2 d^3x = \int \boldsymbol{A}_2 \cdot \boldsymbol{J}_1 d^3x\,. \tag{4.55}$$

Now consider a body with current distribution $\boldsymbol{J}(\boldsymbol{x})$ that is placed in an "external magnetic field" $\boldsymbol{B}^{\text{ext}} = \nabla \times \boldsymbol{A}^{\text{ext}}$ (i.e., $\boldsymbol{A}^{\text{ext}}$ is produced by a "distant current" $\boldsymbol{J}^{\text{ext}}$ that vanishes in the vicinity of the body). The interaction energy is then given by

$$\mathscr{E}^{\text{int}} = \int \boldsymbol{A}^{\text{ext}}(\boldsymbol{x}) \cdot \boldsymbol{J}(\boldsymbol{x}) d^3x\,. \tag{4.56}$$

If we take the body to be located near the origin and if $\boldsymbol{A}^{\text{ext}}$ is slowly varying over the size of the body, we may Taylor expand $\boldsymbol{A}^{\text{ext}}(\boldsymbol{x})$ about $\boldsymbol{x} = 0$. In component notation, we have

$$A_i^{\text{ext}}(\boldsymbol{x}) = A_i^{\text{ext}}|_{\boldsymbol{x}=0} + \sum_j x_j [\partial_j A_i^{\text{ext}}]\big|_{\boldsymbol{x}=0} + \cdots\,. \tag{4.57}$$

Plugging this expansion into eq. (4.56), we obtain

$$\mathscr{E}^{\text{int}} = \boldsymbol{A}^{\text{ext}}(0) \cdot \int \boldsymbol{J}(\boldsymbol{x}) d^3x + \sum_{i,j} [\partial_j A_i^{\text{ext}}]\big|_{\boldsymbol{x}=0} \int x_j J_i(\boldsymbol{x}) d^3x + \cdots\,. \tag{4.58}$$

The first term vanishes by eq. (4.24). To put the second term in a nicer form, we use the definition, eq. (4.34), of the magnetic dipole moment to write

$$\sum_p \epsilon_{pji} \mu_p = \frac{1}{2} \sum_{p,l,n} \epsilon_{pji} \epsilon_{pln} \int x_l J_n d^3x = \int x_j J_i\,, \tag{4.59}$$

where eq. (4.10) and eq. (4.27) have been used. Thus, we obtain

$$\mathscr{E}^{\text{int}} = \sum_{i,j,p} \epsilon_{pji} \mu_p [\partial_j A_i^{\text{ext}}]\big|_{\boldsymbol{x}=0}$$
$$= \boldsymbol{\mu} \cdot \boldsymbol{B}^{\text{ext}}\,. \tag{4.60}$$

By comparison, we found in eq. (2.34) that the the interaction energy of an electric dipole $\boldsymbol{p}$ in an external electric field $\boldsymbol{E}^{\text{ext}}$ is

$$\mathscr{E}^{\text{int}} = -\boldsymbol{p} \cdot \boldsymbol{E}^{\text{ext}} . \tag{4.61}$$

Note the sign difference between eq. (4.60) and eq. (4.61).

The force density exerted by the electromagnetic field on matter in magnetostatics is given by eq. (1.24) with $\boldsymbol{E} = 0$, namely,

$$\boldsymbol{f} = \boldsymbol{J} \times \boldsymbol{B} . \tag{4.62}$$

Here $\boldsymbol{B}$ is the *total* magnetic field. However, a calculation similar to eq. (2.41) shows that the total "self-force" vanishes in magnetostatics (i.e, the contribution to the total force from $\boldsymbol{B} = \nabla \times \boldsymbol{A}$ with $\boldsymbol{A}$ given by eq. (4.22) vanishes; see problem 5 at the end of this chapter). Therefore, if a body with current distribution $\boldsymbol{J}(\boldsymbol{x})$ is placed in an external magnetic field $\boldsymbol{B}^{\text{ext}}$, the total force on the body is given by

$$\boldsymbol{F} = \int \boldsymbol{f}(\boldsymbol{x}) d^3x = \int \boldsymbol{J}(\boldsymbol{x}) \times \boldsymbol{B}(\boldsymbol{x}) d^3x = \int \boldsymbol{J}(\boldsymbol{x}) \times \boldsymbol{B}^{\text{ext}}(\boldsymbol{x}) d^3x . \tag{4.63}$$

As in eq. (4.57) (but now working directly with $\boldsymbol{B}^{\text{ext}}$ instead of $\boldsymbol{A}^{\text{ext}}$), we can Taylor expand $\boldsymbol{B}^{\text{ext}}(\boldsymbol{x})$ about $\boldsymbol{x} = 0$,

$$B_i^{\text{ext}}(\boldsymbol{x}) = B_i^{\text{ext}}|_{\boldsymbol{x}=0} + \sum_j x_j [\partial_j B_i^{\text{ext}}]\big|_{\boldsymbol{x}=0} + \cdots , \tag{4.64}$$

and then plug the result into eq. (4.63). The contribution from the zeroth order term in the Taylor expansion is

$$\left[\int \boldsymbol{J}(\boldsymbol{x}) d^3x \right] \times \boldsymbol{B}^{\text{ext}}(0) = 0, \tag{4.65}$$

where eq. (4.24) has been used. Thus, the leading-order contribution to the total force is

$$F_k = \sum_{m,i,j} \epsilon_{kmi} [\partial_j B_i^{\text{ext}}]\big|_{\boldsymbol{x}=0} \int J_m x_j d^3x . \tag{4.66}$$

Using eq. (4.59), we obtain

$$\begin{aligned} F_k &= \sum_{m,i,j,p} \epsilon_{kmi} \epsilon_{pjm} \mu_p [\partial_j B_i^{\text{ext}}]\big|_{\boldsymbol{x}=0} \\ &= \sum_{i,j,p} \left(\delta_{kj} \delta_{ip} - \delta_{kp} \delta_{ij} \right) \mu_p [\partial_j B_i^{\text{ext}}]\big|_{\boldsymbol{x}=0} \\ &= \sum_i \mu_i [\partial_k B_i^{\text{ext}}]\big|_{\boldsymbol{x}=0} , \end{aligned} \tag{4.67}$$

where $\nabla \cdot \boldsymbol{B}^{\text{ext}} = 0$ was used in the last line. Using $\nabla \times \boldsymbol{B}^{\text{ext}} = 0$ at $\boldsymbol{x} = 0$ (since the current $\boldsymbol{J}^{\text{ext}}$ associated with the external field is "distant"), we can write our expression for

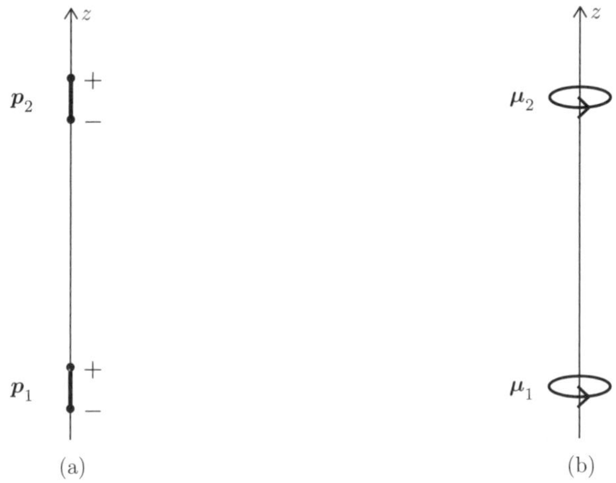

Figure 4.1. Configurations of (a) electrostatic dipoles, p_1 and p_2, and (b) magnetic dipoles, μ_1 and μ_2.

the force in the final form

$$F = (\boldsymbol{\mu} \cdot \nabla)\boldsymbol{B}^{\text{ext}} + \cdots . \tag{4.68}$$

Similarly, the torque, $\boldsymbol{\tau}$, on a body with current distribution $\boldsymbol{J}$ in an external magnetic field $\boldsymbol{B}^{\text{ext}}$ is

$$\boldsymbol{\tau} = \int \boldsymbol{x} \times \boldsymbol{f}(\boldsymbol{x}) d^3x = \int \boldsymbol{x} \times [\boldsymbol{J}(\boldsymbol{x}) \times \boldsymbol{B}(\boldsymbol{x})] d^3x = \int \boldsymbol{x} \times [\boldsymbol{J}(\boldsymbol{x}) \times \boldsymbol{B}^{\text{ext}}(\boldsymbol{x})] d^3x$$
$$= \boldsymbol{\mu} \times \boldsymbol{B}^{\text{ext}} + \cdots . \tag{4.69}$$

We found in eq. (2.44) that the force on an electric dipole $\boldsymbol{p}$ in an external electric field $\boldsymbol{E}^{\text{ext}}$ is

$$\boldsymbol{F} = (\boldsymbol{p} \cdot \nabla)\boldsymbol{E}^{\text{ext}}, \tag{4.70}$$

which is of exactly the same form as eq. (4.68). The fact that eq. (4.68) and eq. (4.70) are of the same form but eq. (4.60) and eq. (4.61) have a sign difference is quite interesting! To see the implications of this, consider the simple example of two charge distributions in electrostatics with vanishing total charge but nonvanishing electrostatic dipoles $\boldsymbol{p}_1$ and $\boldsymbol{p}_2$. We assume that the higher multipole moments of these charge distributions are negligible. A simple case to consider is where both dipoles are aligned in the z-direction and placed on the z-axis, as illustrated in figure 4.1(a). We wish to compare this situation to one in magnetostatics where we have current distributions that produce magnetic dipoles $\boldsymbol{\mu}_1$ and $\boldsymbol{\mu}_2$ and negligible higher multipole moments. We also consider the analogous configuration where both dipoles are aligned in the z-direction and placed on the z-axis, as illustrated in figure 4.1(b). For definiteness, let us view the first of these dipoles ($\boldsymbol{p}_1$ or $\boldsymbol{\mu}_1$) as producing the external field and the second dipole ($\boldsymbol{p}_2$ or, respectively, $\boldsymbol{\mu}_2$) as the body we are considering.

Consider, first, the case of electrostatic charge distributions that produce electric dipoles $\boldsymbol{p}_1$ and $\boldsymbol{p}_2$. According to eq. (4.70), the force between these dipoles is attractive for the configuration shown in figure 4.1(a). Suppose an external agent slowly displaces the dipole $\boldsymbol{p}_2$ toward $\boldsymbol{p}_1$ by $\delta\boldsymbol{x}$, keeping the dipole moments fixed. The work done by the external agent doing this displacement is

$$\delta W = -\boldsymbol{F}\cdot\delta\boldsymbol{x} = -\boldsymbol{p}_2\cdot\delta\boldsymbol{E}_1, \tag{4.71}$$

where $\delta\boldsymbol{E}_1 \equiv (\delta\boldsymbol{x}\cdot\nabla)\boldsymbol{E}_1$ is the change in the external field $\boldsymbol{E}_1$ experienced by the second body. Thus, energy $\boldsymbol{p}_2\cdot\delta\boldsymbol{E}_1$ has been extracted from the system by the external agent. Where did this energy come from? The answer is quite simple: It came from the electromagnetic interaction energy, eq. (4.61), which changes by $-\boldsymbol{p}_2\cdot\delta\boldsymbol{E}_1$. This is consistent with the fact that no energy cost should be associated with maintaining the electric dipole moments $\boldsymbol{p}_1$ and $\boldsymbol{p}_2$ as the second body is moved by the external agent in the field of the first body.

Now let us consider the corresponding situation for magnetic dipoles $\boldsymbol{\mu}_1$ and $\boldsymbol{\mu}_2$. Apart from a factor of $\epsilon_0\mu_0 = 1/c^2$, the force between these dipoles is of exactly the same form as in the electrostatic case, so they also attract each other. Suppose an external agent slowly displaces the dipole $\boldsymbol{\mu}_2$ toward $\boldsymbol{\mu}_1$. For definiteness, suppose that $\boldsymbol{\mu}_2$ and $\boldsymbol{\mu}_1$ are fixed during the displacement, as would be the case if these objects were permanent magnets (see section 4.4). Then, just as in the electrostatic case,

$$\delta W = -\boldsymbol{F}\cdot\delta\boldsymbol{x} = -\boldsymbol{\mu}_2\cdot\delta\boldsymbol{B}_1, \tag{4.72}$$

so the energy $\boldsymbol{\mu}_2\cdot\delta\boldsymbol{B}_1$ is extracted from the system by the external agent. But the sign difference between eq. (4.60) and eq. (4.61) implies that the magnetic interaction energy of the system goes *up* by $\boldsymbol{\mu}_2\cdot\delta\boldsymbol{B}_1$. We have extracted energy from the system and yet the magnetic interaction energy has gotten *bigger*! Where did this energy come from?

The answer is that the energy $\boldsymbol{\mu}_2\cdot\delta\boldsymbol{B}_1$ comes from the rest mass[5]/internal energy of the body with magnetic dipole $\boldsymbol{\mu}_2$, and the energy $\boldsymbol{\mu}_1\cdot\delta\boldsymbol{B}_2 = \boldsymbol{\mu}_2\cdot\delta\boldsymbol{B}_1$ comes from the rest mass/internal energy of the body with dipole $\boldsymbol{\mu}_1$ that produces the "external field." To see this, we have to go beyond magnetostatics to electrodynamics, because even though the external agent may displace the body with dipole $\boldsymbol{\mu}_2$ arbitrarily slowly, the time-integrated dynamical effects cannot be ignored.

As already seen in eq. (1.23) and as will be discussed further in chapter 5, the rate at which energy is transferred from an electromagnetic field to the nonelectromagnetic energy of a body with current distribution $\boldsymbol{J}$ is

$$\frac{d\mathcal{E}_{\text{matter}}}{dt} = \int \boldsymbol{E}\cdot\boldsymbol{J}\,d^3x. \tag{4.73}$$

Note the absence of $\boldsymbol{B}$ in this formula. If we wish to know the change in nonelectromagnetic energy of the second body above, then we replace $\boldsymbol{J}$ by $\boldsymbol{J}_2$ in this formula. The electric field $\boldsymbol{E}$ in this formula is the full electromagnetic field. However, if the electromagnetic field is quasi-static, the self-field contribution $\int \boldsymbol{E}_2\cdot\boldsymbol{J}_2\,d^3x$ merely describes

[5]By the "rest mass" of a body, we mean its total (relativistic) energy in the rest frame of its center of mass. For example, the rest mass of a box of gas will increase if it is heated up. Thus, "rest mass" is synonymous with "internal energy"—provided that the internal energy is the full, relativistic energy.

the exchange of the electromagnetic self-energy of the body with its nonelectromagnetic self-energy.[6] We are not interested in the internal shifting of energy in the second body but only its total change of self-energy, which is thus given by

$$\frac{d\mathcal{E}_2^{\text{self}}}{dt} = \int \boldsymbol{E}_1 \cdot \boldsymbol{J}_2 d^3x = 0 \tag{4.74}$$

since $\boldsymbol{E}_1 = 0$. Thus, if the external field is purely magnetic, it is impossible for the body to exchange energy with the electromagnetic field. Note that this result holds for an arbitrary external magnetic field and a body with an arbitrary current distribution in arbitrary quasi-static motion. It corresponds to the often quoted statement that "a magnetic field can do no work."

Since the external agent interacts only with the second body, and no energy can be exchanged between the second body and the electromagnetic field, it follows that the work done by the external agent must be balanced by a change in the self-energy, $\mathcal{E}_2^{\text{self}}$, of the second body:

$$\delta\mathcal{E}_2^{\text{self}} = \delta W = -\boldsymbol{\mu}_2 \cdot \delta\boldsymbol{B}_1 \,. \tag{4.75}$$

Thus, the self-energy of the second body—which also may be referred to as its "rest mass" or its "internal energy" (see footnote 5)—decreases by $\boldsymbol{\mu}_2 \cdot \delta\boldsymbol{B}_1$, as claimed above.

The analysis of the change of rest mass/internal energy of the body producing the external current distribution with magnetic dipole $\boldsymbol{\mu}_1$ is quite different. Since body 1 interacts only with the electromagnetic field, its change in self-energy is given by

$$\frac{d\mathcal{E}_1^{\text{self}}}{dt} = \int \boldsymbol{E}_2 \cdot \boldsymbol{J}_1 d^3x \,. \tag{4.76}$$

As the dipole $\boldsymbol{\mu}_2$ is moved by the external agent, its magnetic field $\boldsymbol{B}_2(\boldsymbol{x})$ will be time dependent. However, by Maxwell's equation (1.4), this implies that an electric field $\boldsymbol{E}_2$ must be present, whose curl is given by

$$\nabla \times \boldsymbol{E}_2 = -\frac{\partial \boldsymbol{B}_2}{\partial t} \,. \tag{4.77}$$

Taylor expanding $\boldsymbol{E}_2$ in the vicinity of the external current distribution—which, for convenience, we take to be located near $\boldsymbol{x} = 0$—we obtain

$$(E_2)_i(\boldsymbol{x}) = (E_2)_i|_{\boldsymbol{x}=0} + \sum_j x_j[\partial_j(E_2)_i]\big|_{\boldsymbol{x}=0} + \cdots \,. \tag{4.78}$$

Plugging this into eq. (4.76) and using eq. (4.24) and eq. (4.59), we obtain

$$\frac{d\mathcal{E}_1^{\text{self}}}{dt} = \boldsymbol{\mu}_1 \cdot (\nabla \times \boldsymbol{E}_2) = -\boldsymbol{\mu}_1 \cdot \frac{\partial \boldsymbol{B}_2}{\partial t} \tag{4.79}$$

and hence

$$\delta\mathcal{E}_1^{\text{self}} = -\boldsymbol{\mu}_1 \cdot \delta\boldsymbol{B}_2 \tag{4.80}$$

[6] In the non-quasi-static case, this exchange of energy involving the second body can include the emission of electromagnetic radiation, which would not be viewed as part of the self-energy of the body.

independently of the details of how the second body is moved. Thus, the rest mass/internal energy of the body producing the external field also changes by $-\boldsymbol{\mu}_1 \cdot \delta\boldsymbol{B}_2 = -\boldsymbol{\mu}_2 \cdot \delta\boldsymbol{B}_1$, as we desired to show. This energy is extracted from the first body by the electromagnetic field.

It is striking that such different calculations lead to such similar results in eq. (4.75) and eq. (4.80). It is particularly striking that if we did the analysis from the point of view of an observer moving with the second body as it is being displaced, then the roles of the first and second bodies would be reversed in the above analysis, but we would end up with the same answers. It was exactly considerations of this sort (in different examples) that led Einstein to conclude that there must be a fundamental relativistic invariance in electrodynamics, despite the blatant failure of Maxwell's equations to take the same form in a moving frame related by a Galilean transformation. This led directly to the formulation of the theory of special relativity.

It is also worth pointing out that the above analysis resolves what should appear to be a blatant paradox to anyone who has taken an introductory course in electromagnetism. Students are taught that "a magnetic field can do no work"—and this is correct, as we have seen above. Nevertheless, as anyone who has ever held a magnetic body near a magnet has observed, if you let go of that body, it will go flying toward the magnet. The body clearly will gain kinetic energy as it moves toward the magnet. Where does this kinetic energy of the body come from? It cannot come from the energy stored in the magnetic field of the magnet or from the magnetic interaction energy of the body and the magnet (which is of the wrong sign in any case), since there can be no transfer of electromagnetic energy to the body. There is no external agent involved after the body is released, so the kinetic energy of the body cannot come from an external agent. Thus, it might appear that there is a blatant contradiction with conservation of energy. The resolution of this apparent contradiction is that the increased kinetic energy of the body comes from the body's own rest mass. The total energy of the body (i.e., its rest mass plus its kinetic energy) remains unchanged as it flies toward the magnet, in accord with the fact that the magnetic field provided by the magnet has "done no work" on the body.

Finally, note that if we wish to consider the behavior of fundamental continuum charged matter interacting with the electromagnetic field, the dynamics of the full system would be described by adding the Lagrangian of the charged matter to the Lagrangian of the electromagnetic field (see chapter 9). The Lagrangian of the charged matter contains terms involving the electromagnetic field (see eq. (9.34) of chapter 9 for the case of a charged scalar field). For a Dirac field, these matter terms yield a contribution to the matter Hamiltonian of the form $-\boldsymbol{\mu}_S \cdot \boldsymbol{B}^{\text{ext}}$, where $\boldsymbol{\mu}_S$ is the magnetic dipole moment associated with the spin of the charged Dirac field. There also will be matter terms arising from the current of the charged field that make a contribution $-\boldsymbol{\mu}_J \cdot \boldsymbol{B}^{\text{ext}}$ to the matter Hamiltonian in the dipole approximation, where $\boldsymbol{\mu}_J$ is the magnetic dipole moment of the current distribution. Note that the presence of terms of the form $-\boldsymbol{\mu} \cdot \boldsymbol{B}^{\text{ext}}$ in the matter Hamiltonian should be expected from eqs. (4.75) and (4.80) above. If one is studying the dynamics of an electron, the purely electromagnetic part of the Hamiltonian makes no contribution to the equations of motion of the electron, so it is only the matter Hamiltonian that is relevant. Thus, the expression $-\boldsymbol{\mu} \cdot \boldsymbol{B}^{\text{ext}}$ is the correct term to add to the quantum mechanical Hamiltonian describing electrons placed in an external magnetic field—as can be found in quantum mechanics texts—even though $+\boldsymbol{\mu} \cdot \boldsymbol{B}^{\text{ext}}$ is the correct formula for the electromagnetic interaction energy of the magnetic field of a magnetic dipole with an external magnetic field.

4.4 Magnetic Materials

Just as many materials display dielectric properties, many materials display magnetic properties. These materials may be modeled as collections of "atoms" (which need not literally be atoms), each of which has a current distribution that contributes a magnetic dipole moment. In addition, we may also allow for the presence of free currents, $\boldsymbol{J}_f$ (i.e., flows of charge not associated with the current distributions in the atoms).

We can analyze magnetic materials in exactly the same way as dielectrics by performing the kind of averaging described in detail in chapter 3. Since averaging commutes with differentiation, we have

$$\langle \boldsymbol{B} \rangle = \nabla \times \langle \boldsymbol{A} \rangle, \tag{4.81}$$

and

$$\nabla \times \langle \boldsymbol{B} \rangle = \mu_0 \langle \boldsymbol{J} \rangle . \tag{4.82}$$

When we compute $\langle \boldsymbol{J} \rangle$, we obtain an analog of eq. (3.10), with q_j replaced by $\int \boldsymbol{J}_j$ and $\boldsymbol{p}_j$ replaced by an integral involving $(\boldsymbol{x} - \boldsymbol{x}_j)$ times $\boldsymbol{J}_j$. Using eq. (4.24) and eq. (4.27) and then performing a computation analogous to that leading to eq. (3.13), we obtain

$$\langle \boldsymbol{J} \rangle = \langle \boldsymbol{J}_f \rangle + \nabla \times \langle \boldsymbol{M} \rangle, \tag{4.83}$$

where $\boldsymbol{M}$ denotes the magnetic dipole moment density or *magnetization* (analogous to $\boldsymbol{P}$ defined by eq. (3.12)). The quantity $\nabla \times \langle \boldsymbol{M} \rangle$ is referred to as the *magnetization current density*. It is then convenient to define the quantity[7] $\langle \boldsymbol{H} \rangle$ by

$$\langle \boldsymbol{H} \rangle = \frac{1}{\mu_0} \langle \boldsymbol{B} \rangle - \langle \boldsymbol{M} \rangle . \tag{4.84}$$

The full content of the average of Maxwell's equations of magnetostatics in a magnetic material is then expressed by eq. (4.81) together with

$$\nabla \times \langle \boldsymbol{H} \rangle = \langle \boldsymbol{J}_f \rangle . \tag{4.85}$$

At the boundary of a magnetic material, one assumes that $\langle \boldsymbol{M} \rangle$ and $\langle \boldsymbol{J}_f \rangle$ have no δ-function surface contributions (i.e., they cannot be worse than discontinuous at the boundary). It follows that $\langle \boldsymbol{A} \rangle$ (in the gauge $\nabla \cdot \langle \boldsymbol{A} \rangle = 0$) must be continuous, from which it follows that $\hat{\boldsymbol{n}} \cdot \langle \boldsymbol{B} \rangle$ is continuous. In addition, eq. (4.85) implies that the tangential components $\langle \boldsymbol{H}_{\parallel} \rangle$ of $\langle \boldsymbol{H} \rangle$ must be continuous.

As in the case of dielectrics, to proceed further, one must know something more about $\langle \boldsymbol{M} \rangle$ and its relationship to $\langle \boldsymbol{B} \rangle$ and/or $\langle \boldsymbol{H} \rangle$. There are many linear, homogeneous, and isotropic materials for which $\langle \boldsymbol{M} \rangle$ is proportional to $\langle \boldsymbol{B} \rangle$. For historical reasons, this proportionality is usually expressed as a proportionality between $\langle \boldsymbol{M} \rangle$ and $\langle \boldsymbol{H} \rangle$, and is expressed as

$$\langle \boldsymbol{M} \rangle = \chi_m \langle \boldsymbol{H} \rangle, \tag{4.86}$$

where χ_m is called the *magnetic susceptiblity*. Equivalently, this linear relationship may be written as

$$\langle \boldsymbol{B} \rangle = \mu \langle \boldsymbol{H} \rangle, \tag{4.87}$$

[7]Unfortunately, for historical reasons, $\langle \boldsymbol{H} \rangle$ is often referred to as "the magnetic field." When this terminology is used, $\boldsymbol{B}$ and/or $\langle \boldsymbol{B} \rangle$ is then called the "magnetic induction." We will not use this terminology here and, in this book, "magnetic field" will always mean $\boldsymbol{B}$.

where

$$\mu = \mu_0(1 + \chi_m) \tag{4.88}$$

is called the *magnetic permeability*. One difference between magnetic materials and dielectric materials is that it is common to have magnetic materials for which χ_m is negative. Such materials are called *diamagnetic*, whereas materials with $\chi_m > 0$ are called *paramagnetic*. Paramagnetic materials with very large χ_m can be used for magnetic shielding (see problem 9).

One other notable difference from dielectric materials is that for some magnetic materials—most notably iron—the atomic dipole moments are sufficiently large that alignment of the dipoles under their own influence can occur (i.e., a nonvanishing $\langle \boldsymbol{M} \rangle$ can occur even if no external field is applied). Such materials are known as *permanent magnets* (see problem 10). At sufficiently high temperatures, alignment of the dipoles will not be thermodynamically favored, and we will have $\langle \boldsymbol{M} \rangle = 0$ if no external field is applied. At temperatures below which the magnetic dipole alignment is thermodynamically favored, the dipoles will typically be aligned on domains of small scale, but the alignment of the domains will depend on how the system was prepared. Thus, the magnetic field of a permanent magnet will typically depend on the past history of the body. This phenomenon is known as *hysteresis*.

Problems

1. Suppose that the current density $\boldsymbol{J}$ is nonvanishing only in a region that has the topology of a solid torus (i.e., suppose that $\boldsymbol{J}$ is nonvanishing only in a loop of wire, but the wire may have arbitrary, variable thickness and may be bent into an arbitrary shape). The current I flowing through an arbitrary cross section S of the torus is defined by

$$I = \int_S \boldsymbol{J} \cdot \hat{\boldsymbol{n}}\, dA,$$

where $\hat{\boldsymbol{n}}$ is unit normal to S.

(a) Show that I is independent of the choice of cross section S, and thus one may speak of the current I flowing through the loop of wire without having to specify S.

(b) Now suppose that the wire is infinitesimally thin. Show that the magnetic field B of the wire is given by the *Biot-Savart law*

$$\boldsymbol{B}(\boldsymbol{x}) = \frac{\mu_0 I}{4\pi} \oint \frac{d\boldsymbol{l}' \times (\boldsymbol{x} - \boldsymbol{x}')}{|\boldsymbol{x} - \boldsymbol{x}'|^3},$$

where the line integral is taken over the loop of wire.

2. A spherical shell of radius R has a total charge Q uniformly spread over the shell. The shell is now put into uniform rotation about the $\hat{\boldsymbol{z}}$-axis with angular velocity ω. Find the vector potential $\boldsymbol{A}(\boldsymbol{x})$ and magnetic field $\boldsymbol{B}(\boldsymbol{x})$ everywhere (i.e., both inside and outside this shell).

3. (a) Consider an arbitrary connected region $\mathcal{V}$ in which $\boldsymbol{J} = 0$. Suppose that, everywhere in this region, $\boldsymbol{B}$ points in the z-direction (i.e., within $\mathcal{V}$ we

have $\boldsymbol{B}=f(x,y,z)\hat{\boldsymbol{z}}$). Show that, in fact, in $\mathcal{V}$ the magnetic field must be uniform: $\boldsymbol{B}=B_0\hat{\boldsymbol{z}}$, where B_0 is a constant.

(b) Consider a current distribution $\boldsymbol{J}(\boldsymbol{x})$ such that (i) $\boldsymbol{J}$ has translational symmetry in the z-direction (i.e., $\boldsymbol{J}$ depends only on (x,y)); (ii) $\hat{\boldsymbol{z}}\cdot\boldsymbol{J}=0$ (i.e., $\boldsymbol{J}$ points in the x-y plane); and (iii) $\boldsymbol{J}=0$ outside a bounded region in (x,y). An example of such a current distribution is that of an infinite solenoid (where we neglect any current flowing in the z-direction). Show that $\boldsymbol{B}$ points in the z-direction everywhere. [Note: Since $\boldsymbol{J}$ does not go to zero at infinity near the z-axis, the usual argument for uniqueness of solutions does not apply. Nevertheless, it is true (and you may assume) that there is a unique solution for $\boldsymbol{A}$ in the Coulomb gauge such that $|\boldsymbol{A}|\to 0$ as $x^2+y^2\to\infty$.]

(c) Now consider an infinite, straight solenoid of arbitrary cross-sectional shape (i.e., do *not* assume that the cross sections of the solenoid are circular). Show that $\boldsymbol{B}=0$ outside the solenoid and $\boldsymbol{B}$ is uniform inside the solenoid.

4. *Ampère's law* is obtained by applying Stokes's theorem to eq. (4.1):

$$\oint \boldsymbol{B}\cdot d\boldsymbol{l}=\int_S(\nabla\times\boldsymbol{B})\cdot\hat{\boldsymbol{n}}\,dA=\mu_0\int_S\boldsymbol{J}\cdot\hat{\boldsymbol{n}}\,dA\,.$$

Although the differential version, eq. (4.1), is generally much more useful (since it leads to the general solution eq. (4.22)), Ampère's law allows one to easily calculate $\boldsymbol{B}$ in certain situations of high symmetry. Use Ampère's law to obtain $\boldsymbol{B}$ in the following two cases:

(a) An infinite, straight wire with circular cross section of radius a has a uniform current density $\boldsymbol{J}$ in the wire and carries a total current I. Find $\boldsymbol{B}$ everywhere (i.e., both outside and inside the wire).

(b) An infinite solenoid has N turns per unit length of wire carrying a current I. As found in part (c) of problem 3, the magnetic field inside the solenoid is uniform. Find the magnitude $|\boldsymbol{B}|$ of $\boldsymbol{B}$ inside the solenoid.

5. Show that the total magnetic force exerted on a current distribution $\boldsymbol{J}(\boldsymbol{x})$ by the field eq. (4.22) associated with $\boldsymbol{J}$ vanishes (i.e., the total self-force vanishes in magnetostatics).

6. A strip of a semiconducting material of thickness d (in the z-direction) and width w (in the y-direction) has a number density n of charges—each of charge q—that are free to move about the strip. An external voltage is applied in the x-direction, and a uniform magnetic field $\boldsymbol{B}$ is applied in the z-direction. As a result of the external voltage, a current density $\boldsymbol{J}$ flows in the x-direction, as shown in the figure. Assume that n and $\boldsymbol{J}$ are uniform throughout the strip. In this situation, a voltage (i.e., potential difference) V_{H} will develop between the ends of the strip in the y-direction. Show that

$$V_{\mathrm{H}}=-\frac{IB}{nqd}\,,$$

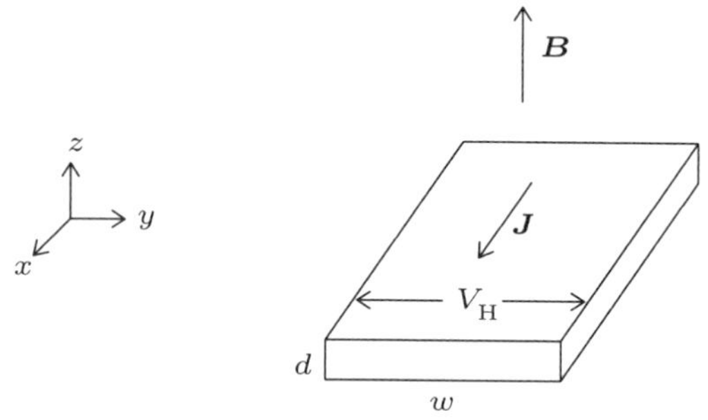

where $I = Jwd$ is the current flowing through the film in the x-direction. The production of such a voltage by this means is known as the (classical) *Hall effect*. [Note: For a thin film in a strong magnetic field, V_H/I is observed to increase with B via a sequence of regularly spaced plateaus rather than the simple linear relation given above. This phenomenon, known as the *quantum Hall effect*, has been—and continues to be—the subject of much theoretical study.]

7. If an electric and magnetic field are both present, the momentum density carried by the electromagnetic field is given by

$$\mathcal{P} = \epsilon_0 \boldsymbol{E} \times \boldsymbol{B}$$

(see eq. (1.16)). Consider a bounded distribution of time-independent charges and currents (i.e., $\rho(\boldsymbol{x})$ and $\boldsymbol{J}(\boldsymbol{x})$ are time independent and vanish when $|\boldsymbol{x}| > R$ for some R).

(a) Show that the total momentum can be written as

$$\boldsymbol{P} \equiv \int \mathcal{P}(\boldsymbol{x}) d^3x = \epsilon_0 \mu_0 \int \phi(\boldsymbol{x}) \boldsymbol{J}(\boldsymbol{x}) d^3x = \frac{1}{c^2} \int \phi(\boldsymbol{x}) \boldsymbol{J}(\boldsymbol{x}) d^3x \,.$$

(b) Give an example of a stationary, bounded charge and current distribution for which $\boldsymbol{P} \neq 0$. (The momentum carried by a stationary electromagnetic field is sometimes referred to as *hidden momentum*.)

8. Consider two disjoint, infinitesimally thin loops of wire. Let $\boldsymbol{A}_1$ be the vector potential corresponding to the case where the first loop carries unit current, $I_1 = 1$, and the second loop carries vanishing current, $I_2 = 0$. Define the *mutual inductance* M_{12} of the loops of wire by

$$M_{12} = \frac{\mu_0}{4\pi} \oint \boldsymbol{A}_1 \cdot d\boldsymbol{l}_2,$$

where the line integral is taken over the second loop of wire. The mutual inductance M_{21} is defined similarly.[8]

(a) Show that M_{12} is gauge invariant.
(b) Show that $M_{12} = M_{21}$.

[8] One also could similarly define the *self-inductance* of each of the loops. However, this is infinite for an infinitesimally thin loop of wire (i.e., one would have to consider a wire of finite thickness to define this notion).

(c) Show that when the loops carry arbitrary currents I_1 and I_2, the magnetic interaction energy of the two loops of wire is

$$\mathscr{E}^{\text{int}} = M_{12} I_1 I_2 \,.$$

9. A material of magnetic permeability μ fills a spherical shell of inner radius a and outer radius b, as shown.

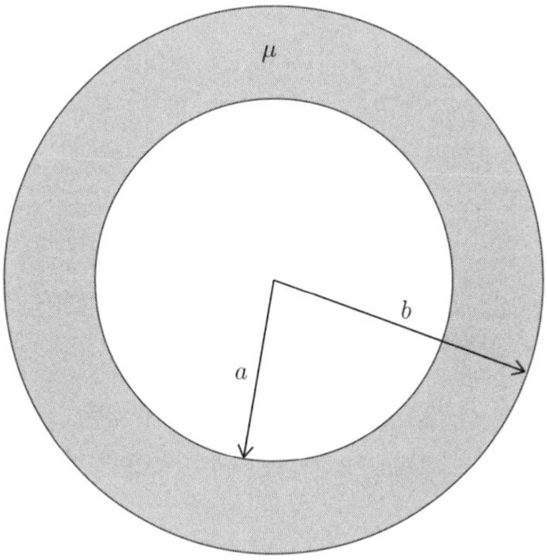

This spherical shell is placed in an originally uniform magnetic field $B_0\hat{\boldsymbol{z}}$. Find the magnetic field in the region $r < a$. Note that for $\mu > \mu_0$, the magnetic field is smaller in magnitude than B_0; a shell made of a material with large μ can therefore be used for *magnetic shielding*. [Hint: One can introduce magnetic scalar potentials for each of the three separate regions $r < a$, $a < r < b$, and $r > b$.]

10. A permanent magnet in the form of a ball of radius R has uniform magnetic dipole moment density $\langle \boldsymbol{M} \rangle = M_0\hat{\boldsymbol{z}}$. Find the magnetization current $\boldsymbol{\nabla} \times \langle \boldsymbol{M} \rangle$. (Note that this current has only a surface contribution.) Obtain $\langle \boldsymbol{B} \rangle$ both inside and outside the magnet. [Hint: One can introduce magnetic scalar potentials for $r < R$ and $r > R$.]

CHAPTER 5

Electrodynamics

In this chapter, we consider the general case where $\phi, \boldsymbol{A}$ and $\rho, \boldsymbol{J}$ are time dependent. The equations of electrodynamics are discussed in section 5.1, and the Lorenz gauge is introduced. We solve for the retarded Green's function in section 5.2, from which the solution to Maxwell's equations for general $\rho, \boldsymbol{J}$ with no incoming radiation can be obtained. Multipole expansions of the electromagnetic field of the retarded solution are obtained in section 5.3. The retarded Green's function is then used in section 5.4 to obtain the solution to Maxwell's equations with prescribed values of the electromagnetic field at an initial time. We discuss plane wave solutions in section 5.5. The chapter concludes with a discussion of electrodynamics in conducting cavities and waveguides in section 5.6.

5.1 The Equations of Electrodynamics

The equations of electrodynamics have been presented in chapter 1. As discussed in section 1.1, the fundamental dynamical variables are the potentials ϕ and $\boldsymbol{A}$, which are considered to be equivalent if and only if they differ by a gauge transformation:

$$\phi \to \phi' = \phi - \frac{\partial \chi}{\partial t}, \qquad \boldsymbol{A} \to \boldsymbol{A}' = \boldsymbol{A} + \nabla \chi . \tag{5.1}$$

The equations of electrodynamics (i.e., Maxwell's equations) are

$$\boldsymbol{E} = -\nabla \phi - \frac{\partial \boldsymbol{A}}{\partial t}, \tag{5.2}$$

$$\boldsymbol{B} = \nabla \times \boldsymbol{A}, \tag{5.3}$$

$$\nabla \cdot \boldsymbol{E} = \frac{\rho}{\epsilon_0}, \tag{5.4}$$

$$\nabla \times \boldsymbol{B} - \frac{1}{c^2}\frac{\partial \boldsymbol{E}}{\partial t} = \mu_0 \boldsymbol{J} . \tag{5.5}$$

The first two of these equations define $\boldsymbol{E}$ and $\boldsymbol{B}$ in terms of ϕ and $\boldsymbol{A}$. They imply that

$$\nabla \cdot \boldsymbol{B} = 0, \tag{5.6}$$

$$\nabla \times \boldsymbol{E} + \frac{\partial \boldsymbol{B}}{\partial t} = 0\,. \tag{5.7}$$

Equations (5.6) and (5.7) are equivalent to eqs. (5.2) and (5.3) in a topologically trivial region[1] (i.e., if eqs. (5.6) and (5.7) hold in a topologically trivial region, they imply the existence of potentials ϕ and $\boldsymbol{A}$ in that region satisfying eqs. (5.2) and (5.3)). Taking the time derivative of eq. (5.4) and adding it to c^2 times the divergence of eq. (5.5)—using the fact that $c^2 = 1/(\epsilon_0\mu_0)$—we obtain the charge-current conservation law:

$$\frac{\partial \rho}{\partial t} + \nabla \cdot \boldsymbol{J} = 0\,. \tag{5.8}$$

It is worth noting that in the source-free case (i.e., when $\rho = \boldsymbol{J} = 0$), except for one sign difference and factors of c, eqs. (5.6) and (5.7) take the same form as eqs. (5.4) and (5.5) with $\boldsymbol{E}$ and $\boldsymbol{B}$ interchanged. It follows immediately that in any source-free, topologically trivial region, Maxwell's equations are invariant under a *duality transformation*:

$$\boldsymbol{E} \to c\boldsymbol{B}, \qquad c\boldsymbol{B} \to -\boldsymbol{E}, \tag{5.9}$$

that is, if $\boldsymbol{E}$, $\boldsymbol{B}$ solve eqs. (5.4)–(5.7) with $\rho = 0$, $\boldsymbol{J} = 0$, then so do $\boldsymbol{E}'$, $\boldsymbol{B}'$ with $\boldsymbol{E}' = c\boldsymbol{B}$, $\boldsymbol{B}' = -\boldsymbol{E}/c$. More generally, for any real number α, the source-free Maxwell equations (5.4)–(5.7) are invariant under the *duality rotation*:

$$\boldsymbol{E} \to \cos\alpha \boldsymbol{E} + \sin\alpha (c\boldsymbol{B}), \qquad \boldsymbol{B} \to \cos\alpha \boldsymbol{B} - \sin\alpha (\boldsymbol{E}/c)\,. \tag{5.10}$$

As already discussed in section 1.2, the energy density, momentum density, and stress tensor of the electromagnetic field are, respectively,

$$\mathcal{E} = \frac{1}{2}\left(\epsilon_0|\boldsymbol{E}|^2 + \frac{1}{\mu_0}|\boldsymbol{B}|^2\right), \tag{5.11}$$

$$\mathcal{P} = \epsilon_0 \boldsymbol{E} \times \boldsymbol{B}, \tag{5.12}$$

$$\Theta_{ij} = \epsilon_0 E_i E_j + \frac{1}{\mu_0} B_i B_j - \frac{1}{2}\delta_{ij}\left(\epsilon_0|\boldsymbol{E}|^2 + \frac{1}{\mu_0}|\boldsymbol{B}|^2\right). \tag{5.13}$$

The total energy and momentum of the electromagnetic field are obtained by integrating eq. (5.11) and eq. (5.12) over all of space.

In special relativity, there is no distinction between a "flow of mass" (i.e., momentum) and a "flow of energy"—apart from a factor of c^2 needed to give these quantities conventional units. Thus, c^2 times the momentum density of the electromagnetic field

---

[1]More precisely, by a "topologically trivial region" in the present context, I mean a region in which any closed 2-dimensional surface S is the boundary of a 3-dimensional (compact) volume, and every closed loop C is the boundary of a 2-dimensional (compact) surface. The necessary and sufficient condition for the existence of a vector potential $\boldsymbol{A}$ such that $\boldsymbol{B} = \nabla \times \boldsymbol{A}$ is that $\int_S \boldsymbol{B} \cdot \hat{\boldsymbol{n}} = 0$ for any closed 2-dimensional surface S. The necessary and sufficient condition for the existence of a scalar potential ϕ satisfying eq. (5.2) is $\int_C [\boldsymbol{E} + \frac{\partial \boldsymbol{A}}{\partial t}] \cdot d\boldsymbol{l} = 0$ for any closed loop C. If eqs. (5.6) and (5.7) hold in a topologically trivial region, then the necessary and sufficient conditions for the existence of $\boldsymbol{A}$ will automatically hold by Gauss's theorem, and the necessary and sufficient conditions for the existence of ϕ will automatically hold by Stokes's theorem.

gives the electromagnetic energy flux,

$$\boldsymbol{\mathcal{S}} = c^2 \boldsymbol{\mathcal{P}} = c^2 \epsilon_0 \boldsymbol{E} \times \boldsymbol{B} = \frac{1}{\mu_0} \boldsymbol{E} \times \boldsymbol{B}, \tag{5.14}$$

where, again, we use the fact that $c^2 = 1/(\epsilon_0 \mu_0)$. $\boldsymbol{\mathcal{S}}$ is known as the *Poynting vector*. For a surface S, the flux $\mathcal{F}$ of electromagnetic energy through S (i.e., the electromagnetic energy that flows through S per unit area per unit time) is given by

$$\mathcal{F} = \boldsymbol{\mathcal{S}} \cdot \hat{\boldsymbol{n}}, \tag{5.15}$$

where $\hat{\boldsymbol{n}}$ is the unit normal to S. The flow of electromagnetic energy out of a volume $\mathcal{V}$ bounded by surface S is thus given by

$$\int_S \mathcal{F} dS = \int_S \boldsymbol{\mathcal{S}} \cdot \hat{\boldsymbol{n}} dS = \int_{\mathcal{V}} \nabla \cdot \boldsymbol{\mathcal{S}} dV. \tag{5.16}$$

If electromagnetic energy were conserved by itself, we would have $\frac{d}{dt} \int_{\mathcal{V}} \mathcal{E} dV = \int_{\mathcal{V}} \frac{\partial \mathcal{E}}{\partial t} dV = -\int_{\mathcal{V}} \nabla \cdot \boldsymbol{\mathcal{S}} dV$, and therefore (since $\mathcal{V}$ is arbitrary), $\partial \mathcal{E}/\partial t = -\nabla \cdot \boldsymbol{\mathcal{S}}$. However, a direct computation yields

$$\begin{aligned}
\frac{\partial \mathcal{E}}{\partial t} + \nabla \cdot \boldsymbol{\mathcal{S}} &= \epsilon_0 \boldsymbol{E} \cdot \frac{\partial \boldsymbol{E}}{\partial t} + \frac{1}{\mu_0} \boldsymbol{B} \cdot \frac{\partial \boldsymbol{B}}{\partial t} + \frac{1}{\mu_0} \nabla \cdot (\boldsymbol{E} \times \boldsymbol{B}) \\
&= \epsilon_0 \boldsymbol{E} \cdot (c^2 \nabla \times \boldsymbol{B} - c^2 \mu_0 \boldsymbol{J}) + \frac{1}{\mu_0} \boldsymbol{B} \cdot (-\nabla \times \boldsymbol{E}) + \frac{1}{\mu_0} \nabla \cdot (\boldsymbol{E} \times \boldsymbol{B}) \\
&= -\boldsymbol{E} \cdot \boldsymbol{J} + \frac{1}{\mu_0} [\boldsymbol{E} \cdot (\nabla \times \boldsymbol{B}) - \boldsymbol{B} \cdot (\nabla \times \boldsymbol{E}) + \nabla \cdot (\boldsymbol{E} \times \boldsymbol{B})] \\
&= -\boldsymbol{E} \cdot \boldsymbol{J},
\end{aligned} \tag{5.17}$$

where eqs. (5.5) and (5.7) were used in the second line, the relation $c^2 \epsilon_0 \mu_0 = 1$ was used in the third line, and the identity

$$\begin{aligned}
\nabla \cdot (\boldsymbol{E} \times \boldsymbol{B}) &= \sum_{i,j,k} \epsilon_{ijk} \partial_i (E_j B_k) = \sum_{i,j,k} \epsilon_{ijk} \left[(\partial_i E_j) B_k + E_j \partial_i B_k \right] \\
&= \boldsymbol{B} \cdot (\nabla \times \boldsymbol{E}) - \boldsymbol{E} \cdot (\nabla \times \boldsymbol{B})
\end{aligned} \tag{5.18}$$

was used to get the last line. This means that if total energy (i.e., the energy of the electromagnetic field plus the energy of matter) is to be conserved, then the electromagnetic field must be adding energy density to the matter at the rate

$$\frac{\partial \mathcal{E}_{\text{matter}}}{\partial t} = \boldsymbol{J} \cdot \boldsymbol{E}. \tag{5.19}$$

Thus $\boldsymbol{J} \cdot \boldsymbol{E}$ is the rate at which energy per unit volume is transferred from the electromagnetic field to matter.

In a similar manner and by a similar calculation, the failure of momentum conservation to hold for the electromagnetic field momentum alone is given by

$$\frac{\partial \mathcal{P}_i}{\partial t} - \sum_{j=1}^{3} \partial_j \Theta_{ij} = - \left[\rho E_i + (\boldsymbol{J} \times \boldsymbol{B})_i \right] . \tag{5.20}$$

If total momentum is to be conserved, then the rate of change of the momentum density of matter must be given by minus the right side of eq. (5.20). In other words, the electromagnetic field must exert a force per unit volume, $\boldsymbol{f}$, on matter given by

$$\boldsymbol{f} = \rho \boldsymbol{E} + \boldsymbol{J} \times \boldsymbol{B} . \tag{5.21}$$

which is referred to as the *Lorentz force*.

It should be emphasized that the entire content of electromagnetic theory is expressed by Maxwell's equations (5.2)–(5.5); equations (5.11)–(5.13) for energy density, momentum density, and stress of the electromagnetic field; and equations (5.19) and (5.21), which express that energy and momentum are conserved for the total system composed of the electromagnetic field and matter.

We now substitute eqs. (5.2) and (5.3) into eqs. (5.4) and (5.5) to write Maxwell's equations purely in terms of ϕ and $\boldsymbol{A}$:

$$-\nabla^2 \phi - \frac{\partial}{\partial t} \nabla \cdot \boldsymbol{A} = \frac{\rho}{\epsilon_0}, \tag{5.22}$$

$$-\nabla^2 \boldsymbol{A} + \nabla(\nabla \cdot \boldsymbol{A}) + \frac{1}{c^2}\frac{\partial}{\partial t} \nabla \phi + \frac{1}{c^2}\frac{\partial^2 \boldsymbol{A}}{\partial t^2} = \mu_0 \boldsymbol{J}, \tag{5.23}$$

where the identity eq. (4.16) was used to get eq. (5.23). A tremendous simplification of these equations can be made by transforming the potentials to a new gauge $(\phi', \boldsymbol{A}')$ such that

$$\frac{1}{c^2}\frac{\partial \phi'}{\partial t} + \nabla \cdot \boldsymbol{A}' = 0 . \tag{5.24}$$

This condition is known as the *Lorenz*[2] *gauge condition*. By eq. (5.1), such a gauge can be chosen if we can find a function χ that satisfies

$$\frac{1}{c^2}\frac{\partial \phi}{\partial t} - \frac{1}{c^2}\frac{\partial^2 \chi}{\partial t^2} + \nabla \cdot \boldsymbol{A} + \nabla^2 \chi = 0, \tag{5.25}$$

that is, if we can solve the equation

$$\Box \chi = -s, \tag{5.26}$$

with $s = (1/c^2)\partial\phi/\partial t + \nabla \cdot \boldsymbol{A}$, where

$$\Box \equiv -\frac{1}{c^2}\frac{\partial^2}{\partial t^2} + \nabla^2 . \tag{5.27}$$

---

[2]Note that there is no "t" in Lorenz. The gauge condition eq. (5.24) is named after Ludvig Lorenz, a nineteenth century Danish physicist/mathematician, not Hendrik Lorentz, the Dutch physicist after whom the Lorentz transformation is named.

The operator $\Box$ is known as the *d'Alembertian* or *wave operator*, and equation (5.26) is known as the *wave equation with source s*. It is well known that this equation can be solved for any smooth s, and indeed, this will follow from results we obtain in section 5.4. Thus, without loss of generality, we may always use the gauge freedom eq. (5.1) to put the potentials ϕ and $\boldsymbol{A}$ in the Lorenz gauge. It is important to note that the Lorenz gauge is *not* unique. If χ satisfies eq. (5.25), then so does $\chi + \psi$, where ψ is any solution to the homogeneous wave equation $\Box\psi = 0$. As we shall see in section 5.4, there are many solutions to this equation.

We now transform our potentials ϕ and $\boldsymbol{A}$ to the Lorenz gauge potentials satisfying eq. (5.24). For notational simplicity, we drop the primes and denote the transformed potentials as ϕ and $\boldsymbol{A}$ rather than ϕ' and $\boldsymbol{A}'$. Maxwell's equations (5.22) and (5.23) then become simply

$$\Box\phi = -\frac{\rho}{\epsilon_0}, \tag{5.28}$$

$$\Box\boldsymbol{A} = -\mu_0\boldsymbol{J}\,. \tag{5.29}$$

Thus, the full content of Maxwell's equations is expressed by the wave equations (5.28) and (5.29) together with the Lorenz gauge condition

$$\frac{1}{c^2}\frac{\partial\phi}{\partial t} + \nabla\cdot\boldsymbol{A} = 0\,. \tag{5.30}$$

5.2 Retarded Green's Function

We see from eqs. (5.28) and (5.29) that the key to being able to solve Maxwell's equations is to be able to solve the wave equation with source

$$\Box\psi = -f\,. \tag{5.31}$$

If we know how to obtain solutions ψ to eq. (5.31) for a given f, then we can immediately solve eqs. (5.28) and (5.29). Of course, we must still solve eq. (5.30) as well, but we will see that this equation is automatically satisfied for the retarded solution to eqs. (5.28) and (5.29) for sources with suitable fall-off.

As in electrostatics, we will be able to obtain a solution to eq. (5.31) if we can find a Green's function, that is, a solution to

$$\Box_{(t,\boldsymbol{x})}G(t,\boldsymbol{x};t',\boldsymbol{x}') = -\delta(\boldsymbol{x}-\boldsymbol{x}')\delta(t-t'), \tag{5.32}$$

where the subscript $(t,\boldsymbol{x})$ on $\Box$ indicates that the derivatives appearing in the d'Alembertian operator $\Box$ are taken with respect to the unprimed variables, not the primed variables. Note that, in contrast to eq. (2.67), G depends on t and t' as well as $\boldsymbol{x}$ and $\boldsymbol{x}'$, and the right side of eq. (5.32) has a delta function in $t-t'$ as well as in $\boldsymbol{x}-\boldsymbol{x}'$. As in electrostatics, given G, a solution ψ to eq. (5.31) can then be obtained via

$$\psi(t,\boldsymbol{x}) = \int G(t,\boldsymbol{x};t',\boldsymbol{x}')f(t',\boldsymbol{x}')d^3x'dt', \tag{5.33}$$

provided that this integral converges. Note that I have used the expressions "*a* Green's function" and "*a* solution," because neither the Green's function (5.32) nor solutions to eq. (5.31) are unique. Indeed, we will explicitly encounter the nonuniqueness below when attempting to solve for G. However, we will see that there is a unique "retarded" Green's function, for which G vanishes for $t < t'$. In section 5.4, we use the retarded Green's function to characterize the general solution to eq. (5.31).

We seek to solve eq. (5.32) by using Fourier transforms. For an integrable function $F : \mathbb{R} \to \mathbb{R}$, we define its Fourier transform $\hat{F}$ by

$$\hat{F}(k) = \frac{1}{\sqrt{2\pi}} \int_{-\infty}^{\infty} F(x) e^{-ikx} dx \,. \tag{5.34}$$

The notion of a Fourier transform can be extended to distributions, in which case the Fourier transform yields a distribution. In particular, the Fourier transform of the delta function $\delta_{x_0}(x) = \delta(x - x_0)$ is well defined and given by the function

$$\hat{\delta}_{x_0}(k) = \frac{1}{\sqrt{2\pi}} e^{-ikx_0}, \tag{5.35}$$

as would be expected by formally replacing $F(x)$ by $\delta(x - x_0)$ in eq. (5.34).

For a smooth function F, with suitably fast fall-off at infinity, it can be shown that the inverse of the Fourier transform is given by

$$F(x) = \frac{1}{\sqrt{2\pi}} \int_{-\infty}^{\infty} \hat{F}(k) e^{+ikx} dk, \tag{5.36}$$

that is, one can recover F from its Fourier transform $\hat{F}$ by the same formula as eq. (5.34) except for the sign change in the exponential. Equation (5.36) also applies to distributions when suitably interpreted. In particular, the delta function is given by

$$\delta_{x_0}(x) = \frac{1}{\sqrt{2\pi}} \int_{-\infty}^{\infty} \hat{\delta}_{x_0}(k) e^{+ikx} dk = \frac{1}{2\pi} \int_{-\infty}^{\infty} e^{-ikx_0} e^{+ikx} dk \,. \tag{5.37}$$

This equation—which appears very commonly in physics texts—may not look very sensible, since the integral on the right side of eq. (5.37) clearly does not converge. However, what this equation is really supposed to mean is that for any smooth function f with suitably fast fall-off at infinity, we have

$$\delta_{x_0}(f) = \frac{1}{2\pi} \int_{-\infty}^{\infty} dk e^{-ikx_0} \int_{-\infty}^{\infty} dx e^{+ikx} f(x) \,. \tag{5.38}$$

This statement is correct, because

$$\frac{1}{2\pi} \int_{-\infty}^{\infty} dk e^{-ikx_0} \int_{-\infty}^{\infty} dx e^{+ikx} f(x) = \frac{1}{\sqrt{2\pi}} \int_{-\infty}^{\infty} dk e^{-ikx_0} \hat{f}(-k) = f(x_0) \,. \tag{5.39}$$

A major reason Fourier transforms are so useful is that differentiation in physical space corresponds to multiplication by ik in Fourier transform space. To see this, let F be a smooth function that falls off sufficiently rapidly at infinity. Then the Fourier transform of dF/dx is given by

$$\begin{aligned}\widehat{\frac{dF}{dx}}(k) &= \frac{1}{\sqrt{2\pi}}\int_{-\infty}^{\infty}\frac{dF}{dx}(x)\, e^{-ikx}dx \\ &= \frac{1}{\sqrt{2\pi}}\int_{-\infty}^{\infty}(ik)F(x)e^{-ikx}dx \\ &= ik\hat{F}(k), \end{aligned} \tag{5.40}$$

where we integrated by parts in the second line, disgarding the boundary term at infinity because of the fall-off of F. On account of eq. (5.40), any partial differential equation with constant coefficients can be converted to an algebraic equation in Fourier transform space.

We now attempt to solve for G, eq. (5.32), by means of Fourier transforms. To simplify the notation, we set $t' = \boldsymbol{x}' = 0$, and we also set $c = 1$. (We will restore t', $\boldsymbol{x}'$, and c at the end of the calculation.) Define the (4-dimensional) Fourier transform of G by

$$\hat{G}(\omega, \boldsymbol{k}) = \frac{1}{(2\pi)^2}\int_{-\infty}^{\infty} G(t, \boldsymbol{x})e^{+i\omega t}e^{-i\boldsymbol{k}\cdot\boldsymbol{x}}dtd^3x\,. \tag{5.41}$$

Note that, by standard convention, the time Fourier transform is defined[3] by integrating with $e^{+i\omega t}$ rather than $e^{-i\omega t}$. Taking the Fourier transform of eq. (5.32) with respect to t and $\boldsymbol{x}$ and using eq. (5.35) (with $x_0 = 0$) and eq. (5.40), we obtain

$$(\omega^2 - k^2)\hat{G}(\omega, \boldsymbol{k}) = -\left(\frac{1}{\sqrt{2\pi}}\right)^4 = -\frac{1}{4\pi^2}, \tag{5.42}$$

where we have written $k = |\boldsymbol{k}|$. One might think that the solution to this equation is simply

$$\hat{G}(\omega, \boldsymbol{k}) = -\frac{1}{4\pi^2}\frac{1}{(\omega^2 - k^2)} = -\frac{1}{4\pi^2}\frac{1}{(\omega - k)}\frac{1}{(\omega + k)}\,. \tag{5.43}$$

However, dividing by $(\omega^2 - k^2)$ is actually an illegal step, because this quantity can be zero. The difficulty arising from this can be seen if we attempt to take the inverse Fourier transform of $\hat{G}$ with respect to ω (but *not* $\boldsymbol{k}$), so as to obtain the Fourier transform of G with respect to space but not time, which we denote as $\tilde{G}$:

$$\tilde{G}(t, \boldsymbol{k}) \equiv \frac{1}{\sqrt{2\pi}}\int_{-\infty}^{\infty}\hat{G}(\omega, \boldsymbol{k})e^{-i\omega t}d\omega\,. \tag{5.44}$$

Then, we have

$$\tilde{G}(t, \boldsymbol{k}) = -\frac{1}{4\pi^2\sqrt{2\pi}}\int_{-\infty}^{\infty}\frac{e^{-i\omega t}}{(\omega - k)(\omega + k)}d\omega\,. \tag{5.45}$$

However, there are logarithmic divergences in the integral on the right side of eq. (5.45) at $\omega = \pm k$, so the right side is ill defined (even as a distribution). This reflects the fact

[3]This is done so that $(\frac{\omega}{c}, \boldsymbol{k})$ are the components of a 4-vector in special relativity (see chapter 8).

that many Green's functions satisfy eq. (5.32), so we cannot be expected to be able to solve for G without providing additional input as to which Green's function we seek.

This difficulty can be dealt with by suitably regularizing eq. (5.45) in such a way that the right side is well defined and eq. (5.42) continues to hold. A simple way of doing this is to infinitesimally displace the poles at $\omega = \pm k$ in eq. (5.45) into the complex ω-plane. The case of most interest for us is to displace both poles infinitesimally into the lower half of the ω-plane, thereby defining the *retarded Green's function*[4] $\tilde{G}_{\text{ret}}$, given by

$$\tilde{G}_{\text{ret}}(t, \boldsymbol{k}) = -\frac{1}{4\pi^2\sqrt{2\pi}} \int_{-\infty}^{\infty} \frac{e^{-i\omega t}}{(\omega - k + i\epsilon)(\omega + k + i\epsilon)} d\omega, \tag{5.46}$$

where $\epsilon > 0$ and the limit $\epsilon \to 0$ is to be taken after the integral is performed. We can view eq. (5.46) as a contour integral in the complex ω-plane. For $t < 0$, the integral is exponentially damped in the upper half of the ω-plane, and we may "close the contour" in that half-plane. The resulting closed contour encloses no singularities—since the poles have been pushed into the lower half-plane—so, by Cauchy's theorem, we obtain

$$\tilde{G}_{\text{ret}}(t, \boldsymbol{k}) = 0, \qquad \text{for } t < 0. \tag{5.47}$$

This condition uniquely characterizes the retarded Green's function. The fact that $\tilde{G}_{\text{ret}}$ vanishes prior to the "turn-on" of the delta function source at $t = 0$ can be interpreted as saying that it is providing the solution with "no incoming radiation." This is the solution of physical relevance in problems where no radiation is present prior to the presence of the source.

For $t > 0$, we can similarly close the contour in the lower half of the ω-plane. However, the resulting closed contour now contains poles at $\omega = \pm k - i\epsilon$. By Cauchy's theorem, we obtain for $t > 0$

$$\begin{aligned}\tilde{G}_{\text{ret}}(t, \boldsymbol{k}) &= +\frac{1}{4\pi^2\sqrt{2\pi}} 2\pi i \left[\frac{e^{-ikt}}{2k} - \frac{e^{+ikt}}{2k} \right] \\ &= \frac{1}{2\pi\sqrt{2\pi}} \frac{\sin kt}{k}, \qquad \text{for } t > 0,\end{aligned} \tag{5.48}$$

where the sign change in the first line as compared with eq. (5.46) results from the contour running the "wrong way."

We now take the inverse Fourier transform of eq. (5.48) with respect to $\boldsymbol{k}$ to obtain the retarded Green's function for $t > 0$ in position space:

$$\begin{aligned}G_{\text{ret}}(t, \boldsymbol{x}) &= \frac{1}{(2\pi)^{3/2}} \frac{1}{2\pi\sqrt{2\pi}} \int \frac{\sin kt}{k} e^{i\boldsymbol{k}\cdot\boldsymbol{x}} d^3k \\ &= \frac{1}{8\pi^3} \int \frac{\sin kt}{k} e^{ik|\boldsymbol{x}|\cos\theta} k^2 \sin\theta \, d\theta \, d\varphi \, dk\end{aligned}$$

[4] As noted below, displacement of both poles into the upper half of the ω-plane yields the advanced Green's function. Displacement of the pole at $\omega = k$ into the lower half-plane and the pole at $\omega = -k$ into the upper half-plane yields the Feynman propagator.

$$
\begin{aligned}
&= \frac{1}{4\pi^2}\int_0^\infty k^2 dk \left[\frac{e^{ik|\boldsymbol{x}|}}{ik|\boldsymbol{x}|} - \frac{e^{-ik|\boldsymbol{x}|}}{ik|\boldsymbol{x}|}\right]\frac{\sin kt}{k} \\
&= -\frac{1}{8\pi^2|\boldsymbol{x}|}\int_0^\infty dk\left(e^{ik|\boldsymbol{x}|} - e^{-ik|\boldsymbol{x}|}\right)\left(e^{ikt} - e^{-ikt}\right) \\
&= -\frac{1}{8\pi^2|\boldsymbol{x}|}\int_{-\infty}^\infty dk\left(e^{ik(|\boldsymbol{x}|+t)} - e^{ik(|\boldsymbol{x}|-t)}\right) \\
&= -\frac{1}{4\pi\,|\boldsymbol{x}|}\left[\delta(t+|\boldsymbol{x}|) - \delta(t-|\boldsymbol{x}|)\right] \\
&= \frac{\delta(t-|\boldsymbol{x}|)}{4\pi\,|\boldsymbol{x}|}, \qquad \text{for } t>0\,.
\end{aligned} \tag{5.49}
$$

Here we have introduced polar coordinates (k,θ,φ) in $\boldsymbol{k}$-space in line 2 and performed the integral over θ and φ in line 3; the four exponential terms in line 4 are compressed to two terms in line 5 by changing the range of integration over k to $-\infty<k<\infty$; eq. (5.37) was used to get to line 6; and $\delta(t+|\boldsymbol{x}|)$ was set to zero in line 7, since we are considering only the case $t>0$ in this calculation.

Reinserting t', $\boldsymbol{x}'$, and c in this result, we find that

$$
G_{\text{ret}}(t,\boldsymbol{x};t',\boldsymbol{x}') = \begin{cases} 0, & \text{for } t<t', \\ \dfrac{\delta(t-t'-|\boldsymbol{x}-\boldsymbol{x}'|/c)}{4\pi\,|\boldsymbol{x}-\boldsymbol{x}'|}, & \text{for } t>t'. \end{cases} \tag{5.50}
$$

The most important feature of G_{ret} is that it is nonvanishing only on the *future light cone* of the source point $(t',\boldsymbol{x}')$; that is, it is nonvanishing only when $|\boldsymbol{x}-\boldsymbol{x}'| = c(t-t')$. In other words, if a field satisfying the wave equation (5.31) vanishes at early times and we then put a delta function source at $(t',\boldsymbol{x}')$, the resulting disturbance to the field propagates away from $(t',\boldsymbol{x}')$ at exactly the speed of light.

It is worth noting that if we had displaced both poles in eq. (5.45) into the upper (rather than lower) half of the ω-plane, we would have obtained the *advanced Green's function*:

$$
G_{\text{adv}}(t,\boldsymbol{x};t',\boldsymbol{x}') = \begin{cases} \dfrac{\delta(t-t'+|\boldsymbol{x}-\boldsymbol{x}'|/c)}{4\pi\,|\boldsymbol{x}-\boldsymbol{x}'|}, & \text{for } t<t', \\ 0, & \text{for } t>t'. \end{cases} \tag{5.51}
$$

The advanced Green's function is characterized by the fact that it vanishes for $t>t'$. Note that, by inspection, we have

$$
G_{\text{adv}}(t,\boldsymbol{x};t',\boldsymbol{x}') = G_{\text{ret}}(t',\boldsymbol{x}';t,\boldsymbol{x})\,. \tag{5.52}
$$

This relation can be shown to hold more generally (e.g., it would continue to hold if we modified the wave equation by addition of a term involving a potential; see problem 7).

The *retarded solution* to eq. (5.31) is the solution obtained from the retarded Green's function via eq. (5.33):

$$
\psi(t,\boldsymbol{x}) = \frac{1}{4\pi}\int \frac{f(t-|\boldsymbol{x}-\boldsymbol{x}'|/c,\boldsymbol{x}')}{|\boldsymbol{x}-\boldsymbol{x}'|}d^3x'\,. \tag{5.53}
$$

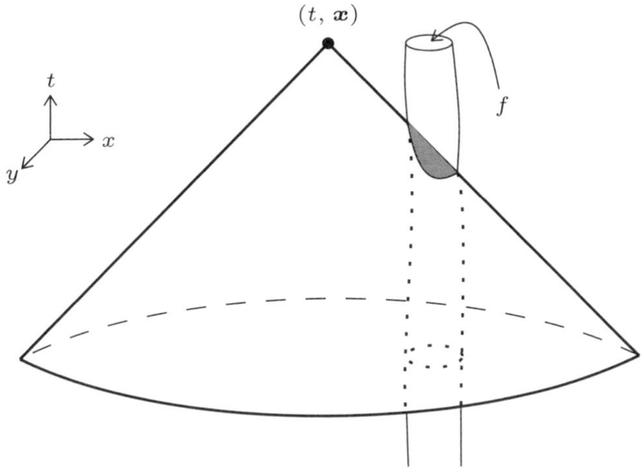

Figure 5.1. A spacetime diagram showing the past light cone of the point $(t, \boldsymbol{x})$ at which the field ψ is being observed. (Only 2 spatial dimensions are illustrated; i.e., the $\hat{z}$-direction is suppressed.) In this figure, the source term f is assumed to be nonvanishing only in the "world tube" shown. The only contribution of f to the integral eq. (5.54) comes from the shaded region corresponding to the intersection of the world tube of f with the past light cone of $(t, \boldsymbol{x})$.

Here we have carried out the integral over t' in eq. (5.33) by integrating the delta function in eq. (5.50). We can write eq. (5.53) in more compact notation as

$$\psi(t, \boldsymbol{x}) = \frac{1}{4\pi} \int \frac{[f(t', \boldsymbol{x}')]_{\text{ret}}}{|\boldsymbol{x} - \boldsymbol{x}'|} d^3x', \tag{5.54}$$

where the notation $[f(t', \boldsymbol{x}')]_{\text{ret}}$ means that $f(t', \boldsymbol{x}')$ is to be evaluated at the *retarded time*

$$t' = t - |\boldsymbol{x} - \boldsymbol{x}'|/c. \tag{5.55}$$

(Thus, in eq. (5.54), t' is not an independent variable but rather is a function of t, $\boldsymbol{x}$, and $\boldsymbol{x}'$.) When written in this form, the retarded solution to the wave equation takes exactly the same form as the solution, eq. (2.20), to Poisson's equation, except that the source is evaluated at the retarded time. However, it should be emphasized that the integral in eq. (5.54) is not actually being taken over all of space at an instant of time—as the "d^3x'" might suggest—but rather is being taken over the past light cone of $(t, \boldsymbol{x})$, as illustrated in figure 5.1. Of course, the solution eq. (5.54) is well defined only if f falls to zero at infinity along the past light cone of $(t, \boldsymbol{x})$ in a sufficiently rapid manner that the integral converges.[5] In the following, we will assume that this is the case.

Thus far in this section, we have been considering only the wave equation (5.31). However, in electromagnetism, ϕ and the Cartesian coordinate components of $\boldsymbol{A}$ satisfy precisely this equation (see eqs. (5.28) and (5.29)). Therefore, in electromagnetism,

[5] If f is such that the integral on the right side of eq. (5.54) does not converge, then there still exist solutions to the wave equation with source f, as can be shown from the results of section 5.4. However, there is no notion of the "retarded solution."

we can immediately write down the retarded solution corresponding to a source with charge density ρ and current density $\boldsymbol{J}$:

$$\phi(t,\boldsymbol{x}) = \frac{1}{4\pi\epsilon_0}\int \frac{[\rho(t',\boldsymbol{x}')]_{\text{ret}}}{|\boldsymbol{x}-\boldsymbol{x}'|}d^3x', \tag{5.56}$$

$$\boldsymbol{A}(t,\boldsymbol{x}) = \frac{\mu_0}{4\pi}\int \frac{[\boldsymbol{J}(t',\boldsymbol{x}')]_{\text{ret}}}{|\boldsymbol{x}-\boldsymbol{x}'|}d^3x'. \tag{5.57}$$

Of course, ϕ and $\boldsymbol{A}$ also must satisfy the Lorenz gauge condition (5.30). However, we have

$$\frac{1}{c^2}\frac{\partial\phi}{\partial t}(t,\boldsymbol{x}) = \frac{1}{4\pi\epsilon_0 c^2}\frac{\partial}{\partial t}\int \frac{[\rho(t',\boldsymbol{x}')]_{\text{ret}}}{|\boldsymbol{x}-\boldsymbol{x}'|}d^3x' = \frac{1}{4\pi\epsilon_0 c^2}\int \frac{[\frac{\partial\rho}{\partial t'}(t',\boldsymbol{x}')]_{\text{ret}}}{|\boldsymbol{x}-\boldsymbol{x}'|}d^3x', \tag{5.58}$$

whereas

$$\begin{aligned}
\nabla\cdot\boldsymbol{A}(t,\boldsymbol{x}) &= \frac{\mu_0}{4\pi}\int \nabla_x\cdot\left(\frac{[\boldsymbol{J}(t',\boldsymbol{x}')]_{\text{ret}}}{|\boldsymbol{x}-\boldsymbol{x}'|}\right)d^3x' \\
&= -\frac{\mu_0}{4\pi}\int \nabla_{x'}\cdot\left(\frac{[\boldsymbol{J}(t',\boldsymbol{x}')]_{\text{ret}}}{|\boldsymbol{x}-\boldsymbol{x}'|}\right)d^3x' + \frac{\mu_0}{4\pi}\int \frac{[\nabla_{x'}\cdot\boldsymbol{J}(t',\boldsymbol{x}')]_{\text{ret}}}{|\boldsymbol{x}-\boldsymbol{x}'|}d^3x' \\
&= \frac{\mu_0}{4\pi}\int \frac{[\nabla_{x'}\cdot\boldsymbol{J}(t',\boldsymbol{x}')]_{\text{ret}}}{|\boldsymbol{x}-\boldsymbol{x}'|}d^3x'.
\end{aligned} \tag{5.59}$$

Here, in the first term of the second line, we use the fact that the $\boldsymbol{x}$-dependence in the integrand appears only in the form $|\boldsymbol{x}-\boldsymbol{x}'|$ to replace ∇_x by $-\nabla_{x'}$; the second term of that line then corrects for the additional explicit $\boldsymbol{x}'$-dependence of $\boldsymbol{J}$. In the last line, we discarded the total divergence, assuming that $\boldsymbol{J}(t',\boldsymbol{x}') \to 0$ sufficiently rapidly as $|\boldsymbol{x}'| \to \infty$ along the past light cone of $(t,\boldsymbol{x})$. Satisfaction of the Lorenz gauge condition (5.30) then follows immediately from charge-current conservation, eq. (5.8), together with the identity $\epsilon_0\mu_0 c^2 = 1$ of SI units.

In summary, the solution to Maxwell's equations with charge density ρ and current density $\boldsymbol{J}$ corresponding to the absence of incoming radiation is given by eqs. (5.56) and (5.57).

5.3 Multipole Expansion

In this section, we perform a multipole expansion of the retarded solution in electrodynamics, eqs. (5.56) and (5.57), analogous to the multipole expansions we performed in electrostatics in section 2.4 and in magnetostatics in section 4.2.

5.3.1 CARTESIAN MULTIPOLE EXPANSION OF THE RADIATION FIELD FOR A NONRELATIVISTIC SOURCE

In electrostatics and magnetostatics, we expanded $1/|\boldsymbol{x}-\boldsymbol{x}'|$ in powers of $1/|\boldsymbol{x}|$ (see eq. (2.51)) and substituted the result in the Green's function expressions, eq. (2.50) and eq. (4.22), for ϕ and $\boldsymbol{A}$, respectively. We then found that the monopole moment of the source—which is absent in magnetostatics—contributes a term that decays at large $|\boldsymbol{x}|$

as $1/|\boldsymbol{x}|$, the dipole moment of the source contributes a term that decays as $1/|\boldsymbol{x}|^2$, and so forth.

Formulas (5.56) and (5.57) are quite similar to eqs. (2.50) and (4.22), with the only difference being that the sources in eqs. (5.56) and (5.57) are evaluated at the retarded time. However, as we shall see below, for a time-dependent source, all angular multipole moments of the source will now contribute at order $1/|\boldsymbol{x}|$. We will refer to the electric and magnetic fields at order $1/|\boldsymbol{x}|$ as the *radiation field*, and will focus our attention in this section on the calculation of the radiation field. (This restriction is not made in section 5.3.2.) Note that the flux of energy, eq. (5.14), carried to infinity by the electromagnetic field is determined by the radiation field.

Let us begin with the calculation of ϕ to order $1/|\boldsymbol{x}|$. To this order, we may replace $1/|\boldsymbol{x}-\boldsymbol{x}'|$ in the integrand of eq. (5.56) by $1/|\boldsymbol{x}|$ and pull it out of the integral. Thus, we obtain

$$\phi(t,\boldsymbol{x}) = \frac{1}{4\pi\epsilon_0|\boldsymbol{x}|}\int \rho\left(t-\frac{1}{c}|\boldsymbol{x}-\boldsymbol{x}'|,\boldsymbol{x}'\right)d^3x' + O\left(\frac{1}{|\boldsymbol{x}|^2}\right). \tag{5.60}$$

Next we expand $|\boldsymbol{x}-\boldsymbol{x}'|$ around $\boldsymbol{x}'=0$ as

$$\begin{aligned}
|\boldsymbol{x}-\boldsymbol{x}'| &= |\boldsymbol{x}|\left|\hat{\boldsymbol{x}}-\frac{\boldsymbol{x}'}{|\boldsymbol{x}|}\right| \\
&= |\boldsymbol{x}|\left[1-\frac{\hat{\boldsymbol{x}}\cdot\boldsymbol{x}'}{|\boldsymbol{x}|}+O\left(\frac{1}{|\boldsymbol{x}|^2}\right)\right] \\
&= |\boldsymbol{x}|-\hat{\boldsymbol{x}}\cdot\boldsymbol{x}'+O\left(\frac{1}{|\boldsymbol{x}|}\right).
\end{aligned} \tag{5.61}$$

Thus, to order $1/|\boldsymbol{x}|$, we have

$$\phi(t,\boldsymbol{x}) = \frac{1}{4\pi\epsilon_0|\boldsymbol{x}|}\int \rho\left(t-\frac{|\boldsymbol{x}|}{c}+\frac{\hat{\boldsymbol{x}}\cdot\boldsymbol{x}'}{c},\boldsymbol{x}'\right)d^3x'. \tag{5.62}$$

Similarly, to order $1/|\boldsymbol{x}|$, we have

$$\boldsymbol{A}(t,\boldsymbol{x}) = \frac{\mu_0}{4\pi|\boldsymbol{x}|}\int \boldsymbol{J}\left(t-\frac{|\boldsymbol{x}|}{c}+\frac{\hat{\boldsymbol{x}}\cdot\boldsymbol{x}'}{c},\boldsymbol{x}'\right)d^3x'. \tag{5.63}$$

If ρ and $\boldsymbol{J}$ are localized near $\boldsymbol{x}'=0$ and vary sufficiently slowly with time, it is useful to Taylor expand them in the t variable (but *not* the $\boldsymbol{x}'$ variable) about the time $t-|\boldsymbol{x}|/c$. The formal Taylor series expressions are

$$\rho\left(t-\frac{|\boldsymbol{x}|}{c}+\frac{\hat{\boldsymbol{x}}\cdot\boldsymbol{x}'}{c},\boldsymbol{x}'\right) = \sum_{n=0}^{\infty}\frac{1}{n!}\frac{\partial^n\rho}{\partial t^n}\bigg|_{(t-\frac{|\boldsymbol{x}|}{c},\boldsymbol{x}')}\left[\frac{\hat{\boldsymbol{x}}\cdot\boldsymbol{x}'}{c}\right]^n, \tag{5.64}$$

$$\boldsymbol{J}\left(t-\frac{|\boldsymbol{x}|}{c}+\frac{\hat{\boldsymbol{x}}\cdot\boldsymbol{x}'}{c},\boldsymbol{x}'\right) = \sum_{n=0}^{\infty}\frac{1}{n!}\frac{\partial^n\boldsymbol{J}}{\partial t^n}\bigg|_{(t-\frac{|\boldsymbol{x}|}{c},\boldsymbol{x}')}\left[\frac{\hat{\boldsymbol{x}}\cdot\boldsymbol{x}'}{c}\right]^n. \tag{5.65}$$

If ρ and $\boldsymbol{J}$ are assumed to be analytic functions of time, then, by definition, these series will have a finite radius of convergence. However, what we really have in mind is approximating ρ and $\boldsymbol{J}$ by taking a finite number of terms in these series. This will provide a good approximation if ρ and $\boldsymbol{J}$ are sufficiently differentiable and sufficiently slowly varying in time, with no need to make any assumptions about analyticity. Substituting these expansions into eq. (5.62) and eq. (5.63), we obtain

$$\phi(t,\boldsymbol{x}) = \frac{1}{4\pi\epsilon_0|\boldsymbol{x}|}\sum_{n=0}^{\infty}\sum_{j_1\ldots j_n=1}^{3}\frac{1}{n!}\hat{\boldsymbol{x}}_{j_1}\cdots\hat{\boldsymbol{x}}_{j_n}\frac{1}{c^n}\frac{d^n}{dt^n}\int\rho\left(t-\frac{|\boldsymbol{x}|}{c},\boldsymbol{x}'\right)x'_{j_1}\cdots x'_{j_n}\,d^3x', \tag{5.66}$$

$$A_i(t,\boldsymbol{x}) = \frac{\mu_0}{4\pi|\boldsymbol{x}|}\sum_{n=0}^{\infty}\sum_{j_1\ldots j_n=1}^{3}\frac{1}{n!}\hat{\boldsymbol{x}}_{j_1}\cdots\hat{\boldsymbol{x}}_{j_n}\frac{1}{c^n}\frac{d^n}{dt^n}\int J_i\left(t-\frac{|\boldsymbol{x}|}{c},\boldsymbol{x}'\right)x'_{j_1}\cdots x'_{j_n}\,d^3x'. \tag{5.67}$$

Let us now make a rough estimation of the relative sizes of the terms appearing in eq. (5.66) and eq. (5.67). Since the integrand in the nth term contains n powers of x', the integral appearing in the nth term should scale as R^n, where R denotes the spatial size of the source. Since n derivatives with respect to t are taken, this term also should scale as $1/T^n$, where T denotes the timescale over which the source varies significantly. Thus, the nth term on the right side of eq. (5.66) and eq. (5.67) should be of order

$$n\text{th term} \propto \frac{1}{n!}\left(\frac{R}{cT}\right)^n. \tag{5.68}$$

Now, if the source is nonrelativistic, one would expect the timescale over which the source varies significantly to be much larger than the light travel time across the source,

$$T \gg R/c. \tag{5.69}$$

If that is the case, then the terms with small n will dominate.

If we keep only the $n=0,1$ terms in eq. (5.66), we obtain

$$\begin{aligned}\phi(t,\boldsymbol{x}) &= \frac{1}{4\pi\epsilon_0|\boldsymbol{x}|}\int\rho\left(t-\frac{|\boldsymbol{x}|}{c},\boldsymbol{x}'\right)d^3x' + \frac{1}{4\pi\epsilon_0 c|\boldsymbol{x}|}\frac{d}{dt}\int\rho\left(t-\frac{|\boldsymbol{x}|}{c},\boldsymbol{x}'\right)\hat{\boldsymbol{x}}\cdot\boldsymbol{x}'\,d^3x' \\ &= \frac{q}{4\pi\epsilon_0|\boldsymbol{x}|} + \frac{1}{4\pi\epsilon_0 c|\boldsymbol{x}|}\hat{\boldsymbol{x}}\cdot\left.\frac{d\boldsymbol{p}}{dt}\right|_{\text{ret}}.\end{aligned} \tag{5.70}$$

Here, q is the charge of the source (which is independent of time by charge conservation), and $\boldsymbol{p}$ is the electric dipole moment of the source, with the time derivative of $\boldsymbol{p}$ in eq. (5.70) being evaluated at the retarded time $t-|\boldsymbol{x}|/c$. Similarly, keeping only the $n=0$ term in eq. (5.67), we obtain

$$\boldsymbol{A}(t,\boldsymbol{x}) = \frac{\mu_0}{4\pi|\boldsymbol{x}|}\int\boldsymbol{J}\left(t-\frac{|\boldsymbol{x}|}{c},\boldsymbol{x}'\right)d^3x'. \tag{5.71}$$

In magnetostatics, we have $\nabla\cdot\boldsymbol{J}=0$, which implies $\int\boldsymbol{J}(\boldsymbol{x})d^3x=0$ (see eq. (4.24)). In electrodynamics, we have $\nabla\cdot\boldsymbol{J}=-\partial\rho/\partial t$. The computation analogous to eq. (4.26) now yields

$$0=\int \nabla \cdot (x_j \boldsymbol{J}) d^3x = \int \left[x_j(\nabla \cdot \boldsymbol{J}) + \boldsymbol{J} \cdot \nabla x_j\right] d^3x$$
$$= -\int x_j \frac{\partial \rho}{\partial t} d^3x + \sum_i \int J_i \partial_i x_j d^3x = -\frac{d}{dt}\int x_j \rho d^3x + \int J_j d^3x. \qquad (5.72)$$

Thus, in electrodynamics, we have

$$\int \boldsymbol{J}(\boldsymbol{x}) d^3x = \frac{d\boldsymbol{p}}{dt}, \qquad (5.73)$$

and the $n=0$ contribution to $\boldsymbol{A}$ is

$$\boldsymbol{A}(t,\boldsymbol{x}) = \frac{\mu_0}{4\pi |\boldsymbol{x}|} \left.\frac{d\boldsymbol{p}}{dt}\right|_{\text{ret}}. \qquad (5.74)$$

Consequently, for nonrelativistic motion, the dominant contribution to the electromagnetic potentials at order $1/|\boldsymbol{x}|$ is given by eq. (5.70) and eq. (5.74), and it is determined by the charge and by the time derivative of the electric dipole moment of the source at the retarded time of the observation.

The electric field associated with the potentials, eq. (5.70) and eq. (5.74), is

$$\boldsymbol{E}(t,\boldsymbol{x}) = -\nabla\phi - \frac{\partial \boldsymbol{A}}{\partial t}$$
$$= \frac{1}{4\pi\epsilon_0 c^2 |\boldsymbol{x}|}\left(\hat{\boldsymbol{x}} \cdot \left.\frac{d^2\boldsymbol{p}}{dt^2}\right|_{\text{ret}}\right)\hat{\boldsymbol{x}} - \frac{\mu_0}{4\pi|\boldsymbol{x}|}\left.\frac{d^2\boldsymbol{p}}{dt^2}\right|_{\text{ret}} + O\left(\frac{1}{|\boldsymbol{x}|^2}\right)$$
$$= \frac{\mu_0}{4\pi|\boldsymbol{x}|}\left[\left(\hat{\boldsymbol{x}} \cdot \frac{d^2\boldsymbol{p}}{dt^2}\right)\hat{\boldsymbol{x}} - \frac{d^2\boldsymbol{p}}{dt^2}\right]\Bigg|_{\text{ret}} + O\left(\frac{1}{|\boldsymbol{x}|^2}\right). \qquad (5.75)$$

Here we have used the fact that

$$\frac{\partial}{\partial t}\left[\left.\frac{d\boldsymbol{p}}{dt}\right|_{\text{ret}}\right] = \frac{\partial}{\partial t}\left[\frac{d\boldsymbol{p}}{dt}\left(t-\frac{1}{c}|\boldsymbol{x}|\right)\right] = \frac{d^2\boldsymbol{p}}{dt^2}\left(t-\frac{1}{c}|\boldsymbol{x}|\right) = \left.\frac{d^2\boldsymbol{p}}{dt^2}\right|_{\text{ret}}, \qquad (5.76)$$

whereas

$$\frac{\partial}{\partial x_i}\left[\left.\frac{d\boldsymbol{p}}{dt}\right|_{\text{ret}}\right] = -\frac{x_i}{c|\boldsymbol{x}|}\left.\frac{d^2\boldsymbol{p}}{dt^2}\right|_{\text{ret}}. \qquad (5.77)$$

Note that $\nabla(q/|\boldsymbol{x}|) = O(1/|\boldsymbol{x}|^2)$, so the Coulomb term in ϕ does not contribute to $\boldsymbol{E}$ at order $1/|\boldsymbol{x}|$. Similarly, we obtain

$$\boldsymbol{B}(t,\boldsymbol{x}) = \nabla \times \boldsymbol{A} = -\frac{\mu_0}{4\pi c|\boldsymbol{x}|}\hat{\boldsymbol{x}} \times \left.\frac{d^2\boldsymbol{p}}{dt^2}\right|_{\text{ret}} + O\left(\frac{1}{|\boldsymbol{x}|^2}\right). \qquad (5.78)$$

Note that, at order $1/|\boldsymbol{x}|$, $\boldsymbol{E}$ and $\boldsymbol{B}$ are orthogonal to each other, and both are orthogonal to $\hat{\boldsymbol{x}}$. Furthermore, at this order, we have $|\boldsymbol{E}| = c|\boldsymbol{B}|$.

The energy flux is given by the Poynting vector, eq. (5.14). To order $1/|\boldsymbol{x}|^2$, we obtain

$$\boldsymbol{\mathcal{S}} = \frac{1}{\mu_0}\boldsymbol{E}\times\boldsymbol{B} = \frac{\mu_0}{16\pi^2 c|\boldsymbol{x}|^2}\left[\left|\frac{d^2\boldsymbol{p}}{dt^2}\right|^2_{\text{ret}} - \left(\hat{\boldsymbol{x}}\cdot\frac{d^2\boldsymbol{p}}{dt^2}\right)^2_{\text{ret}}\right]\hat{\boldsymbol{x}}. \tag{5.79}$$

Thus, the energy flux points radially outward at all observation angles, and its magnitude varies with observation angle as $\sin^2\theta$, where θ is the angle between the observation point $\boldsymbol{x}$ and $d^2\boldsymbol{p}/dt^2$ at the corresponding retarded time. The radiated power per unit solid angle is given by

$$\frac{dP}{d\Omega} \equiv \lim_{|\boldsymbol{x}|\to\infty} |\boldsymbol{x}|^2 \boldsymbol{\mathcal{S}}\cdot\hat{\boldsymbol{x}} = \frac{\mu_0}{16\pi^2 c}\left|\frac{d^2\boldsymbol{p}}{dt^2}\right|^2_{\text{ret}} \sin^2\theta\,. \tag{5.80}$$

The total power $P = d\mathcal{E}/dt$ radiated to infinity is obtained by integrating $dP/d\Omega$ over all angles, yielding *Larmor's formula*:

$$\begin{aligned} P &= \int \frac{dP}{d\Omega}d\Omega = \frac{\mu_0}{16\pi^2 c}\left|\frac{d^2\boldsymbol{p}}{dt^2}\right|^2_{\text{ret}} \int \sin^2\theta \sin\theta\, d\theta\, d\varphi = \frac{\mu_0}{6\pi c}\left|\frac{d^2\boldsymbol{p}}{dt^2}\right|^2_{\text{ret}} \\ &= \frac{1}{6\pi\epsilon_0 c^3}\left|\frac{d^2\boldsymbol{p}}{dt^2}\right|^2_{\text{ret}}. \end{aligned} \tag{5.81}$$

Thus, eq. (5.81) yields the dominant contribution to the electromagnetic energy radiated to infinity for a nonrelativistic source. This leading contribution is referred to as *electric dipole radiation*, since it is determined by the time variation of the electric dipole moment of the source.

For a nonrelativistic source, (i.e., a source that satisfies eq. (5.69)), the next leading correction to the above formulas arises from including both the $n=2$ term in eq. (5.66) and the $n=1$ term in eq. (5.67). This gives rise to terms in ϕ and $\boldsymbol{A}$ proportional to $d^2 Q_{ij}/dt^2$ and $d\boldsymbol{\mu}/dt$, where Q_{ij} and $\boldsymbol{\mu}$ are, respectively, the electric quadrupole moment and magnetic dipole moment of the source, evaluated at the retarded time. These terms are referred to, respectively, as *electric quadrupole radiation* and *magnetic dipole radiation*. They give rise to contributions to $\boldsymbol{E}$ and $\boldsymbol{B}$ proportional to $d^3 Q_{ij}/dt^3$ and $d^2\boldsymbol{\mu}/dt^2$, and a corresponding radiated power proportional to $|d^3 Q_{ij}/dt^3|^2$ and $|d^2\boldsymbol{\mu}/dt^2|^2$.

5.3.2 GENERAL MULTIPOLE EXPANSION FOR A RELATIVISTIC SOURCE

Here we describe how to obtain a systematic multipole expansion that does not require that the source be nonrelativistic and is not restricted to the calculation of the radiation field at order $1/|\boldsymbol{x}|$.

Let us first consider the inhomogeneous scalar wave equation (5.31). The first step toward obtaining a systematic multipole expansion of the retarded solution to this equation is to take the Fourier transform of this equation in time but *not* in space:

$$[\nabla^2 + \frac{\omega^2}{c^2}]\hat{\psi}(\omega, \boldsymbol{x}) = -\hat{f}(\omega, \boldsymbol{x}), \tag{5.82}$$

where[6]

$$\hat{\psi}(\omega,\boldsymbol{x}) = \frac{1}{\sqrt{2\pi}}\int_{-\infty}^{\infty} e^{i\omega t}\psi(t,\boldsymbol{x})dt\,. \tag{5.83}$$

Equation (5.82) with $\hat{f}=0$ is known as the *Helmholtz equation*. The solution of the inhomogeneous Helmholtz equation (5.82) corresponding to the retarded solution of the inhomogeneous wave equation (5.31) can be obtained by taking the Fourier transform of the retarded solution, eq. (5.53):

$$\hat{\psi}(\omega,\boldsymbol{x}) = \frac{1}{4\pi\sqrt{2\pi}}\int_{-\infty}^{\infty} dt e^{i\omega t}\int \frac{f(t-|\boldsymbol{x}-\boldsymbol{x}'|/c,\boldsymbol{x}')}{|\boldsymbol{x}-\boldsymbol{x}'|}d^3x'\,. \tag{5.84}$$

Making the change of integration variable $t\to\tilde{t}=t-|\boldsymbol{x}-\boldsymbol{x}'|/c$, we obtain

$$\hat{\psi}(\omega,\boldsymbol{x}) = \frac{1}{4\pi}\int \frac{e^{i\omega|\boldsymbol{x}-\boldsymbol{x}'|/c}}{|\boldsymbol{x}-\boldsymbol{x}'|}\hat{f}(\omega,\boldsymbol{x}')d^3x'\,. \tag{5.85}$$

Thus, we see that the Green's function $G^{\mathrm{H}}(\boldsymbol{x},\boldsymbol{x}')$ for the Helmholtz equation that corresponds to the retarded Green's function for the wave equation is given by

$$G^{\mathrm{H}}(\boldsymbol{x},\boldsymbol{x}') = \frac{1}{4\pi}\frac{e^{i\omega|\boldsymbol{x}-\boldsymbol{x}'|/c}}{|\boldsymbol{x}-\boldsymbol{x}'|}\,. \tag{5.86}$$

We now can obtain a multipole expansion of the Helmholtz Green's function eq. (5.86) in exact parallel with the derivation of eq. (2.82). In parallel with eq. (2.69), we expand $G^{\mathrm{H}}(\boldsymbol{x},\boldsymbol{x}')$ in spherical harmonics in the $\boldsymbol{x}$ variable:

$$G^{\mathrm{H}}(\boldsymbol{x},\boldsymbol{x}') = \sum_{\ell,m} G^{\mathrm{H}}_{\ell m}(r,\boldsymbol{x}')Y_{\ell m}(\theta,\varphi)\,. \tag{5.87}$$

In exact parallel with the electrostatic case, we obtain

$$G^{\mathrm{H}}_{\ell m}(r,\boldsymbol{x}') = g^{\mathrm{H}}_{\ell m}(r,r')Y^*_{\ell m}(\theta',\varphi'), \tag{5.88}$$

where $g^{\mathrm{H}}_{\ell m}(r,r')$ satisfies

$$\frac{d}{dr}\left(r^2\frac{dg^{\mathrm{H}}_{\ell m}(r,r')}{dr}\right) + \left(\frac{\omega^2}{c^2}r^2 - \ell(\ell+1)\right)g^{\mathrm{H}}_{\ell m}(r,r') = -\delta(r-r')\,. \tag{5.89}$$

Note that the left side of this equation differs from the left side of eq. (2.74) only by the presence of the additional term $\omega^2 r^2/c^2$ in the factor multiplying $g^{\mathrm{H}}_{\ell m}$. The general solution to this equation in the region $r>r'$ or in the region $r<r'$ is of the form

$$g^{\mathrm{H}}_{\ell m}(r,r') = a^{\mathrm{H}}_{\ell m}(r')n_\ell(\omega r/c) + b^{\mathrm{H}}_{\ell m}(r')j_\ell(\omega r/c), \tag{5.90}$$

[6]Note that in section 5.3.2, the "hat" denotes the Fourier transform with respect to the time variable only.

where $a^{\mathrm{H}}_{\ell m}(r')$ and $b^{\mathrm{H}}_{\ell m}(r')$ are arbitrary functions of r'; and j_ℓ and n_ℓ are, respectively, the *spherical Bessel* and *spherical Neumann* functions, defined by

$$j_\ell(x) = (-x)^\ell \left(\frac{1}{x}\frac{d}{dx}\right)^\ell \left(\frac{\sin x}{x}\right), \tag{5.91}$$

$$n_\ell(x) = -(-x)^\ell \left(\frac{1}{x}\frac{d}{dx}\right)^\ell \left(\frac{\cos x}{x}\right). \tag{5.92}$$

The spherical Bessel function is regular at $x = 0$, and as $x \to 0$, it behaves as

$$j_\ell(x) = \frac{2^\ell \ell!}{(2\ell+1)!} x^\ell + O(x^{\ell+2}). \tag{5.93}$$

As $x \to 0$, the Neumann function n_ℓ behaves as

$$n_\ell(x) = -\frac{(2\ell-1)!}{2^\ell(\ell-1)!}\frac{1}{x^{\ell+1}} + O(\frac{1}{x^{\ell-1}}) \tag{5.94}$$

and thus is singular at $x = 0$.

Since $n_\ell(x)$ is singular at $x = 0$, the only acceptable solution for $r < r'$ is

$$g^{\mathrm{H}}_{\ell m}(r, r') = b^{\mathrm{H}}_{\ell m}(r') j_\ell(\omega r/c) \qquad \text{for } r < r'. \tag{5.95}$$

In contrast, as $x \to \infty$, the asymptotic behavior of j_ℓ and n_ℓ is given by

$$j_\ell = \frac{1}{x}\sin(x - \pi\ell/2) + O\left(\frac{1}{x^2}\right), \qquad n_\ell = -\frac{1}{x}\cos(x - \pi\ell/2) + O\left(\frac{1}{x^2}\right), \tag{5.96}$$

so they both decay as $1/x$ in an oscillatory manner. Thus, in contrast with solutions to Laplace's equation, both $j_\ell(x)$ and $n_\ell(x)$ are regular as $x \to \infty$. This reflects the fact that there are many Green's functions for the wave equation (5.31) and the Helmholtz equation (5.82) that give well-behaved solutions at infinity, whereas a unique such Green's function exists for Poisson's equation (2.3). However, our interest is in obtaining the Green's function (5.86) for the Helmholtz equation that corresponds to the retarded Green's function for the wave equation. As can be seen from eq. (5.86), this corresponds to having behavior $e^{i\omega|\mathbf{x}|/c}/|\mathbf{x}|$ as $|\mathbf{x}| \to \infty$. Since the corresponding time dependence is $e^{-i\omega t}$, this yields oscillatory behavior in spacetime that is referred to as an *outgoing wave*. (The requirement that a solution behave as an outgoing wave near infinity is usually referred to as the *Sommerfeld radiation condition*.) The linear combination of n_ℓ and j_ℓ that corresponds to an outgoing wave is

$$h^{(1)}_\ell(x) = j_\ell(x) + i n_\ell(x) \tag{5.97}$$

and is called a *spherical Hankel function of the first kind*. The asymptotic behavior of $h^{(1)}_\ell(x)$ as $x \to \infty$ is

$$h^{(1)}_\ell(x) = (-i)^{\ell+1}\frac{e^{ix}}{x} + O(1/x^2). \tag{5.98}$$

Thus, the solution for $r > r'$ that yields the Green's function for the Helmholtz equation that corresponds to the retarded Green's function for the wave equation is obtained by requiring $g^{\mathrm{H}}_{\ell m}$ to be an outgoing wave near infinity:

$$g^{\mathrm{H}}_{\ell m}(r, r') = a^{\mathrm{H}}_{\ell m}(r') h^{(1)}_{l}(\omega r/c), \qquad \text{for } r > r' . \tag{5.99}$$

The unknown functions $a^{\mathrm{H}}_{\ell m}(r')$ and $b^{\mathrm{H}}_{\ell m}(r')$ can now be determined by imposing exactly the same jump/matching conditions as led from eq. (2.76) to eq. (2.82). Carrying through these steps,[7] we obtain our final expression:

$$\begin{aligned} G^{\mathrm{H}}(\boldsymbol{x}, \boldsymbol{x}') &= \frac{e^{i\omega|\boldsymbol{x}-\boldsymbol{x}'|/c}}{4\pi|\boldsymbol{x}-\boldsymbol{x}'|} \\ &= \begin{cases} \sum_{\ell,m} \frac{i\omega}{c} j_\ell(\omega r/c) h^{(1)}_\ell(\omega r'/c) Y^*_{\ell m}(\theta', \varphi') Y_{\ell m}(\theta, \varphi), & \text{if } r < r' \\ \sum_{\ell,m} \frac{i\omega}{c} h^{(1)}_\ell(\omega r/c) j_\ell(\omega r'/c) Y^*_{\ell m}(\theta', \varphi') Y_{\ell m}(\theta, \varphi), & \text{if } r > r'. \end{cases} \end{aligned} \tag{5.100}$$

Let us now return to the wave equation (5.31) and consider the case where the source term $f(t, \boldsymbol{x})$ is nonvanishing only in a ball of radius R at all times. We may use eq. (5.100) to obtain a multipole expansion of the retarded solution ψ to the wave equation for $r > R$ as follows. The time Fourier transform, $\hat{\psi}$, of ψ is given in terms of $\hat{f}$ by

$$\hat{\psi}(\omega, \boldsymbol{x}) = \int G^{\mathrm{H}}(\boldsymbol{x}, \boldsymbol{x}') \hat{f}(\omega, \boldsymbol{x}) d^3x' . \tag{5.101}$$

Substituting the series expansion for the $r > r'$ case of eq. (5.100), we obtain

$$\hat{\psi}(\omega, \boldsymbol{x}) = \frac{i\omega}{c} \sum_{\ell,m} \tilde{q}_{\ell m}(\omega) h^{(1)}_\ell(\omega r/c) Y_{\ell m}(\theta, \varphi), \tag{5.102}$$

where the source multipole moments, $\tilde{q}_{\ell m}(\omega)$, at frequency ω are given by

$$\tilde{q}_{\ell m}(\omega) = \int j_\ell(\omega r'/c) Y^*_{\ell m}(\theta', \varphi') \hat{f}(\omega, \boldsymbol{x}') d^3x' . \tag{5.103}$$

The retarded solution $\psi(t, \boldsymbol{x})$ can then be obtained by taking the inverse Fourier transform of $\hat{\psi}(\omega, \boldsymbol{x})$:

$$\psi(t, \boldsymbol{x}) = \frac{1}{\sqrt{2\pi}} \int_{-\infty}^{\infty} e^{-i\omega t} \hat{\psi}(\omega, \boldsymbol{x}) d\omega . \tag{5.104}$$

Note that the solution eq. (5.102) is valid at all $r > R$, not just to leading order in $1/r$. We may also obtain the retarded solution for $r < R$ by using the $r < r'$ case of eq. (5.100) to account for the contribution from the portion of the source lying outside the radius at which ψ is being evaluated (see eq. (2.83) for the analogous expression in electrostatics).

[7] To calculate the jump condition, we use the fact that $j_\ell(x) \frac{dn_\ell}{dx}(x) - n_\ell(x) \frac{dj_\ell}{dx}(x) = \frac{1}{x^2}$.

We now turn our attention to the electromagnetic field. Since the potential ϕ satisfies the wave equation with source ρ, the above analysis applies to it, and we can write down multipole expansion formulas for ϕ exactly as in the previous paragraph. The Cartesian components of $\boldsymbol{A}$ also satisfy the wave equation with the source given by the corresponding Cartesian components of $\boldsymbol{J}$, so the formulas of the previous paragraph also apply to it. Thus, we thereby immediately obtain an exact multipole expansion for the retarded solution for the electromagnetic field for an arbitrary charge-current source without having to do any additional work.

However, as discussed in section 4.2, the Cartesian basis is not rotationally invariant, and the current conservation equation and Lorenz gauge condition couple terms of different values of ℓ and m in the spherical harmonic expansion of the Cartesian components of $\boldsymbol{J}$ and $\boldsymbol{A}$. Consequently, it is much more useful to expand $\boldsymbol{A}$ and $\boldsymbol{J}$ in vector spherical harmonics, eq. (4.38). When one does so, terms with different values of ℓ and m decouple from each other. Terms of different parity also decouple. Furthermore, terms with different frequencies also decouple. Thus, we can separately consider solutions with a given ω, ℓ, and m and parity type. We then sum the result over ℓ, m, and parity type and integrate over ω to get the general solution. We now explain in more detail how this works and what the results are.

We begin by taking the Fourier transform of all quantities with respect to time:

$$\hat{\rho}(\omega,\boldsymbol{x}) = \frac{1}{\sqrt{2\pi}}\int_{-\infty}^{\infty}\rho(t,\boldsymbol{x})e^{i\omega t}dt\,, \qquad \hat{\boldsymbol{J}}(\omega,\boldsymbol{x}) = \frac{1}{\sqrt{2\pi}}\int_{-\infty}^{\infty}\boldsymbol{J}(t,\boldsymbol{x})e^{i\omega t}dt, \tag{5.105}$$

$$\hat{\phi}(\omega,\boldsymbol{x}) = \frac{1}{\sqrt{2\pi}}\int_{-\infty}^{\infty}\phi(t,\boldsymbol{x})e^{i\omega t}dt\,, \qquad \hat{\boldsymbol{A}}(\omega,\boldsymbol{x}) = \frac{1}{\sqrt{2\pi}}\int_{-\infty}^{\infty}\boldsymbol{A}(t,\boldsymbol{x})e^{i\omega t}dt\,. \tag{5.106}$$

We then expand the Fourier transformed quantities in scalar and vector spherical harmonics. Because of the decoupling mentioned in the previous paragraph, we may consider each ω, ℓ, m, and parity type separately.

Consider, first, a magnetic parity charge-current distribution of a given ω, ℓ, m. Since any scalar quantity such as $\hat{\rho}$ is automatically of electric parity, we have $\hat{\rho} = 0$. Furthermore, as in eq. (4.39), a magnetic parity current distribution must take the form

$$\hat{\boldsymbol{J}} = j^{\mathrm{M}}_{\omega\ell m}(r)\,\boldsymbol{r}\times\nabla Y_{\ell m}\,. \tag{5.107}$$

Similarly, the corresponding potentials in the magnetic parity case are $\hat{\phi} = 0$ and $\hat{\boldsymbol{A}}$ of the form

$$\hat{\boldsymbol{A}} = a^{\mathrm{M}}_{\omega\ell m}(r)\,\boldsymbol{r}\times\nabla Y_{\ell m}\,. \tag{5.108}$$

The only difference between the magnetostatic equation (4.20) and the electrodynamic equation (5.29) is the additional second time derivative term appearing in eq. (5.29). Thus, plugging eq. (5.107) and eq. (5.108) into eq. (5.29), we obtain

$$\frac{d^2a^{\mathrm{M}}_{\omega\ell m}}{dr^2} + \frac{2}{r}\frac{da^{\mathrm{M}}_{\omega\ell m}}{dr} - \frac{\ell(\ell+1)}{r^2}a^{\mathrm{M}}_{\omega\ell m} + \frac{\omega^2}{c^2}a^{\mathrm{M}}_{\omega\ell m} = -\mu_0 j^{\mathrm{M}}_{\omega\ell m}\,, \tag{5.109}$$

which differs from eq. (4.41) only by the addition of the last term on the left side. The operator appearing on the left side of eq. (5.109) is just the radial Helmholtz operator of eq. (5.89). In parallel with eq. (4.42), the solution to eq. (5.109) that is regular at the origin and corresponds to an outgoing wave at infinity is

$$a^{\mathrm{M}}_{\omega\ell m}(r) = \frac{i\omega\mu_0}{c}\left[h^{(1)}_\ell(\tfrac{\omega r}{c})\int_0^r j^{\mathrm{M}}_{\omega\ell m}(r')j_\ell(\tfrac{\omega r'}{c})r'^2 dr' \right.$$
$$\left. + j_\ell(\tfrac{\omega r}{c})\int_r^\infty j^{\mathrm{M}}_{\omega\ell m}(r')h^{(1)}_\ell(\tfrac{\omega r'}{c})r'^2 dr'\right]. \qquad (5.110)$$

In particular, if the current distribution vanishes for $r > R$, then for $r > R$, we have

$$a^{\mathrm{M}}_{\omega\ell m}(r) = -i\mu_0 M_{\omega\ell m} h^{(1)}_\ell(\tfrac{\omega r}{c}), \qquad (5.111)$$

where the *magnetic parity multipole moments* $M_{\omega\ell m}$ are defined by

$$M_{\omega\ell m} \equiv -\frac{\omega}{c}\int_0^\infty j^{\mathrm{M}}_{\omega\ell m}(r)j_\ell(\tfrac{\omega r}{c})r^2 dr = -\frac{\omega}{c\ell(\ell+1)}\int j_\ell(\tfrac{\omega r}{c})\left[\boldsymbol{r}\times\nabla Y^*_{\ell m}\right]\cdot\hat{\boldsymbol{J}}\, d^3x$$
$$= \frac{\omega}{c\ell(\ell+1)}\int j_\ell(\tfrac{\omega r}{c})Y^*_{\ell m}(\theta,\varphi)\,\boldsymbol{r}\cdot(\nabla\times\hat{\boldsymbol{J}})\, d^3x\,. \qquad (5.112)$$

The electric and magnetic fields can then be computed from $\hat{\boldsymbol{E}} = i\omega\hat{\boldsymbol{A}}$ and $\hat{\boldsymbol{B}} = \nabla\times\hat{\boldsymbol{A}}$. Note that in the magnetic parity case, it follows immediately from eq. (5.108) that $\boldsymbol{r}\cdot\hat{\boldsymbol{E}} = 0$ at all $\boldsymbol{x}$.

In the electric parity case, we will get nontrivial contributions at a given ω, ℓ, m from $\hat{\rho}$ as well as two terms in $\hat{\boldsymbol{J}}$, as given in eq. (4.45). These quantities will be related by charge-current conservation. Similarly, the potentials will get contributions from $\hat{\phi}$ and from the two electric parity spherical harmonic terms in $\hat{\boldsymbol{A}}$, with the Lorenz gauge condition relating these three quantities. However, as in the magnetostatic case, we can avoid the complications arising from having many unknown variables and relations between them by working directly with the magnetic field, which takes the magnetic parity form

$$\hat{\boldsymbol{B}} = b^{\mathrm{E}}_{\omega\ell m}(r)\,\boldsymbol{r}\times\nabla Y_{\ell m}\,. \qquad (5.113)$$

By problem 1 (at the end of this chapter), $\hat{\boldsymbol{B}}$ satisfies

$$\nabla^2\hat{\boldsymbol{B}} + \frac{\omega^2}{c^2}\hat{\boldsymbol{B}} = -\mu_0\nabla\times\hat{\boldsymbol{J}}\,. \qquad (5.114)$$

Furthermore, if $\hat{\boldsymbol{J}}$ is of electric parity, then $\nabla\times\hat{\boldsymbol{J}}$ is of magnetic parity, so $\nabla\times\hat{\boldsymbol{J}}$ must be of the general magnetic parity form

$$\nabla\times\hat{\boldsymbol{J}} = f^{\mathrm{E}}_{\omega\ell m}(r)\,\boldsymbol{r}\times\nabla Y_{\ell m}\,. \qquad (5.115)$$

Consequently, in the electric parity case, $b^{\mathrm{E}}_{\omega\ell m}$ is related to $f^{\mathrm{E}}_{\omega\ell m}$ in exactly the same way that $a^{\mathrm{M}}_{\omega\ell m}$ is related to $j^{\mathrm{M}}_{\omega\ell m}$ in the magnetic parity case analyzed earlier in this section. Thus, eq. (5.110) holds with these substitutions. In particular, if the charge-current distribution vanishes for $r > R$, then for $r > R$, we have

$$b^{\mathrm{E}}_{\omega\ell m}(r) = -\omega\mu_0 Q_{\omega\ell m} h^{(1)}_\ell(\tfrac{\omega r}{c}), \tag{5.116}$$

where the *electric parity multipole moments* $Q_{\omega\ell m}$ are defined by[8]

$$Q_{\omega\ell m} \equiv -\frac{i}{c}\int_0^\infty f^{\mathrm{E}}_{\omega\ell m}(r) j_\ell(\tfrac{\omega r}{c}) r^2 dr = -\frac{i}{c\ell(\ell+1)}\int j_\ell(\tfrac{\omega r}{c})\left[\boldsymbol{r}\times\nabla Y^*_{\ell m}\right]\cdot(\nabla\times\hat{\boldsymbol{J}})d^3x$$

$$= \frac{i}{c\ell(\ell+1)}\int j_\ell(\tfrac{\omega r}{c}) Y^*_{\ell m}(\theta,\varphi)\,\boldsymbol{r}\cdot[\nabla\times(\nabla\times\hat{\boldsymbol{J}})]\,d^3x. \tag{5.117}$$

The electric field can then be obtained from $\frac{i\omega}{c^2}\hat{\boldsymbol{E}} = -\nabla\times\hat{\boldsymbol{B}} + \mu_0\hat{\boldsymbol{J}}$. Note that in the electric parity case, it follows immediately from eq. (5.113) that $\boldsymbol{r}\cdot\hat{\boldsymbol{B}} = 0$ for all $\boldsymbol{x}$.

Putting the above results together, we see that if we have a charge-current distribution that is nonvanishing only when $r < R$, then for $r > R$, the magnetic field is given by

$$\boldsymbol{B}(t,\boldsymbol{x}) = -\mu_0\sum_{\ell,m}\frac{1}{\sqrt{2\pi}}\int_{-\infty}^{\infty} d\omega e^{-i\omega t}\Big[icM_{\omega\ell m}\nabla\times\Big(h^{(1)}_\ell(\tfrac{\omega r}{c})\boldsymbol{r}\times\nabla Y_{\ell m}\Big)$$

$$+\,\omega Q_{\omega\ell m} h^{(1)}_\ell(\tfrac{\omega r}{c})\boldsymbol{r}\times\nabla Y_{\ell m}\Big] \tag{5.118}$$

where $M_{\omega\ell m}$ and $Q_{\omega\ell m}$ are given by eqs. (5.112) and (5.117), respectively. Since $\boldsymbol{J}$ vanishes for $r > R$, the corresponding electric field is obtained by replacing the quantity in square brackets in the integrand of section 5.3.2 by ic^2/ω times its curl. We obtain

$$\boldsymbol{E}(t,\boldsymbol{x}) = \mu_0 c\sum_{\ell,m}\frac{1}{\sqrt{2\pi}}\int_{-\infty}^{\infty} d\omega e^{-i\omega t}\Big[\omega M_{\omega\ell m} h^{(1)}_\ell(\tfrac{\omega r}{c})\boldsymbol{r}\times\nabla Y_{\ell m}$$

$$-\,icQ_{\omega\ell m}\nabla\times\Big(h^{(1)}_\ell(\tfrac{\omega r}{c})\boldsymbol{r}\times\nabla Y_{\ell m}\Big)\Big] \tag{5.119}$$

where we used eq. (4.16) together with $\nabla\cdot(h^{(1)}_\ell(\frac{\omega r}{c})\boldsymbol{r}\times\nabla Y_{\ell m}) = 0$, and $[\nabla^2 + \frac{\omega^2}{c^2}](h^{(1)}_\ell(\frac{\omega r}{c})\boldsymbol{r}\times\nabla Y_{\ell m}) = 0$. The electromagnetic field for $r < R$ can be obtained by using eq. (5.110) and the similar equation for $b^{\mathrm{E}}_{\omega\ell m}$ rather than eqs. (5.111) and (5.116), but we shall not write out the formulas here. The leading-order nonrelativistic results obtained in section 5.3.1 correspond to considering the $\ell = 1$ electric parity contribution in the limit of low frequencies (see problem 10).

In summary, we have obtained a series expansion for the electromagnetic field of the retarded solution for an arbitrary relativistic charge-current distribution. In particular, if we want to find the electromagnetic field outside the source, we need only calculate the multipole moments eqs. (5.112) and (5.117) at each frequency of the time Fourier transform of the current. The electric and magnetic fields are then given by eqs. (5.118) and (5.119). These formulas are exact.

[8] A useful alternative formula for $Q_{\omega\ell m}$ is given in problem 9 (at the end of this chapter).

5.4 The Initial Value Formulation for Maxwell's Equations

In this section, we show how the electromagnetic field at all times is uniquely determined by appropriate initial data for the electromagnetic field at one time. This will characterize in a precise way the independent degrees of freedom of the electromagnetic field. Furthermore, the formula determining the electromagnetic field from its initial data also will show in a precise way that the electromagnetic field propagates at the speed of light.

We begin by analyzing the initial value formulation of the wave equation (5.31). The key tool that we need is a version of Green's theorem of section 2.5 for the wave equation rather than for Poisson's equation. This version of Green's theorem will involve integration over spacetime rather than over space. To formulate it, it is very convenient to introduce the notation of special relativity. Write $x^0 = ct$, and write the coordinates on spacetime as $x^\mu = (x^0, x^1, x^2, x^3)$. We also write

$$\partial_\mu = (\frac{\partial}{\partial x^0}, \frac{\partial}{\partial x^1}, \frac{\partial}{\partial x^2}, \frac{\partial}{\partial x^3}) . \tag{5.120}$$

Thus, ∂_μ is the spacetime generalization of $\boldsymbol{\nabla}$. Define

$$\eta^{\mu\nu} = \begin{pmatrix} -1 & 0 & 0 & 0 \\ 0 & 1 & 0 & 0 \\ 0 & 0 & 1 & 0 \\ 0 & 0 & 0 & 1 \end{pmatrix} . \tag{5.121}$$

As we shall see in chapter 8, $\eta^{\mu\nu}$ is the inverse spacetime metric. Then we have

$$\Box = \sum_{\mu,\nu} \eta^{\mu\nu} \partial_\mu \partial_\nu . \tag{5.122}$$

Now, let ψ_1 and ψ_2 satisfy

$$\Box\psi_1 = -f_1, \tag{5.123}$$

$$\Box\psi_2 = -f_2 . \tag{5.124}$$

Then, by the same type of calculation performed in eq. (2.105), we have

$$\begin{aligned} \sum_{\mu,\nu} \partial_\mu \left[\eta^{\mu\nu} (\psi_1 \partial_\nu \psi_2 - \psi_2 \partial_\nu \psi_1)\right] &= \psi_1 \Box \psi_2 - \psi_2 \Box \psi_1 \\ &= -(\psi_1 f_2 - \psi_2 f_1) . \end{aligned} \tag{5.125}$$

The left side of eq. (5.125) is a total divergence. Thus, if we integrate it over a *spacetime* (i.e., 4-dimensional) "volume," we can use a 4-dimensional version of Gauss's theorem to convert it to a (3-dimensional) "surface" integral over the boundary of this volume. Although the 4-dimensional version of Gauss's theorem holds quite generally (see the boxed side comment in section 2.1), for our purposes, we need only consider a rectangular spacetime volume, in which case, the integral of a total divergence is easily computed directly. Specifically, let $v^\mu = (v^0, v^1, v^2, v^3)$ be an arbitrary (differentiable) vector field on spacetime. Let $\mathscr{R}$ be the rectangular region in spacetime defined by $c_0^\mu \leq x^\mu \leq c_1^\mu$ for $\mu = 0, 1, 2, 3$, where c_0^μ, c_1^μ are constants. Then we have

$$\int_{\mathscr{R}} \sum_\mu \partial_\mu v^\mu d^4x = \int_{\mathscr{R}} \left[\frac{\partial v^0}{\partial x^0} + \frac{\partial v^1}{\partial x^1} + \frac{\partial v^2}{\partial x^2} + \frac{\partial v^3}{\partial x^3} \right] dx^0 dx^1 dx^2 dx^3$$
$$= \left[\int v^0 dx^1 dx^2 dx^3 \right] \Bigg|_{x^0=c_0^0}^{x^0=c_1^0} + \left[\int v^1 dx^0 dx^2 dx^3 \right] \Bigg|_{x^1=c_0^1}^{x^1=c_1^1}$$
$$+ \left[\int v^2 dx^0 dx^1 dx^3 \right] \Bigg|_{x^2=c_0^2}^{x^2=c_1^2} + \left[\int v^3 dx^0 dx^1 dx^2 \right] \Bigg|_{x^3=c_0^3}^{x^3=c_1^3}, \tag{5.126}$$

where the integrals on the right side of eq. (5.126) are taken over the indicated faces of the rectangle $\mathscr{R}$. We now apply this formula, choosing

$$v^\mu = \sum_\nu \eta^{\mu\nu} \left(\psi_1 \partial_\nu \psi_2 - \psi_2 \partial_\nu \psi_1 \right). \tag{5.127}$$

Integrating eq. (5.125) over $\mathscr{R}$, we find that the left side is given by eq. (5.126), whereas the right side is simply $-\int_{\mathscr{R}} (\psi_1 f_2 - \psi_2 f_1)$. This is our desired version of Green's theorem.

Now let ψ satisfy the homogeneous wave equation,

$$\Box \psi = 0. \tag{5.128}$$

We apply Green's theorem, choosing $\psi_1 = \psi$ and $\psi_2(x^\mu) = G_{\rm adv}(x^\mu, x'^\mu)$—where the advanced Green's function is given by eq. (5.51)—with x'^μ chosen to be an arbitrary point in spacetime with $x'^0 > 0$. We choose the spacetime rectangle $\mathscr{R}$ to have its bottom face at $x^0 = c_0^0 = 0$, and we choose its top face to be at $x^0 = c_1^0 > x'^0$. We choose its side faces so that for $\mu = 1, 2, 3$, we have $c_0^\mu < x'^\mu - x'^0$ and $c_1^\mu > x'^\mu + x'^0$. This choice ensures that $\mathscr{R}$ not only encloses the spacetime point x'^μ but that the side faces of $\mathscr{R}$ are far enough away from x'^μ that the past light cone of x'^μ does not intersect these faces, as illustrated in figure 5.2.

Let us now integrate eq. (5.125) over $\mathscr{R}$ with the above choices of ψ_1 and ψ_2, and evaluate the left side using eq. (5.126). When we do so, we get integrals involving $G_{\rm adv}(x^\mu, x'^\mu)$, where x^μ lies on one of the faces of $\mathscr{R}$. However, $G_{\rm adv}(x^\mu, x'^\mu) = 0$ unless x^μ lies on the past light cone of x'^μ. Consequently, only the bottom face contributes to the right side of eq. (5.126), and so we obtain

$$\int_{\mathscr{R}} \sum_{\mu,\nu} \partial_\mu \left[\eta^{\mu\nu} \left(\psi(x^\mu) \partial_\nu G_{\rm adv}(x^\mu, x'^\mu) - G_{\rm adv}(x^\mu, x'^\mu) \partial_\nu \psi(x^\mu) \right) \right] d^4x$$
$$= -\int_{x^0=0} v^0 d^3x, \tag{5.129}$$

where

$$v^0 = -\left[\psi(x^\mu) \partial_0 G_{\rm adv}(x^\mu, x'^\mu) - G_{\rm adv}(x^\mu, x'^\mu) \partial_0 \psi(x^\mu) \right], \tag{5.130}$$

and we have used the fact that $\eta^{0\nu} = -\delta^{0\nu}$. The limits of integration over x^1, x^2, x^3 on the right side of eq. (5.129) can be taken to be over all space (rather than over just the

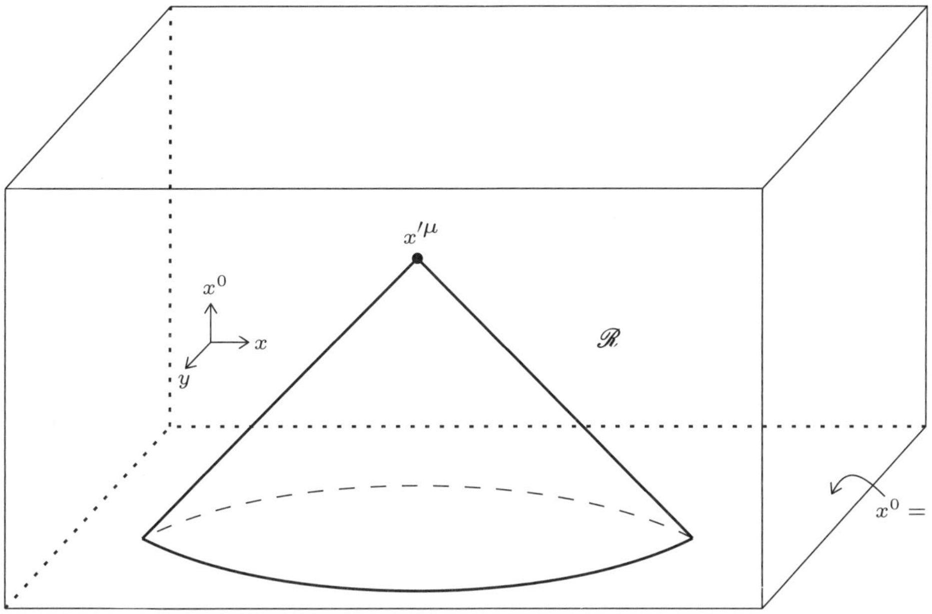

Figure 5.2. A spacetime diagram showing the spacetime rectangle $\mathscr{R}$, chosen so that the bottom face is at $x^0 = 0$, the top face lies above x'^0, and the sides are sufficiently far away that the past light cone of x'^{μ} does not intersect the sides.

bottom face of $\mathscr{R}$), since v^0 is nonvanishing only on the past light cone of x'^{μ}. The integral of the right side of eq. (5.125) over $\mathscr{R}$ yields

$$\int_{\mathscr{R}} \left[\psi(x^{\mu})\Box G_{\mathrm{adv}}(x^{\mu}, x'^{\mu}) - G_{\mathrm{adv}}(x^{\mu}, x'^{\mu})\Box\psi(x^{\mu})\right] d^4x$$
$$= -\int_{\mathscr{R}} \psi(x^{\mu})\delta(t-t')\delta(\boldsymbol{x}-\boldsymbol{x}')c\,dt\,d^3x = -c\psi(x'^{\mu}), \tag{5.131}$$

where we have used $\Box G_{\mathrm{adv}}(x^{\mu}, x'^{\mu}) = -\delta(t-t')\delta(\boldsymbol{x}-\boldsymbol{x}')$, and $\Box\psi = 0$. Thus, we obtain

$$\psi(x'^{\mu}) = -\frac{1}{c}\int_{x^0=0} \left[\psi(x^{\mu})\partial_0 G_{\mathrm{adv}}(x^{\mu}, x'^{\mu}) - G_{\mathrm{adv}}(x^{\mu}, x'^{\mu})\partial_0\psi(x^{\mu})\right] d^3x. \tag{5.132}$$

Reversing the roles of x^{μ} and x'^{μ} and using $G_{\mathrm{adv}}(x'^{\mu}, x^{\mu}) = G_{\mathrm{ret}}(x^{\mu}, x'^{\mu})$ (see eq. (5.52)), we may rewrite this equation as

$$\psi(x^{\mu}) = -\frac{1}{c}\int_{x'^0=0} \left[\psi(x'^{\mu})\partial'_0 G_{\mathrm{ret}}(x^{\mu}, x'^{\mu}) - G_{\mathrm{ret}}(x^{\mu}, x'^{\mu})\partial'_0\psi(x'^{\mu})\right] d^3x'. \tag{5.133}$$

Using the explicit form, eq. (5.50), of the retarded Green's function and also switching back from the notation x^{μ} to $(t, \boldsymbol{x})$ for labeling spacetime events, we find that for all $t > 0$, we have

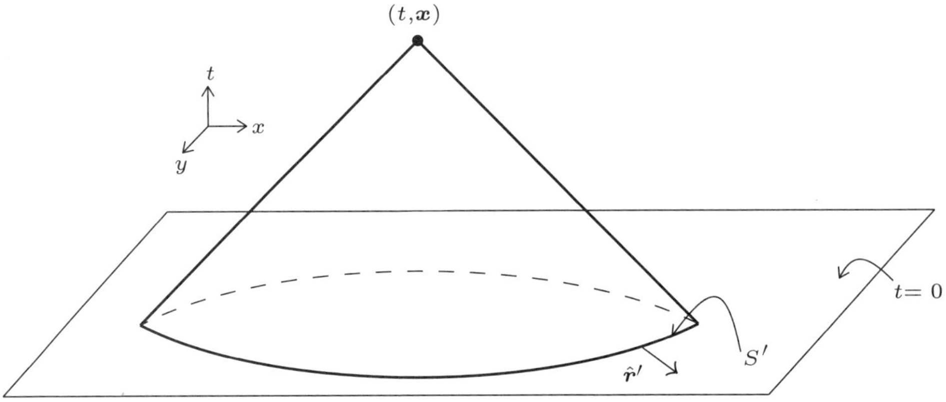

Figure 5.3. A spacetime diagram showing the sphere S' obtained by taking the intersection of the past light cone of (t, x) with the hypersurface $t = 0$. The unit outward pointing normal to S' in this hypersurface is $\hat{r}'$.

$$\psi(t, \boldsymbol{x}) = \frac{1}{4\pi}\int_{S'}\left[\frac{1}{r'^2}\psi(\theta',\varphi') + \frac{1}{r'}\left(\frac{1}{c}\frac{\partial\psi}{\partial t}(\theta',\varphi') + \hat{\boldsymbol{r}}'\cdot\nabla\psi(\theta',\varphi')\right)\right] r'^2 \sin\theta' d\theta' d\varphi'. \tag{5.134}$$

Here the integral is taken over the sphere S' that is obtained by intersecting the past light cone of $(t, \boldsymbol{x})$ with the hypersurface $t = 0$, as illustrated in figure 5.3. We have introduced polar coordinates, θ', φ' on this sphere, r' denotes the radius of the sphere, and $\hat{\boldsymbol{r}}'$ denotes the unit, outward-pointing normal to S' lying within the hypersurface $t = 0$.

Equation (5.134) shows that if we want to calculate $\psi(t, \boldsymbol{x})$ for $t > 0$, all we need to know is the value of ψ and its first derivatives on the sphere S' at time $t = 0$. This formula has been derived for $t > 0$, but a similar formula (with $t \to -t$) also holds for $t < 0$, with S' now taken to be the intersection of the future light cone of $(t, \boldsymbol{x})$ with the hypersurface $t = 0$. This shows that a solution ψ to the homogeneous wave equation (5.128) is uniquely determined everywhere if one is given ψ and $\partial\psi/\partial t$ for all $\boldsymbol{x}$ at $t = 0$, since this provides enough information to compute the right side of eq. (5.134) at any t and $\boldsymbol{x}$. Furthermore, given an arbitrary smooth specification of ψ and $\partial\psi/\partial t$ for all $\boldsymbol{x}$ at $t = 0$, it can be shown that if we define $\psi(t, \boldsymbol{x})$ for all t by eq. (5.134) and the similar formula for $t < 0$, then ψ will satisfy the homogeneous wave equation (5.128).

If ψ satisfies the inhomogeneous wave equation (5.31) with source $f(t, \boldsymbol{x})$, the above derivation and results continue to apply, with the only change being that an additional spacetime volume integral term appears in eq. (5.131), resulting from the fact that we now have $\Box\psi = -f$. This gives rise to the additional term $\int G_{\text{ret}}(x^\mu, x'^\mu) f(x'^\mu) d^4x'$ in eq. (5.133) for $\psi(x^\mu)$, where the integral is taken only over the region $t \geq 0$ (since $\mathscr{R}$ includes only $t \geq 0$). The final result in the inhomogeneous case corresponding to eq. (5.134) in the homogeneous case is

$$\begin{aligned}\psi(t, \boldsymbol{x}) = {} & \frac{1}{4\pi}\int_{S'}\left[\frac{1}{r'^2}\psi(\theta',\varphi') + \frac{1}{r'}\left(\frac{1}{c}\frac{\partial\psi}{\partial t}(\theta',\varphi') + \hat{\boldsymbol{r}}'\cdot\nabla\psi(\theta',\varphi')\right)\right] r'^2 \sin\theta' d\theta' d\varphi' \\ & + \int_{|\boldsymbol{x}-\boldsymbol{x}'|\leq ct} \frac{[f(t', \boldsymbol{x}')]_{\text{ret}}}{4\pi|\boldsymbol{x}-\boldsymbol{x}'|} d^3x'. \end{aligned} \tag{5.135}$$

Thus, ψ is uniquely determined if the source term $f(t, \boldsymbol{x})$ is specified everywhere on spacetime, and ψ and $\partial\psi/\partial t$ are given for all $\boldsymbol{x}$ at an initial time. Furthermore, if we define ψ in terms of its initial values and f by eq. (5.135), then ψ will satisfy the inhomogeneous wave equation (5.31) with source $f(t, \boldsymbol{x})$. Thus, we have the following existence and uniqueness theorem.

> ***Theorem (Initial Value Theorem for the Wave Equation):*** *Let $f(t, \boldsymbol{x})$ be an arbitrary smooth function on spacetime, and let $\chi_1(\boldsymbol{x})$ and $\chi_2(\boldsymbol{x})$ be arbitrary smooth functions on space. Then there exists a unique smooth solution $\psi(t, \boldsymbol{x})$ of $\Box\psi = -f$ such that $\psi(0, \boldsymbol{x}) = \chi_1(\boldsymbol{x})$ and $\frac{\partial\psi}{\partial t}(0, \boldsymbol{x}) = \chi_2(\boldsymbol{x})$.*

This theorem shows that the freedom in specifying a solution ψ to the wave equation (5.31) with a given source $f(t, \boldsymbol{x})$ consists of an arbitrary specification of the value of ψ and its first time derivative $\partial\psi/\partial t$ at an initial time. This is closely analogous to the situation in particle mechanics, where the freedom in specifying a solution to particle motion in a given force field consists of an arbitrary specification of the position and momentum of the particle at an initial time. In particle mechanics, these free initial data comprise the *phase space* of the system, which can be viewed as describing the independent dynamical degrees of freedom of the system. One can similarly think of ψ and $\partial\psi/\partial t$ at a given time as describing the independent dynamical degrees of freedom of solutions to the wave equation. Since the specification of ψ and $\partial\psi/\partial t$ as initial data for the wave equation allows arbitrary functions of $\boldsymbol{x}$—as opposed to finitely many parameters, as needed to specify initial data in the particle case—the solutions to the wave equation have infinitely many independent dynamical degrees of freedom.

Another extremely important feature of equations (5.134) and (5.135) is as follows. The initial data at $t = 0$ that is relevant to determining the solution at $(t, \boldsymbol{x})$ is only the initial data on S', that is, the initial data lying on the past light cone of $(t, \boldsymbol{x})$. This provides a precise statement of the notion that solutions to the wave equation propagate at the speed of light. This fact is often referred to as *Huygen's principle*. It should be noted that the above initial value theorem holds quite generally for partial differential equations whose highest derivative terms are similar to that of the wave equation. However, the Huygen's principle property is very special to the specific form of the wave equation. If we were to consider a modification of the wave equation involving lower derivative terms (such as the massive Klein-Gordon equation, $(\Box + m^2)\psi = -f$), then the the above initial value theorem would continue to hold, but the solution ψ at $(t, \boldsymbol{x})$ would depend on the initial data interior to S' as well as on S'. Furthermore, in odd spacetime dimensions, Huygen's principle fails even for the unmodified wave equation.

Let us now apply the above results to the electromagnetic case. Since ϕ and the Cartesian components of $\boldsymbol{A}$ satisfy the wave equations (5.28) and (5.29), all of the above formulas—in particular, eq. (5.135)—apply to these quantities. The only significant complication is that to get a solution to Maxwell's equation, we must ensure that our solutions to eqs. (5.28) and (5.29) also satisfy the Lorenz gauge condition, eq. (5.30). To analyze what is needed for this, we write

$$\Psi = \frac{1}{c^2}\frac{\partial\phi}{\partial t} + \nabla \cdot \boldsymbol{A}, \tag{5.136}$$

so that the Lorenz gauge condition is $\Psi = 0$. Then, if ϕ and $\boldsymbol{A}$ satisfy eqs. (5.28) and (5.29), we have

$$\Box\Psi = \frac{1}{c^2}\frac{\partial}{\partial t}(\Box\phi) + \nabla\cdot(\Box \boldsymbol{A}) = -\frac{1}{c^2\epsilon_0}\frac{\partial\rho}{\partial t} - \mu_0\nabla\cdot\boldsymbol{J} = -\mu_0\left[\frac{\partial\rho}{\partial t} + \nabla\cdot\boldsymbol{J}\right] = 0, \tag{5.137}$$

where we have used the charge-current conservation eq. (5.8). Thus, if ϕ and $\boldsymbol{A}$ satisfy eqs. (5.28) and (5.29), then Ψ satisfies the homogeneous wave equation. By the above initial value theorem, we will have $\Psi = 0$ everywhere (i.e., the Lorenz gauge condition will hold) if and only if the initial data for ϕ and $\boldsymbol{A}$ are chosen so that $\Psi = 0$ and $\partial\Psi/\partial t = 0$ at $t = 0$. We can try to satisfy these initial conditions by using the gauge freedom, $\phi \to \phi - \frac{1}{c}\frac{\partial\chi}{\partial t}$, $\boldsymbol{A} \to \boldsymbol{A} + \nabla\chi$. It is easy to see that we can use this gauge freedom to set $\Psi = 0$ at $t = 0$. However, we have

$$\frac{\partial\Psi}{\partial t} = \frac{1}{c^2}\frac{\partial^2\phi}{\partial t^2} + \nabla\cdot\frac{\partial\boldsymbol{A}}{\partial t} = \frac{\rho}{\epsilon_0} + \nabla^2\phi + \nabla\cdot\frac{\partial\boldsymbol{A}}{\partial t} = \frac{\rho}{\epsilon_0} - \nabla\cdot\boldsymbol{E}, \tag{5.138}$$

where eq. (5.28) has been used. The right side of this equation is gauge invariant, so we cannot set it to zero by a gauge transformation. Thus, we cannot get a solution to eqs. (5.28) and (5.29) that also satisfies the Lorenz gauge condition unless the initial data for eqs. (5.28) and (5.29) are chosen so that this quantity vanishes. Thus, the initial data for the electromagnetic field at $t = 0$ must satisfy the *initial value constraint*:

$$\nabla\cdot\boldsymbol{E} = \rho/\epsilon_0\,. \tag{5.139}$$

Of course, this is one of Maxwell's equations, so it is not surprising that the initial data must satisfy this condition, since it clearly would not be possible to obtain a solution to Maxwell's equations for all t if this equation did not hold at $t = 0$.

The above analysis shows that to solve Maxwell's equations (5.28), (5.29), and (5.30), we need only solve eqs. (5.28) and (5.29), with the initial data for ϕ and $\boldsymbol{A}$ chosen to satisfy $\frac{1}{c^2}\frac{\partial\phi}{\partial t} + \nabla\cdot\boldsymbol{A} = 0$ and $\nabla\cdot\boldsymbol{E} = \rho/\epsilon_0$ at $t = 0$. The first condition is merely a gauge condition, but the second condition is a genuine physical restriction on the allowed initial data.

It is convenient to reformulate these results in terms of the gauge invariant field strengths $\boldsymbol{E}$ and $\boldsymbol{B}$. Given initial values of ϕ, $\boldsymbol{A}$, and their first time derivatives over all of space at $t = 0$, we can compute $\boldsymbol{E}$ and $\boldsymbol{B}$ at $t = 0$. Conversely, given $\boldsymbol{E}$ and $\boldsymbol{B}$ over all of space at $t = 0$ with $\boldsymbol{B}$ satisfying $\nabla\cdot\boldsymbol{B} = 0$, we can determine ϕ, $\boldsymbol{A}$, and their first time derivatives at $t = 0$ up to a gauge transformation. Thus, we can replace the specification of initial data for the potentials by the specification of initial values of $\boldsymbol{E}$ and $\boldsymbol{B}$. We thereby obtain the following theorem.

Theorem (Initial Value Theorem for Maxwell's Equations): *Let $\rho(t,\boldsymbol{x})$ and $\boldsymbol{J}(t,\boldsymbol{x})$ be arbitrary, smooth specifications of the charge density and current density on spacetime, subject to charge-current conservation, eq.* (5.8). *Let $\boldsymbol{E}_0(\boldsymbol{x})$ and $\boldsymbol{B}_0(\boldsymbol{x})$ be arbitrary, smooth vector fields on space satisfying*

$$\nabla\cdot\boldsymbol{E}_0(\boldsymbol{x}) = \frac{1}{\epsilon_0}\rho(t=0,\boldsymbol{x})\,, \qquad \nabla\cdot\boldsymbol{B}_0(\boldsymbol{x}) = 0\,. \tag{5.140}$$

Then there exists a unique, smooth solution $\boldsymbol{E}(t,\boldsymbol{x})$, $\boldsymbol{B}(t,\boldsymbol{x})$ to Maxwell's equations in the form of eqs. (5.4)–(5.7) with the property that $\boldsymbol{E}(t=0,\boldsymbol{x}) = \boldsymbol{E}_0(\boldsymbol{x})$, and $\boldsymbol{B}(t=0,\boldsymbol{x}) = \boldsymbol{B}_0(\boldsymbol{x})$.

The initial data $\boldsymbol{E}_0(\boldsymbol{x})$ and $\boldsymbol{B}_0(\boldsymbol{x})$ subject to eq. (5.140) thus provides a gauge invariant way of characterizing the independent dynamical degrees of freedom of solutions to Maxwell's equations. Again, for a specified source $\rho(t,\boldsymbol{x})$, $\boldsymbol{J}(t,\boldsymbol{x})$, and specified initial data $\boldsymbol{E}_0(\boldsymbol{x})$ and $\boldsymbol{B}_0(\boldsymbol{x})$, the solution $\boldsymbol{E}$, $\boldsymbol{B}$, at a point $(t,\boldsymbol{x})$ for $t>0$ is given by a formula similar in character to eq. (5.135), involving an integral over S' of quantities computable directly from the initial data and an integral of the sources over the portion of the past light cone of $(t,\boldsymbol{x})$ with $t>0$. Thus, solutions to Maxwell's equations also satisfy Huygen's principle.

5.5 Plane Waves

In this section, we consider the general, source-free solution to Maxwell's equations. Maxwell's equations with $\rho=\boldsymbol{J}=0$ in Lorenz gauge are

$$\Box\phi=0, \tag{5.141}$$

$$\Box\boldsymbol{A}=0, \tag{5.142}$$

$$\frac{1}{c^2}\frac{\partial\phi}{\partial t}+\nabla\cdot\boldsymbol{A}=0\,. \tag{5.143}$$

As already noted near the end of section 5.1, we still have the residual *restricted gauge freedom*

$$\phi\to\phi'=\phi-\frac{\partial\chi}{\partial t}, \qquad \boldsymbol{A}\to\boldsymbol{A}'=\boldsymbol{A}+\nabla\chi, \tag{5.144}$$

where, to preserve the Lorenz gauge condition eq. (5.143), χ must satisfy

$$\Box\chi=0\,. \tag{5.145}$$

From the initial value theorem for the wave equation of section 5.4, we can specify χ and $\partial\chi/\partial t$ arbitrarily as a function of $\boldsymbol{x}$ at $t=0$ and then obtain a unique solution to eq. (5.145) with these initial values. We choose this initial data for χ as follows:

$$\left.\frac{\partial\chi}{\partial t}\right|_{t=0}(\boldsymbol{x})=\phi(t=0,\boldsymbol{x}), \tag{5.146}$$

$$\left.\nabla^2\chi\right|_{t=0}(\boldsymbol{x})=-\frac{1}{c^2}\frac{\partial\phi}{\partial t}(t=0,\boldsymbol{x})\,. \tag{5.147}$$

We may always solve eq. (5.147) for $\chi|_{t=0}$, but if $\frac{\partial\phi}{\partial t}(t=0,\boldsymbol{x})\to 0$ suitably rapidly as $|\boldsymbol{x}|\to\infty$, we may additionally demand that $\chi|_{t=0}\to 0$ as $|\boldsymbol{x}|\to\infty$, in which case eq. (5.147) has a unique solution, and eqs. (5.145)–(5.147) uniquely determine χ for all of spacetime.

We now make the gauge transformation eq. (5.144) with the choice of χ given by eqs. (5.145)–(5.147). In the new gauge, we have $\Box\phi'=0$. In addition, by eq. (5.146), we have $\phi'|_{t=0}=0$, and by eq. (5.147) and eq. (5.145), we have $\frac{\partial\phi'}{\partial t}|_{t=0}=0$. Consequently the initial value theorem for the wave equation implies that

$$\phi'(t,\boldsymbol{x})=0 \tag{5.148}$$

for all $(t, \boldsymbol{x})$. Since $\phi' = 0$, the Lorenz gauge condition eq. (5.143) yields $\nabla \cdot \boldsymbol{A}' = 0$, that is, in this gauge, both the Lorenz and Coulomb gauge conditions hold (see eq. (4.21)). In this gauge, Maxwell's equations are simply

$$\Box \boldsymbol{A}' = 0, \tag{5.149}$$

and

$$\nabla \cdot \boldsymbol{A}' = 0\,. \tag{5.150}$$

In the following, we work in this gauge but drop the prime on $\boldsymbol{A}'$.

We now take the Fourier transform of $\boldsymbol{A}$ in space but *not* in time:[9]

$$\hat{\boldsymbol{A}}(t, \boldsymbol{k}) = \frac{1}{(2\pi)^{3/2}} \int e^{-i\boldsymbol{k}\cdot\boldsymbol{x}} \boldsymbol{A}(t, \boldsymbol{x}) d^3x\,. \tag{5.151}$$

Reality of $\boldsymbol{A}(t, \boldsymbol{x})$ requires that

$$\hat{\boldsymbol{A}}^*(t, \boldsymbol{k}) = \hat{\boldsymbol{A}}(t, -\boldsymbol{k}), \tag{5.152}$$

where $*$ denotes complex conjugation. Equations (5.149) and (5.150) imply that $\hat{\boldsymbol{A}}(t, \boldsymbol{k})$ satisfies

$$\left(-\frac{1}{c^2}\frac{\partial^2}{\partial t^2} - k^2\right)\hat{\boldsymbol{A}}(t, \boldsymbol{k}) = 0, \tag{5.153}$$

$$\boldsymbol{k} \cdot \hat{\boldsymbol{A}}(t, \boldsymbol{k}) = 0, \tag{5.154}$$

where we have written $k = |\boldsymbol{k}|$. The general solution to eq. (5.153) is

$$\hat{\boldsymbol{A}}(t, \boldsymbol{k}) = \boldsymbol{c}_1(\boldsymbol{k}) e^{-i\omega t} + \boldsymbol{c}_2(\boldsymbol{k}) e^{+i\omega t}, \tag{5.155}$$

where $\boldsymbol{c}_1$ and $\boldsymbol{c}_2$ are arbitrary constant (complex) vectors (which depend on $\boldsymbol{k}$) and

$$\omega = kc\,. \tag{5.156}$$

Equation (5.154) implies that

$$\boldsymbol{k} \cdot \boldsymbol{c}_1(\boldsymbol{k}) = \boldsymbol{k} \cdot \boldsymbol{c}_2(\boldsymbol{k}) = 0\,. \tag{5.157}$$

The general solution to the source-free Maxwell equations is obtained by taking the inverse Fourier transform of $\hat{\boldsymbol{A}}(t, \boldsymbol{k})$:

$$\boldsymbol{A}(t, \boldsymbol{x}) = \frac{1}{(2\pi)^{3/2}} \int e^{+i\boldsymbol{k}\cdot\boldsymbol{x}} \hat{\boldsymbol{A}}(t, \boldsymbol{k}) d^3k\,. \tag{5.158}$$

Taking account of the reality condition, eq. (5.152), we may write the general solution in the form

$$\boldsymbol{A}(t, \boldsymbol{x}) = \int \boldsymbol{C}(\boldsymbol{k}) e^{-i\omega t} e^{i\boldsymbol{k}\cdot\boldsymbol{x}} d^3k + \text{c.c.}, \tag{5.159}$$

[9] Note that in this section, the hat denotes the Fourier transform with respect to $\boldsymbol{x}$ but not t, whereas in section 5.3.2, we used the hat to denote the Fourier transform in time but not in space.

where "c.c." denotes the complex conjugate of the preceding term, and $\boldsymbol{C}(\boldsymbol{k})$ is a complex vector that does not depend on $(t, \boldsymbol{x})$ and is such that

$$\boldsymbol{k} \cdot \boldsymbol{C}(\boldsymbol{k}) = 0\,. \tag{5.160}$$

A solution of the form $\phi = 0$ and

$$\boldsymbol{A}(t, \boldsymbol{x}) = \boldsymbol{C} e^{-i\omega t} e^{i\boldsymbol{k}\cdot\boldsymbol{x}}, \tag{5.161}$$

with $\omega = kc$ and with $\boldsymbol{C}$ a (complex) constant vector field on spacetime satisfying $\boldsymbol{k} \cdot \boldsymbol{C} = 0$, is called a *plane electromagnetic wave*. We refer to $\boldsymbol{k}$ as the *wave vector* of the plane wave. Note that since $\boldsymbol{C}$ must be orthogonal to $\boldsymbol{k}$, $\boldsymbol{C}$ has two independent components. This corresponds to the two *polarization* degrees of freedom of the plane wave. We refer to the unit vector $\hat{\boldsymbol{C}}$, pointing in the $\boldsymbol{C}$-direction, as the *polarization vector* of the plane wave. Equation (5.159) shows that every solution to the source-free Maxwell equations can be obtained by taking the real part of a superposition of plane waves. Thus, we can understand a great deal about the general solution to Maxwell's equations by studying the properties of plane wave solutions.

The (complex) electric and magnetic fields of a plane wave solution are given by

$$\boldsymbol{E} = -\frac{\partial \boldsymbol{A}}{\partial t} = i\omega \boldsymbol{C}(\boldsymbol{k}) e^{-i\omega t} e^{i\boldsymbol{k}\cdot\boldsymbol{x}}, \tag{5.162}$$

$$\boldsymbol{B} = \nabla \times \boldsymbol{A} = i\boldsymbol{k} \times \boldsymbol{C}(\boldsymbol{k}) e^{-i\omega t} e^{i\boldsymbol{k}\cdot\boldsymbol{x}}\,. \tag{5.163}$$

Thus $\boldsymbol{E}$ and $\boldsymbol{B}$ of the plane wave solution eq. (5.161) are each orthogonal to $\boldsymbol{k}$, and they are orthogonal to each other. Furthermore, they are of constant magnitude on spacetime, with $|\boldsymbol{E}| = c|\boldsymbol{B}|$. To see the properties of the real solution to Maxwell's equations obtained by adding a plane wave to its complex conjugate, let us orient the z-axis so that $\boldsymbol{k} = k\hat{\boldsymbol{z}}$ and write the components of $\boldsymbol{C}$ as

$$2i\omega \boldsymbol{C} = (\alpha_x e^{i\beta_x}, \alpha_y e^{i\beta_y}, 0), \tag{5.164}$$

where $\alpha_x, \alpha_y, \beta_x, \beta_y$ are real, and we have used the fact that the z-component of $\boldsymbol{C}$ vanishes by eq. (5.160). Then the real solution corresponding to this plane wave has electric and magnetic fields given by

$$\boldsymbol{E} = \alpha_x \cos(\omega t - kz - \beta_x)\hat{\boldsymbol{x}} + \alpha_y \cos(\omega t - kz - \beta_y)\hat{\boldsymbol{y}}, \tag{5.165}$$

$$c\boldsymbol{B} = -\alpha_y \cos(\omega t - kz - \beta_y)\hat{\boldsymbol{x}} + \alpha_x \cos(\omega t - kz - \beta_x)\hat{\boldsymbol{y}}\,. \tag{5.166}$$

If $\beta_x = \beta_y$, then at fixed z, $\boldsymbol{E}$ and $\boldsymbol{B}$ oscillate in phase with each other about the fixed directions $\boldsymbol{\alpha}$ and $\hat{\boldsymbol{z}} \times \boldsymbol{\alpha}$, respectively. Such a plane wave is said to be *linearly polarized*. In contrast, if $\alpha_x = \alpha_y$ and $\beta_x = \beta_y \pm \pi/2$, then at fixed z, $\boldsymbol{E}$ and $\boldsymbol{B}$ each maintain a constant magnitude but rotate in the x-y plane, where the direction of rotation is determined by the $\pm$. The plane wave is then said to be right- or left-handed *circularly polarized*, depending on whether the direction of rotation is in the positive or negative φ-direction relative to the z-axis. In the general case where neither the linear nor circular polarization conditions hold, the plane wave is said to be *elliptically polarized*. It is easy to see that the general plane wave solution, eqs. (5.165) and (5.166) with $\boldsymbol{k} = k\hat{z}$, is a linear combination of a linearly polarized plane wave with polarization in the x-direction (i.e., $\alpha_y = 0$) and a linearly polarized plane wave with polarization in the y-direction (i.e., $\alpha_x = 0$). Similarly, a general plane wave solution can be written as a linear combination of right- and left-handed circularly polarized plane waves.

Finally, we note that a plane wave occupies all of space[10] at all times, so it is not obvious how to meaningfully assign a "velocity" to a plane wave. Nevertheless, for a general plane wave of the form of eqs. (5.165) and (5.166), $\boldsymbol{E}$ and $\boldsymbol{B}$ are constant on surfaces of constant values of $\omega t - kz$. Thus, the *phase velocity* of the wave is $z/t = \omega/k = c$ (i.e., the pattern of the wave propagates in the z-direction at the speed of light). Since this phase velocity is the same[11] for all k, by taking a superposition of plane waves in the z-direction for different values of k, one can produce wave packets that maintain their profile in z and propagate in the z-direction at the speed of light (see problem 11). The general solution eq. (5.159) to the source-free Maxwell's equations can be viewed as a superposition of such wave packets propagating in different directions at the speed of light. This is a manifestation of the fact that electromagnetic radiation propagates at the speed of light. A more mathematically precise formulation of this notion was given in section 5.4.

5.6 Conducting Cavities and Waveguides

This section is an introduction to electrodynamics inside conducting cavities and waveguides. In addition to having important practical applications, these topics provide a very good illustration of the utility of the mathematical technique of eigenfunction expansions.

By a *conducting cavity*, we mean a bounded region of space, $\mathcal{V}$, surrounded by a conductor, as we considered in electrostatics in section 2.5. By a *waveguide*, we mean a region of space enclosed by a conductor that is bounded in two dimensions but has translational symmetry (and thus is infinite) in the third dimension. We consider "perfect conductors" here, for which $\boldsymbol{E} = 0$ inside the conductor, so the electric field in the region enclosed by the conductor satisfies the boundary condition

$$\hat{\boldsymbol{n}} \times \boldsymbol{E}\big|_S = 0, \tag{5.167}$$

where S is the boundary of the enclosed region, and $\hat{\boldsymbol{n}}$ denotes the unit normal to S. In the more realistic case of finite conductivity, the electric field will penetrate somewhat into the conductor (see section 6.3.2), resulting in some energy loss due to dissipation (see problem 9 of chapter 6). We make some brief remarks about dissipation at the end of section 5.6.1, but we shall not otherwise consider dissipative effects arising from imperfect conductors. Throughout this section, we take the interior of the cavity or waveguide to be vacuum. However, if it were filled with a homogeneous medium with dielectric constant ϵ and magnetic permeability μ, the corresponding results could be obtained straightforwardly by using the substitutions $\epsilon_0 \to \epsilon$, $\mu_0 \to \mu$, $c \to c/n$ (see eq. (6.12) in chapter 6).[12]

---

[10] It is possible to construct "beam" solutions that have similar characteristics to plane waves but are localized in two directions of space (see problem 12).

[11] We analyze the case where the phase velocity varies with k in section 6.2.

[12] Since we consider only source-free solutions, the only relevant substitution is $c \to c/n$, where n is the index of refraction of the medium.

5.6.1 CONDUCTING CAVITIES

As usual, it is useful to first consider a real scalar field ψ satisfying

$$\Box\psi = 0 \tag{5.168}$$

in a bounded region $\mathcal{V}$ of space. A close mathematical analog for the scalar field of the perfect conductor boundary condition eq. (5.167) is the Dirichlet condition

$$\psi|_S = 0\,. \tag{5.169}$$

The solutions to eq. (5.168) with the boundary condition eq. (5.169) can be constructed as follows.

The operator $-\nabla^2$ on functions on $\mathcal{V}$ with the boundary condition eq. (5.169) can be shown to be a positive, self-adjoint operator with a discrete spectrum, so it has an orthonormal basis of eigenvectors (see the boxed side comment in section 2.4). Let $\{\xi_n(\boldsymbol{x})\}$ denote this basis of eigenvectors with corresponding eigenvalues $\lambda_n \geq 0$. Since we can expand any function on $\mathcal{V}$ in this basis, the general solution to eq. (5.168) can be written as

$$\psi(t,\boldsymbol{x}) = \sum_n f_n(t)\xi_n(\boldsymbol{x})\,. \tag{5.170}$$

Plugging this into eq. (5.168), we find that f_n satisfies

$$\frac{1}{c^2}\frac{d^2 f_n}{dt^2} = -\lambda_n f_n\,. \tag{5.171}$$

Thus, the general solution to eq. (5.168) is

$$\psi(t,\boldsymbol{x}) = \sum_n c_n e^{-i\omega_n t}\xi_n(\boldsymbol{x}) + \text{c.c.} \tag{5.172}$$

where $\omega_n = c\sqrt{\lambda_n}$, c_n is an arbitrary complex constant, and c.c. denotes the complex conjugate of the preceding term. Thus, the general solution is a superposition of modes, $\xi_n(\boldsymbol{x})$, where each mode simply oscillates in time with frequency ω_n.

We turn now to the electromagnetic case. Since the boundary condition eq. (5.167) is given in terms of $\boldsymbol{E}$, it is convenient to formulate the equations in terms of $\boldsymbol{E}$. The interior of $\mathcal{V}$ is source free, so we have

$$\nabla\cdot\boldsymbol{E} = 0 \tag{5.173}$$

in $\mathcal{V}$. It follows from Maxwell's equations with $\rho = \boldsymbol{J} = 0$ that

$$\Box\boldsymbol{E} = 0 \tag{5.174}$$

(see problem 1). We wish to solve eqs. (5.173) and (5.174) in $\mathcal{V}$ with the boundary condition eq. (5.167). In parallel with the scalar case, the general solution can be obtained as follows.

The operator $-\nabla^2$ with the boundary condition eq. (5.167) is a positive, self-adjoint operator on the space of divergence-free vector fields on $\mathcal{V}$ (see problem 13) and has an orthonormal basis of eigenvectors. We denote this basis by $\{\boldsymbol{\zeta}_n(\boldsymbol{x})\}$ and denote the corresponding eigenvalues by $\Lambda_n \geq 0$. Thus, each $\boldsymbol{\zeta}_n$ satisfies $\nabla \cdot \boldsymbol{\zeta}_n = 0$ and $\nabla^2 \boldsymbol{\zeta}_n = -\Lambda_n \boldsymbol{\zeta}_n$. We may expand $\boldsymbol{E}$ as

$$\boldsymbol{E} = \sum_n F_n(t) \boldsymbol{\zeta}_n(\boldsymbol{x}) . \tag{5.175}$$

Substituting in eq. (5.174), we find that the general solution for $\boldsymbol{E}$ inside a conducting cavity is

$$\boldsymbol{E}(t, \boldsymbol{x}) = \sum_n C_n e^{-i\omega_n t} \boldsymbol{\zeta}_n(\boldsymbol{x}) + \text{c.c.}, \tag{5.176}$$

where $\omega_n = c\sqrt{\Lambda_n}$, and C_n is an arbitrary complex constant. By eq. (5.7), the magnetic field $\boldsymbol{B}$ is then given by

$$\boldsymbol{B}(t, \boldsymbol{x}) = \sum_n \frac{C_n}{i\omega_n} e^{-i\omega_n t} \nabla \times \boldsymbol{\zeta}_n(\boldsymbol{x}) + \text{c.c.} \tag{5.177}$$

Again, the general solution is a sum of modes described by $\boldsymbol{\zeta}_n(\boldsymbol{x})$ that simply oscillate in time.

A very significant difference between solutions to Maxwell's equations in a conducting cavity and those in free space is that the cavity solutions only exist at the discrete frequencies $\omega_n = c\sqrt{\Lambda_n}$, as opposed to the continuum of frequencies of plane waves found in section 5.5. The eigenmodes $\boldsymbol{\zeta}_n(\boldsymbol{x})$ and eigenvalues Λ_n depend on the size and shape of the cavity $\mathcal{V}$. For a rectangular cavity, the eigenmodes can be expressed in terms of products of Fourier modes in x, y, and z. For a spherical cavity, they can be expressed in terms of products of vector spherical harmonics and spherical Bessel functions of r (see problem 14). For a cylindrical cavity, they can be written in terms of products of Fourier modes in z, Fourier modes in φ, and cylindrical Bessel functions of the cylindrical radial coordinate. For a general cavity with no symmetries, one would need to determine $\boldsymbol{\zeta}_n(\boldsymbol{x})$ and Λ_n by numerical means.

The following is an interesting consequence of the fact that source-free solutions inside a conducting cavity exist only at the discrete frequencies $\omega_n = c\sqrt{\Lambda_n}$. Suppose one has an atom in an excited state capable of making an electromagnetic transition at frequency ω to another state. Now, suppose that one puts the atom in a conducting cavity, so that none of the cavity frequencies ω_n are close to ω. (The linear size of the cavity would have to be comparable to or smaller than c/ω for this to be the case.) Then this transition would be greatly suppressed.

If the conductor is not perfect, then as already mentioned at the beginning of this section, there will be energy loss due to dissipation (see problem 9 of chapter 6). Instead of merely oscillating at frequency ω_n, each mode will also exponentially decay due to the dissipation, so its time dependence will be of the form $e^{-\alpha_n t - i\omega_n t}$. The ratio $Q_n = \omega_n / 2\alpha_n$ is known as the *quality factor* of the cavity for the nth mode.

5.6.2 WAVEGUIDES

In section 5.6.1, we considered a region of space enclosed by a conductor that was bounded in all directions. Let us now consider a region of space—also enclosed by a conductor—that is bounded in x and y but infinite in extent in z. More precisely, for a waveguide, the region bounded by a conductor is obtained by restricting x and y to a bounded region $\mathcal{A}$ in the x-y plane but allowing z to take any value. Of course, a real waveguide won't be infinite in extent or perfectly straight, but these idealizations allow us to analyze the essential features of real waveguides in a simple, mathematically clear manner. As we shall see, in a waveguide, electromagnetic waves can propagate in the z-direction. Waveguides are thereby very useful devices for transmitting electromagnetic signals, since they avoid the signal spreading in x and y that would occur in free space, and they also provide shielding of the signal from external influences.

Our analysis proceeds by focusing attention on the transverse electric field

$$\boldsymbol{E}^T \equiv (E_x, E_y)\,. \tag{5.178}$$

We obtain a basis expansion of $\boldsymbol{E}^T$ and solve the equations satisfied by $\boldsymbol{E}^T$ alone. We then can determine E_z by using eq. (5.173) and determine $\boldsymbol{B}$ by using eq. (5.7).

At any fixed t and z, $\boldsymbol{E}^T$ is a vector field on the cross section $\mathcal{A}$ of the waveguide. We denote the (1-dimensional) boundary of $\mathcal{A}$ by $\mathcal{C}$. By eq. (5.167), we have

$$\hat{\boldsymbol{n}} \times \boldsymbol{E}^T\big|_{\mathcal{C}} = 0, \tag{5.179}$$

where the unit normal $\hat{\boldsymbol{n}}$ to the boundary surface S of the conductor in $\mathcal{V}$ coincides with the unit normal to $\mathcal{C}$ in $\mathcal{A}$, since the z-direction is tangent to the conductor. Our analysis of $\boldsymbol{E}^T$ will be based on the following general theorem[13] on vector fields $\boldsymbol{V}$ on a 2-dimensional bounded region $\mathcal{A}$ of the plane with boundary $\mathcal{C}$.

Theorem (Decomposition of a Vector Field in 2 Dimensions): *Let $\mathcal{A}$ be an arbitrary bounded region of the x-y plane with smooth boundary $\mathcal{C}$. Let $\boldsymbol{V}$ be a smooth vector field on $\mathcal{A}$ that is tangent to the x-y plane and is such that on $\mathcal{C}$, the component of $\boldsymbol{V}$ tangential to $\mathcal{C}$ vanishes, $\boldsymbol{V}_\parallel|_{\mathcal{C}} = 0$. Then $\boldsymbol{V}$ can be uniquely written as the following orthogonal decomposition:*

$$\boldsymbol{V} = \nabla\alpha + \hat{\boldsymbol{z}} \times \nabla\beta + \boldsymbol{\eta}\,. \tag{5.180}$$

Here $\alpha = \alpha(x,y)$ is a function on $\mathcal{A}$ with $\alpha|_{\mathcal{C}} = 0$; $\beta = \beta(x,y)$ is a function on $\mathcal{A}$ with $(\hat{\boldsymbol{n}} \cdot \nabla\beta)|_{\mathcal{C}} = 0$; and $\boldsymbol{\eta}$ is tangent to $\mathcal{A}$ and satisfies

$$\nabla \cdot \boldsymbol{\eta} = 0\,, \qquad \nabla \times \boldsymbol{\eta} = 0\,, \qquad \boldsymbol{\eta}_\parallel|_{\mathcal{C}} = 0\,. \tag{5.181}$$

In particular, $\boldsymbol{\eta}$ satisfies Laplace's equation $\nabla^2\boldsymbol{\eta} = 0$ (i.e., $\boldsymbol{\eta}$ is harmonic).

Proof. Let α satisfy

$$\nabla^2\alpha = \nabla \cdot \boldsymbol{V} \tag{5.182}$$

with the Dirichlet boundary condition $\alpha|_{\mathcal{C}} = 0$. By a 2-dimensional version of the existence and uniqueness theorem of section 2.5, a unique solution for α exists. Similarly, let β satisfy

$$\nabla^2\beta = -\nabla \cdot (\hat{\boldsymbol{z}} \times \boldsymbol{V}), \tag{5.183}$$

[13] This theorem is a version of the much more general *Hodge decomposition theorem*, which holds for an arbitrary p-form on an arbitrary n-dimensional manifold with boundary. Our version is adapted to the boundary conditions relevant for waveguides.

with the Neumann boundary condition $(\hat{\boldsymbol{n}} \cdot \nabla \beta)|_{\mathcal{C}} = 0$. By a 2-dimensional version of the existence and uniqueness theorem of section 2.5, a unique (up to the addition of a constant) solution for β also exists.[14] Define $\boldsymbol{\eta}$ by

$$\boldsymbol{\eta} = \boldsymbol{V} - \nabla \alpha - \hat{\boldsymbol{z}} \times \nabla \beta \,. \tag{5.184}$$

Then it can be straightforwardly checked that $\boldsymbol{\eta}$ satisfies all the properties in eq. (5.181). It also is straightforward to check that the terms appearing in eq. (5.180) are orthogonal as elements of the Hilbert space of square integrable vector fields on $\mathcal{A}$. Furthermore, we have

$$\nabla^2 \boldsymbol{\eta} = -\nabla \times (\nabla \times \boldsymbol{\eta}) + \nabla(\nabla \cdot \boldsymbol{\eta}) = 0, \tag{5.185}$$

so $\boldsymbol{\eta}$ satisfies Laplace's equation. □

In fact, it is not difficult to see that if $\mathcal{A}$ has the topology of a disk, then there does not exist a nontrivial $\boldsymbol{\eta}$ satisfying the conditions in eq. (5.181). Namely, since $\mathcal{A}$ is simply connected, the condition $\nabla \times \boldsymbol{\eta} = 0$ implies that $\boldsymbol{\eta}$ is of the form $\boldsymbol{\eta} = \nabla \chi$ for some function χ. The condition $\nabla \cdot \boldsymbol{\eta} = 0$ then implies that $\nabla^2 \chi = 0$, and the condition $\eta_{\parallel}|_{\mathcal{C}} = 0$ implies that χ is locally constant on $\mathcal{C}$. Since the boundary $\mathcal{C}$ of a disk is connected, it follows that χ is constant on $\mathcal{C}$, in which case, the unique solution of Laplace's equation is $\chi = \text{const.}$ throughout $\mathcal{A}$, and thus $\boldsymbol{\eta} = 0$. Thus, if $\mathcal{A}$ has the topology of a disk, then the decomposition eq. (5.180) holds with $\boldsymbol{\eta} = 0$.

In contrast, if $\mathcal{A}$ has the topology of an annulus, the above argument fails both because $\mathcal{A}$ is not simply connected and because its boundary $\mathcal{C}$ is disconnected. In fact, since $\eta_{\parallel}|_{\mathcal{C}} = 0$, it follows that $\boldsymbol{\eta}$ must still be of the form $\boldsymbol{\eta} = \nabla \chi$. However, a nontrivial solution of Laplace's equation for χ can be obtained by assigning different constant values to χ on the inner and outer portions of the boundary $\mathcal{C}$. For the case where $\mathcal{A}$ is the region between two concentric circles, the solution for $\boldsymbol{\eta}$ is

$$\boldsymbol{\eta} = \frac{C}{\rho} \hat{\boldsymbol{\rho}}, \tag{5.186}$$

where C is a constant, and ρ is the polar radial coordinate. As we shall see below, the existence of a harmonic contribution $\boldsymbol{\eta}$ in eq. (5.180) significantly improves the transmission capabilities of a waveguide. This makes coaxial cable far superior to hollow cable for transmitting electromagnetic signals.

We now have all the machinery needed to expand $\boldsymbol{E}^T$ in modes and obtain the general solution to Maxwell's equations in a waveguide. At each fixed t and z, we write $\boldsymbol{E}^T$ in the form eq. (5.180). Then we expand α in an orthonormal basis $\alpha_n(x, y)$ of eigenfunctions of the Laplacian operator on $\mathcal{A}$ with the Dirichlet boundary condition $\alpha_n|_{\mathcal{C}} = 0$. Let $-\kappa_n$ denote the corresponding eigenvalues, so that

$$\nabla^2 \alpha_n = -\kappa_n \alpha_n \,. \tag{5.187}$$

Note that for all n, we have[15]

$$\kappa_n = \frac{\int_{\mathcal{A}} |\nabla \alpha_n|^2}{\int_{\mathcal{A}} |\alpha_n|^2} > 0 \,. \tag{5.188}$$

[14] The analog of the required additional condition eq. (2.96) can be seen to hold using our assumption $V_{\parallel}|_{\mathcal{C}} = 0$.

[15] We cannot have $\kappa_n = 0$, since $\nabla \alpha_n = 0$ together with $\alpha_n|_{\mathcal{C}} = 0$ implies that $\alpha_n = 0$.

Similarly, we expand β in an orthonormal basis of eigenfunctions $\beta_n(x,y)$ of the Laplacian operator on $\mathcal{A}$ with the Neumann boundary condition $(\hat{\boldsymbol{n}}\cdot\nabla\beta_n)|_{\mathcal{C}}=0$. Let $-\mu_n$ denote the corresponding eigenvalues

$$\nabla^2\beta_n=-\mu_n\beta_n\,. \tag{5.189}$$

In the case of Neumann conditions, there is an eigenfunction $\beta_0=\text{const.}$ that has eigenvalue $\mu_0=0$. However, this eigenfunction does not contribute to $\boldsymbol{E}^T$ via eq. (5.180), so we ignore it. All other eigenfunctions have $\mu_n>0$.

It follows from all of the above results that $\boldsymbol{E}^T(t,\boldsymbol{x})$ can be written as the following mode sum:

$$\boldsymbol{E}^T(t,\boldsymbol{x})=\sum_n a_n(t,z)\nabla\alpha_n(x,y)+\sum_m b_m(t,z)\hat{\boldsymbol{z}}\times\nabla\beta_m(x,y)+\sum_i h_i(t,z)\boldsymbol{\eta}_i(x,y)\,. \tag{5.190}$$

Here the first two sums run over the infinite number of eigenfunctions of the Dirichlet/Neumann Laplacians. The last term is a finite sum that runs over a basis of modes $\boldsymbol{\eta}_i$ satisfying eq. (5.181). As we have seen, there are no such modes if the waveguide is a hollow conducting tube, and there is only one such mode for a coaxial cable.

We now plug eq. (5.190) into eq. (5.174). To satisfy eq. (5.174), each mode in this expansion will have to separately satisfy this equation. Thus, we find that each of a_n, b_m, and h_i satisfies a 2-dimensional wave equation of the form

$$-\frac{1}{c^2}\frac{\partial^2 f}{\partial t^2}+\frac{\partial^2 f}{\partial z^2}-\gamma f=0\,. \tag{5.191}$$

In the case $f=a_n$, we have $\gamma=\kappa_n>0$; for $f=b_m$, we have $\gamma=\mu_m>0$; and for $f=h_i$, we have $\gamma=0$. We are mainly interested in solutions f that oscillate in time as $e^{-i\omega t}$, as would occur if one were to put a source that oscillates with frequency ω at one end of a (finite-length) waveguide. In that case, writing $f(t,z)=F(z)e^{-i\omega t}$, we have

$$\frac{\partial^2 F}{\partial z^2}+\left(\frac{\omega^2}{c^2}-\gamma\right)F=0\,. \tag{5.192}$$

Thus, if $\omega<c\sqrt{\gamma}$, the mode will be exponentially damped in z and will not propagate along the waveguide. Such modes are said to be *evanescent*. However, when $\omega>c\sqrt{\gamma}$, the solutions for F are of the form $e^{\pm ikz}$ with $k=\sqrt{(\omega/c)^2-\gamma}$. Wave packets composed of such frequencies will propagate along the waveguide. If $\gamma>0$, the relationship between k and ω is not linear, and such wave packets will undergo dispersion (see section 6.2 and problem 7 of chapter 6 for further discussion of the behavior of such wave packets).

Given $\boldsymbol{E}^T(t,\boldsymbol{x})$, the z-component E_z of $\boldsymbol{E}$ is determined by eq. (5.173) up to the addition of solutions of the wave equation (5.174) that are independent of z. Such z-independent solutions are not of interest, and we shall ignore them. Given $\boldsymbol{E}$, the magnetic field $\boldsymbol{B}$ is determined by eq. (5.7) up to the addition of a static magnetic field. Such static magnetic fields are also not of interest, and we shall ignore them. Thus, the solutions of interest to Maxwell's equations in a waveguide are determined by $\boldsymbol{E}^T(t,\boldsymbol{x})$ and may be classified as follows.

Transverse electric (TE) modes: These modes are associated with $\{\beta_m\}$. If we assume time dependence $e^{-i\omega t}$, the transverse electric field is given by

$$\boldsymbol{E}^T(t,\boldsymbol{x}) = e^{ik_m z}e^{-i\omega t}\hat{\boldsymbol{z}} \times \boldsymbol{\nabla}\beta_m, \tag{5.193}$$

where $k_m^2 = (\omega/c)^2 - \mu_m$. The z-component of the electric field satisfies

$$\frac{\partial E_z}{\partial z} = -\boldsymbol{\nabla}\cdot\boldsymbol{E}^T(t,\boldsymbol{x}) = -e^{ik_m z}e^{-i\omega t}\boldsymbol{\nabla}\cdot(\hat{\boldsymbol{z}}\times\boldsymbol{\nabla}\beta_m) = 0, \tag{5.194}$$

with solution

$$E_z = 0, \tag{5.195}$$

which gives rise to the name "transverse electric." The magnetic field is then given by

$$\boldsymbol{B} = \frac{1}{i\omega}\boldsymbol{\nabla}\times\boldsymbol{E}. \tag{5.196}$$

Note that the z-component of the magnetic field is given by

$$\begin{aligned}
B_z &= \frac{1}{i\omega}\hat{\boldsymbol{z}}\cdot(\boldsymbol{\nabla}\times\boldsymbol{E}^T) = \frac{1}{i\omega}e^{ik_m z}e^{-i\omega t}\hat{\boldsymbol{z}}\cdot(\boldsymbol{\nabla}\times[\hat{\boldsymbol{z}}\times\boldsymbol{\nabla}\beta_m]) \\
&= \frac{1}{i\omega}e^{ik_m z}e^{-i\omega t}\nabla^2\beta_m \\
&= -\frac{\mu_m}{i\omega}e^{ik_m z}e^{-i\omega t}\beta_m
\end{aligned} \tag{5.197}$$

and thus is nonvanishing.

Transverse magnetic (TM) modes: These modes are associated with $\{\alpha_n\}$. If we assume time dependence $e^{-i\omega t}$, the transverse electric field is given by

$$\boldsymbol{E}^T(t,\boldsymbol{x}) = e^{ik_n z}e^{-i\omega t}\boldsymbol{\nabla}\alpha_n, \tag{5.198}$$

where $k_n^2 = (\omega/c)^2 - \kappa_n$. The z-component of the electric field satisfies

$$\frac{\partial E_z}{\partial z} = -\boldsymbol{\nabla}\cdot\boldsymbol{E}^T(t,\boldsymbol{x}) = -e^{ik_n z}e^{-i\omega t}\nabla^2\alpha_n = \kappa_n e^{ik_n z}e^{-i\omega t}\alpha_n \tag{5.199}$$

with solution

$$E_z = \frac{\kappa_n}{ik_n}e^{ik_n z}e^{-i\omega t}\alpha_n. \tag{5.200}$$

The magnetic field is then given by

$$\boldsymbol{B} = \frac{1}{i\omega}\boldsymbol{\nabla}\times\boldsymbol{E}. \tag{5.201}$$

Note that

$$B_z = \frac{1}{i\omega}\hat{\boldsymbol{z}}\cdot(\boldsymbol{\nabla}\times\boldsymbol{E}^T) = \frac{1}{i\omega}e^{ik_m z}e^{-i\omega t}\hat{\boldsymbol{z}}\cdot(\boldsymbol{\nabla}\times\boldsymbol{\nabla}\alpha_n) = 0, \tag{5.202}$$

which gives rise to the name "transverse magnetic."

Transverse electromagnetic (TEM) modes: These modes are associated with $\{\boldsymbol{\eta}_i\}$. If we assume time dependence $e^{-i\omega t}$, the transverse electric field is given by

$$\boldsymbol{E}^T(t, \boldsymbol{x}) = e^{ikz} e^{-i\omega t} \boldsymbol{\eta}_i, \tag{5.203}$$

where $k^2 = \omega^2/c^2$, and hence, TEM waves propagate along the waveguide without dispersion. Since $\nabla \cdot \boldsymbol{\eta}_i = 0 = \nabla \times \boldsymbol{\eta}_i$, these modes have both $E_z = 0$ and $B_z = 0$. There are no TEM modes for a hollow waveguide. For the case of a coaxial cable—where $\mathcal{A}$ is an annulus bounded by concentric circles—the unique harmonic $\boldsymbol{\eta}$ is given by eq. (5.186). Thus, in this case, the TEM modes take the form

$$\boldsymbol{E} = e^{ikz} e^{-i\omega t} \frac{\hat{\boldsymbol{\rho}}}{\rho}, \tag{5.204}$$

$$c\boldsymbol{B} = \frac{c}{i\omega} \nabla \times \boldsymbol{E} = \pm \hat{\boldsymbol{z}} \times \frac{\hat{\boldsymbol{\rho}}}{\rho} e^{ikz} e^{-i\omega t} = \pm e^{ikz} e^{-i\omega t} \frac{\hat{\boldsymbol{\varphi}}}{\rho}, \tag{5.205}$$

where $\hat{\boldsymbol{\varphi}}$ is a unit vector in the φ-direction, and the $\pm$ depends on whether $k = \pm\omega$.

Problems

1. Show that for an arbitrary solution of Maxwell's equations, the electric and magnetic fields, $\boldsymbol{E}$ and $\boldsymbol{B}$, satisfy

$$\Box \boldsymbol{E} = \frac{1}{\epsilon_0} \nabla \rho + \mu_0 \frac{\partial \boldsymbol{J}}{\partial t}, \qquad \Box \boldsymbol{B} = -\mu_0 \nabla \times \boldsymbol{J},$$

where $\Box = \nabla^2 - \frac{1}{c^2}\frac{\partial^2}{\partial t^2}$. Show that these equations do not imply Maxwell's equations (i.e., find a solution to the above equations that is not a solution to Maxwell's equations).

2. The angular momentum density of the electromagnetic field is given by

$$\boldsymbol{l} = \boldsymbol{x} \times \boldsymbol{\mathcal{P}} = \epsilon_0 \boldsymbol{x} \times (\boldsymbol{E} \times \boldsymbol{B}).$$

Consider a source-free ($\rho = 0$, $\boldsymbol{J} = 0$) solution to Maxwell's equations with $\boldsymbol{E}$ and $\boldsymbol{B}$ vanishing rapidly as $|\boldsymbol{x}| \to \infty$, so the total angular momentum

$$\boldsymbol{L} = \int \boldsymbol{l}\, d^3x$$

is well defined. Show that $\boldsymbol{L}$ is conserved (i.e., independent of time).

3. A particle of charge q_1 moves with velocity v in a circular orbit of radius R about the origin in the x-y plane, such that its φ coordinate varies as $\varphi = \omega t$, with $\omega = v/R$. Assume that $v \ll c$. Another particle of charge q_2 is at rest at point $\boldsymbol{x}$, where $|\boldsymbol{x}| \gg R$. To order $1/|\boldsymbol{x}|$, find the force $\boldsymbol{F}$ on the particle of charge q_2 at time t.

4. An antenna is a segment of conducting wire in which a current flows (driven by an external power supply). Suppose an antenna of length L is placed on the

z-axis between $z=0$ and $z=L$, and suppose that the current in the antenna is

$$\boldsymbol{J}(t,z)=I_0\sin(\pi z/L)\cos(\omega t)\delta(x)\delta(y)\hat{\boldsymbol{z}}.$$

(a) Find the charge density $\rho(t,z)$ in the antenna.
(b) Assume that $\omega L \ll c$. Find the electric and magnetic fields, $\boldsymbol{E}(t,\boldsymbol{x})$ and $\boldsymbol{B}(t,\boldsymbol{x})$, at large distances from the antenna (valid to order $1/|\boldsymbol{x}|$).
(c) Find the total radiated power, time averaged over 1 period of oscillation.

5. A square of side d lies in the $z=0$ plane, with the center of the square lying at the origin, $x=y=z=0$. Charges are placed on the corners of the square in the two different manners, (i) and (ii), shown below. The square rotates with angular velocity Ω about the z-axis, where $\Omega d \ll c$. In each case, the radiated power P will scale with q, d, and Ω as

$$P \propto q^{n_1} d^{n_2} \Omega^{n_3}$$

for some integers n_1, n_2, and n_3. Find n_1, n_2, and n_3 for case (i) and for case (ii).

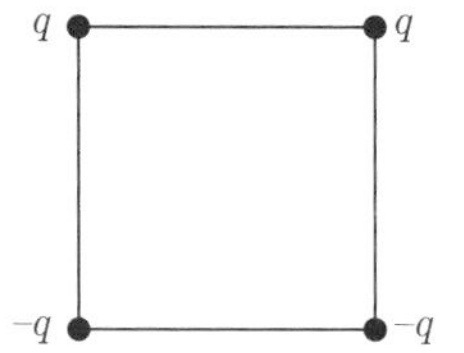

(i)

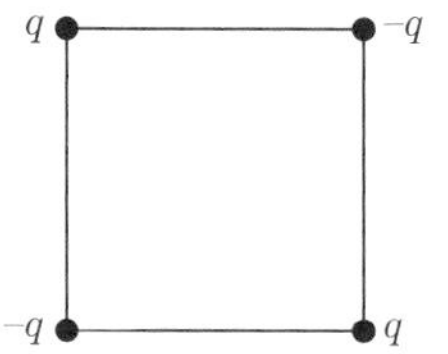

(ii)

6. A point charge of charge q and mass m is placed on the end of a spring with spring constant k. The charge is displaced in the z-direction by an amount α away from its equilibrium position and is then released to oscillate. Assume that the resulting motion is nonrelativistic, $v \ll c$.

(a) Assume that the charge oscillates harmonically with amplitude α. To order $1/r$ in distance from the charge and to leading order in v/c, what are the resulting electromagnetic potentials $\phi, \boldsymbol{A}$?
(b) What is the radiated power?
(c) As a result of the radiation of electromagnetic energy, the maximum amplitude of oscillation, α, will, in fact, slowly decay with time. Find $\alpha(t)$.

7. It can be easily seen from the explicit expressions eq. (5.50) and eq. (5.51) that the retarded and advanced Green's functions for the wave equation are related by $G_{\text{ret}}(t,\boldsymbol{x};t',\boldsymbol{x}')=G_{\text{adv}}(t',\boldsymbol{x}';t,\boldsymbol{x})$. This is not an accident. Consider the wave equation with some arbitrary given "potential" $V(t,\boldsymbol{x})$:

$$\Box\psi + V\psi = -f.$$

Although one may not be able to find their form explicitly, it is possible to prove by general arguments that unique retarded and advanced Green's functions exist for this equation, satisfying the "support properties" that $G_{\text{ret}}(t,\boldsymbol{x};t'\boldsymbol{x}')=0$ for $t<t'+|\boldsymbol{x}-\boldsymbol{x}'|/c$, and $G_{\text{adv}}(t,\boldsymbol{x};t',\boldsymbol{x}')=0$ for

$t > t' - |\boldsymbol{x} - \boldsymbol{x}'|/c$. (Note that, in general, these Green's functions will be nonvanishing inside the light cone—not just on the light cone.)

(a) Show that a Green's identity holds for any two solutions ψ_1 and ψ_2 of this equation of exactly the same form, eq. (5.125), as for the wave equation without a potential.
(b) Prove that for the wave equation with a potential, $G_{\text{ret}}(t, \boldsymbol{x}; t'\boldsymbol{x}') = G_{\text{adv}}(t', \boldsymbol{x}'; t, \boldsymbol{x})$. [Hint: Integrate Green's identity over a suitable spacetime cube for suitable choices of ψ_1 and ψ_2.]

8. Show that e^{ikz} can be expanded in spherical harmonics as

$$e^{ikz} = \sum_{\ell=0}^{\infty} i^\ell \sqrt{4\pi(2\ell+1)}\, j_\ell(kr) Y_{\ell 0}(\theta, \varphi) = \sum_{\ell=0}^{\infty} i^\ell (2\ell+1) j_\ell(kr) P_\ell(\cos\theta).$$

Note that this formula allows one to write a plane wave solution of the scalar wave equation in terms of spherical waves. [Hint: First show that

$$e^{-ikz} = \lim_{\lambda\to\infty} \lambda e^{-ik\lambda} \frac{e^{ik|\boldsymbol{x}-\lambda\hat{\boldsymbol{z}}|}}{|\boldsymbol{x}-\lambda\hat{\boldsymbol{z}}|}.$$

Then use the expansion eq. (5.100), the asymptotic form eq. (5.98) of $h_\ell^{(1)}(x)$, as well as $Y_{\ell m}(0,0) = \sqrt{(2\ell+1)/4\pi}\,\delta_{m0}$.]

9. Show that the electric multipole moment $Q_{\omega\ell m}$ defined by eq. (5.117) can be re-expressed as

$$Q_{\omega\ell m} = \frac{\omega}{c\ell(\ell+1)} \int \left[\hat{\rho}\frac{d}{dr}[r j_\ell(\tfrac{\omega r}{c})] + \frac{i\omega}{c^2}(\boldsymbol{r}\cdot\hat{\boldsymbol{J}}) j_\ell(\tfrac{\omega r}{c})\right] Y^*_{\ell m}(\theta,\varphi) d^3x.$$

10. Suppose that ρ and $\boldsymbol{J}$ are nonvanishing only in the region $r < R$.

(a) Show that at low frequencies ($\omega R/c \ll 1$), the multipole moments $M_{\omega\ell m}$ and $Q_{\omega\ell m}$ are approximately given by

$$M_{\omega\ell m} = \frac{2^\ell(\ell-1)!}{(2\ell+1)!}\left(\frac{\omega}{c}\right)^{\ell+1} m_{\ell m}, \qquad Q_{\omega\ell m} = \frac{2^\ell(\ell-1)!}{(2\ell+1)!}\left(\frac{\omega}{c}\right)^{\ell+1} q_{\ell m},$$

where $m_{\ell m}$ and $q_{\ell m}$ are defined by eqs. (4.44) and (2.87), respectively (with ρ and $\boldsymbol{J}$ in those equations now replaced by $\hat{\rho}$ and $\hat{\boldsymbol{J}}$). [Hint: To show this result for $Q_{\omega\ell m}$, it will be helpful to use the result of problem 9.]
(b) Compute the contribution to $\boldsymbol{E}$ and $\boldsymbol{B}$ in the general solution, eqs. (5.118) and (5.119), that arises from $\ell = 1$ electric parity radiation under the assumption that $\omega R/c \ll 1$. Show that this agrees with eqs. (5.75) and (5.78) obtained in section 5.3.1 to leading order by making a nonrelativistic approximation.

11. Show that $\phi = 0$ and

$$\boldsymbol{A} = f(t - z/c)\hat{\boldsymbol{y}}$$

satisfy the source-free Maxwell's equations (5.141)–(5.143) for any function f. Note that this solution represents a wave with initial profile $f(z/c)$ that propagates in the z-direction at the speed of light. Explicitly write this solution as a superposition of plane waves (i.e., in the form of eq. (5.159)) for the cases where (i) f is a Gaussian,

$$f(x) = e^{-\alpha x^2},$$

where $\alpha > 0$ is a constant; and (ii) f is a step function of the form

$$f(x) = \begin{cases} 0, & \text{if } |x| > 1 \\ \beta, & \text{if } |x| \leq 1. \end{cases}$$

where β is a constant.

12. A solution ψ to the wave equation (5.128) with frequency ω is referred to as a *beam* traveling in the z-direction if it is of the form

$$\psi(t, \boldsymbol{x}) = \Psi(x, y, z) e^{ikz - i\omega t}$$

(where $k \equiv \omega/c$) such that Ψ is highly localized in x and y and is slowly varying in z in the sense that it satisfies the *paraxial condition*

$$\left| \frac{\partial^2 \Psi}{\partial z^2} \right| \ll k \left| \frac{\partial \Psi}{\partial z} \right| .$$

(a) Show that when the paraxial condition holds, Ψ satisfies

$$i \frac{\partial \Psi}{\partial z} = -\frac{1}{2k} \left(\frac{\partial^2}{\partial x^2} + \frac{\partial^2}{\partial y^2} \right) \Psi .$$

(Note that this is of the same form as the Schrodinger equation in 2 dimensions.)

(b) Show that

$$\Psi(x, y, z) = \frac{1}{w_0^2 + 2iz/k} \exp\left(-\frac{x^2 + y^2}{w_0^2 + 2iz/k} \right)$$

is a solution to the paraxial wave equation found in part (a), where w_0 is a constant. Such a solution is referred to as a *Gaussian beam*.

(c) Show that when $w_0 k \gg 1$, the paraxial condition of part (a) is satisfied for $|z| < k w_0^2$ in the region of x and y where the beam is not negligibly small. Thus, a Gaussian beam is a good approximation to a solution to the wave equation in this regime.

(d) Show that $\phi = 0$ and

$$\boldsymbol{A} = \left[\Psi \hat{\boldsymbol{x}} + \frac{i}{k} \frac{\partial \Psi}{\partial x} \hat{\boldsymbol{z}} \right] e^{ikz - i\omega t}$$

approximately satisfies Maxwell's equations if Ψ satisfies the equation of part (a) and its variation with z is small compared with k (that is, $|\partial \Psi / \partial z| \ll k |\Psi|$ and similar relations hold with Ψ replaced by derivatives of Ψ). In particular, for Ψ as in part (b), we thereby obtain a Gaussian electromagnetic beam.

13. Let $\mathcal{V}$ be a bounded region with boundary S. Let $\boldsymbol{v}$ and $\boldsymbol{w}$ be vector fields in $\mathcal{V}$ that are divergence free, $\nabla \cdot \boldsymbol{v} = \nabla \cdot \boldsymbol{w} = 0$, throughout $\mathcal{V}$ and satisfy the boundary condition eq. (5.167) on S, namely, $\hat{\boldsymbol{n}} \times \boldsymbol{v}|_S = \hat{\boldsymbol{n}} \times \boldsymbol{w}|_S = 0$.

    (a) Show that

    $$\int_{\mathcal{V}} \boldsymbol{w} \cdot \nabla^2 \boldsymbol{v}\, d^3x = \int_{\mathcal{V}} (\nabla^2 \boldsymbol{w}) \cdot \boldsymbol{v}\, d^3x .$$

    (b) Show that

    $$\int_{\mathcal{V}} \boldsymbol{v} \cdot \nabla^2 \boldsymbol{v}\, d^3x \le 0 .$$

    [Hint: It will be useful to obtain a relation that holds for divergence-free vector fields that can be obtained by starting with

    $$0 = \int_{\mathcal{V}} (\nabla \cdot \boldsymbol{v})(\nabla \cdot \boldsymbol{w}) d^3x$$

    and integrating by parts twice.]

14. Find the modes $\boldsymbol{\zeta}$ and mode frequencies ω (see eq. (5.176)) of a spherical conducting cavity of radius R as follows. By rotational symmetry and parity symmetry, without loss of generality, we may assume that each mode $\boldsymbol{\zeta}$ is given by vector spherical harmonics of fixed ℓ, m, and parity type.

    (a) For a magnetic parity mode (referred to as a "transverse electric (TE)" mode), this implies that $\boldsymbol{\zeta}$ takes the form

    $$\boldsymbol{\zeta}^{\mathrm{TE}}_{n\ell m} = h_{n\ell m}(r)\, \boldsymbol{r} \times \nabla Y_{\ell m}(\theta, \varphi)$$

    (see eq. (4.38)). (The index n labels the different possible magnetic parity modes of the given ℓ, m.) Since we automatically have $\nabla \cdot \boldsymbol{\zeta}^{\mathrm{TE}}_{n\ell m} = 0$ by virtue of this magnetic parity form, we need only solve $\nabla^2 \boldsymbol{\zeta}^{\mathrm{TE}}_{n\ell m} = -(\omega^{\mathrm{TE}}_{n\ell m}/c)^2 \boldsymbol{\zeta}^{\mathrm{TE}}_{n\ell m}$ together with the boundary condition eq. (5.167). Show that $h_{n\ell m}$ satisfies the radial Helmholtz equation, and thus, the regular solution is $h_{n\ell m}(r) \propto j_\ell(\omega^{\mathrm{TE}}_{n\ell m} r/c)$, where j_ℓ is the spherical Bessel function eq. (5.91). Obtain the condition that must be satisfied by $\omega^{\mathrm{TE}}_{n\ell m}$ for the boundary condition eq. (5.167) to be satisfied.

    (b) For an electric parity mode (referred to as a "transverse magnetic (TM)" mode), one can initially work with the corresponding modes $\boldsymbol{\beta}$ for the magnetic field $\boldsymbol{B}$ (or equivalently, one can work with $\nabla \times \boldsymbol{\zeta}$, where $\boldsymbol{\zeta}$ is a mode of the electric field). Show that all results of part (a) apply to the TM modes $\boldsymbol{\beta}^{\mathrm{TM}}_{n\ell m}$ of $\boldsymbol{B}$ except for the boundary condition. Obtain the condition that must be satisfied by $\omega^{\mathrm{TM}}_{n\ell m}$ for the boundary condition eq. (5.167) to be satisfied. Obtain an expression for the electric field modes $\boldsymbol{\zeta}^{\mathrm{TM}}_{n\ell m}$ in terms of $\boldsymbol{\beta}^{\mathrm{TM}}_{n\ell m}$.

15. Consider a rectangular waveguide of sides a and b. Explicitly find all TE and TM modes, that is, obtain all eigenfunctions β_n and eigenvalues μ_n of the TE modes (see eq. (5.189)) and all eigenfunctions α_n and eigenvalues κ_n of the TM modes (see eq. (5.187)). [Hint: Since d^2/dx^2 and d^2/dy^2 commute, the eigenfunctions of $\nabla^2 = d^2/dx^2 + d^2/dy^2$ with given boundary conditions can be taken to be simultaneous eigenfunctions of d^2/dx^2 and d^2/dy^2 with the corresponding boundary conditions.]

CHAPTER 6

Electrodynamics in Media

In this chapter, we consider electrodynamics in a medium. We assume that the medium consists of electric and magnetic dipoles as well as free charges and free currents. As in chapters 3 and 4, we take the spatial average of quantities over length scale L via eq. (3.2) and denote the resulting averaged quantities with angle brackets $\langle\,\rangle$. As in chapters 3 and 4, we define

$$\langle \boldsymbol{D}\rangle = \epsilon_0 \langle \boldsymbol{E}\rangle + \langle \boldsymbol{P}\rangle, \tag{6.1}$$

$$\langle \boldsymbol{H}\rangle = \frac{1}{\mu_0}\langle \boldsymbol{B}\rangle - \langle \boldsymbol{M}\rangle\,. \tag{6.2}$$

Maxwell's equations in a medium are obtained by spatially averaging Maxwell's equations (5.4)–(5.7). The spatial average of eqs (5.6) and (5.7) straightforwardly yields

$$\nabla\cdot\langle \boldsymbol{B}\rangle = 0, \tag{6.3}$$

$$\nabla\times\langle \boldsymbol{E}\rangle + \frac{\partial\langle \boldsymbol{B}\rangle}{\partial t} = 0\,. \tag{6.4}$$

By following exactly the same steps as led to eq. (3.17), the spatial average of eq. (5.4) again yields

$$\nabla\cdot\langle \boldsymbol{D}\rangle = \langle \rho_f\rangle\,. \tag{6.5}$$

However, in magnetostatics, the derivation of eq. (4.83) made use of $\int \boldsymbol{J} d^3x = 0$ (see eq. (4.24)), which, in turn, followed from $\nabla\cdot\boldsymbol{J} = 0$. In electrodynamics, we have $\nabla\cdot\boldsymbol{J} = -\partial\rho/\partial t$, and, as previously found in eq. (5.73), the corresponding result in electrodynamics is $\int \boldsymbol{J} d^3x = d\boldsymbol{p}/dt$, where $\boldsymbol{p}$ is the electric dipole moment of the source. Parallelling the steps leading to eq. (4.83) with this change, in electrodynamics we obtain

$$\langle \boldsymbol{J}\rangle = \frac{\partial\langle \boldsymbol{P}\rangle}{\partial t} + \langle \boldsymbol{J}_f\rangle + \nabla\times\langle \boldsymbol{M}\rangle\,. \tag{6.6}$$

The spatial average of eq. (5.5) then yields

$$\nabla\times\langle \boldsymbol{H}\rangle - \frac{\partial\langle \boldsymbol{D}\rangle}{\partial t} = \langle \boldsymbol{J}_f\rangle\,. \tag{6.7}$$

Equations (6.3), (6.4), (6.5), and (6.7) correspond to the form of Maxwell's equations commonly found on T-shirts.

It is important to clarify the meaning of "free charges and currents," since this terminology is used somewhat differently in electrostatics and electrodynamics. In electrostatics in a medium, there is a fundamental distinction between the behavior of charges that are bound and charges that are unbound. Bound charges in a static electric field will be displaced by at most a finite amount, but unbound charges would continue to move indefinitely, until they are stopped by the surface of the body, as in a conductor. In electrostatics, the free charge density $\langle \rho_f \rangle$ is the macroscopic average of the net charge of the individual atoms. As stated below eq. (3.5), in electrostatics the "free charge" in a dielectric is assumed *not* to be free to move about the dielectric. However, in electrodynamics, if the electromagnetic field is oscillating in time at a nonzero frequency, there is no fundamental distinction between the behavior of bound and unbound charges, since both will undergo oscillations. Thus, there is no difficulty in allowing unbound charges in the medium. If we declare the unbound charges to be free, then we directly count their contribution to $\langle \rho \rangle$ and $\langle \boldsymbol{J} \rangle$ as $\langle \rho_f \rangle$ and $\langle \boldsymbol{J}_f \rangle$, but we do not consider them as making any contribution to $\langle \boldsymbol{P} \rangle$. Alternatively, we can treat the unbound charges on the same footing as the bound charges. In that case, they do not contribute to $\langle \rho_f \rangle$ or $\langle \boldsymbol{J}_f \rangle$, but they contribute to $\langle \rho \rangle$ via $-\nabla \cdot \langle \boldsymbol{P} \rangle$, and they contribute to $\langle \boldsymbol{J} \rangle$ via $\partial \langle \boldsymbol{P} \rangle / \partial t$. For oscillating charges, these descriptions are equivalent, that is, they lead to the same $\langle \rho \rangle$ and $\langle \boldsymbol{J} \rangle$ and the same fields $\langle \boldsymbol{E} \rangle$ and $\langle \boldsymbol{B} \rangle$. (Note, however, that they would yield a different value of $\langle \boldsymbol{D} \rangle$.) In many references, the charges and currents of unbound charges are treated as free. This is particularly convenient to do at low frequencies, where the bound charges contribute to a (nearly) real dielectric constant, and the unbound charges contribute to a (nearly) real conductivity. However, in the simple model of a medium that we shall consider in section 6.3, it will be convenient to treat the unbound and bound charges on the same footing and consider neither of them to be free. In contrast, when we treat magnetohydrodynamics in section 6.4, it will be convenient to treat all of the charges in the plasma as free. Maxwell's equations in the above forms will hold whether or not we treat the unbound charges as free.

To proceed beyond the above results, we must make additional assumptions about the medium. We shall do so in the following sections.

6.1 Linear, Homogeneous, Isotropic Medium with an Instantaneous Response

The simplest assumption one can make about the medium is that—as in electrostatics and magnetostatics—we have the linear relations

$$\langle \boldsymbol{D} \rangle = \epsilon \langle \boldsymbol{E} \rangle , \qquad \langle \boldsymbol{B} \rangle = \mu \langle \boldsymbol{H} \rangle, \tag{6.8}$$

where ϵ and μ are constants. In electrodynamics, these relations would require an instantaneous response of the electric and magnetic dipoles in the medium to the electric and magnetic fields, which would not be realistic if the electric and magnetic fields were changing rapidly with time (see sections 6.2 and 6.3). However, eq. (6.8) is a good approximation to low-frequency behavior in many materials.

In a medium where eq. (6.8) holds, Maxwell's equations (6.5) and (6.7) become

$$\nabla \cdot \langle \boldsymbol{E} \rangle = \frac{1}{\epsilon} \langle \rho_f \rangle, \tag{6.9}$$

$$\nabla \times \langle \boldsymbol{B} \rangle - \mu\epsilon \frac{\partial \langle \boldsymbol{E} \rangle}{\partial t} = \mu \langle \boldsymbol{J}_f \rangle\,. \tag{6.10}$$

We define the *index of refraction* of the medium by

$$n = \frac{\sqrt{\epsilon\mu}}{\sqrt{\epsilon_0\mu_0}} = c\sqrt{\epsilon\mu}\,. \tag{6.11}$$

By inspection, it can be immediately seen that under the substitutions $\boldsymbol{E} \to \langle \boldsymbol{E} \rangle$, $\boldsymbol{B} \to \langle \boldsymbol{B} \rangle$, $\rho \to \langle \rho_f \rangle$, $\boldsymbol{J} \to \langle \boldsymbol{J}_f \rangle$, as well as

$$\epsilon_0 \to \epsilon, \qquad \mu_0 \to \mu, \qquad c = \frac{1}{\sqrt{\epsilon_0\mu_0}} \to \frac{1}{\sqrt{\epsilon\mu}} = \frac{c}{n}, \tag{6.12}$$

Maxwell's equations (5.4)–(5.7) for the exact fields and sources are transformed to Maxwell's equations (6.3), (6.4), (6.9), and (6.10) for the averaged fields and "free sources" in a medium where eq. (6.8) holds. Consequently, for any result obtained from Maxwell's equations (5.4)–(5.7), there is an analogous result that is valid in a medium where eq. (6.8) holds. In particular, it follows that electromagnetic disturbances in the medium propagate at velocity c/n. If the medium is infinite in extent, we may introduce plane wave solutions, which have all the properties discussed in section 5.5 under these substitutions.

We can rewrite Maxwell's equations in terms of the averaged potentials $\langle \phi \rangle$ and $\langle \boldsymbol{A} \rangle$. To do so, it is very convenient to impose the modified Lorenz gauge condition:

$$\frac{n^2}{c^2}\frac{\partial \langle \phi \rangle}{\partial t} + \nabla \cdot \langle \boldsymbol{A} \rangle = 0\,. \tag{6.13}$$

Maxwell's equations in the medium then become simply

$$\tilde{\Box} \langle \phi \rangle = -\frac{\langle \rho_f \rangle}{\epsilon}, \tag{6.14}$$

$$\tilde{\Box} \langle \boldsymbol{A} \rangle = -\mu \langle \boldsymbol{J}_f \rangle, \tag{6.15}$$

together with the modified Lorenz gauge condition (6.13), where the modified wave operator $\tilde{\Box}$ is given by

$$\tilde{\Box} = -\frac{n^2}{c^2}\frac{\partial^2}{\partial t^2} + \nabla^2\,. \tag{6.16}$$

Equations (6.13)–(6.15) are identical to the exact Maxwell's equations (5.28)–(5.30) in the Lorenz gauge with $\phi \to \langle \phi \rangle$, $\boldsymbol{A} \to \langle \boldsymbol{A} \rangle$, and the substitutions defined in eq. (6.12).

If the medium is of finite extent, we must consider the matching conditions across the boundary. By the same arguments as we used in electrostatics and magnetostatics, we must have continuity of $\langle \boldsymbol{E}_\parallel \rangle$ and $\hat{\boldsymbol{n}} \cdot \langle \boldsymbol{B} \rangle$, and if we assume that there are no δ-function surface contributions to $\langle \rho_f \rangle$ and $\langle \boldsymbol{J}_f \rangle$, then we also must have continuity of $\hat{\boldsymbol{n}} \cdot \langle \boldsymbol{D} \rangle$ and $\langle \boldsymbol{H}_\parallel \rangle$.

If we have a semi-infinite medium with a planar boundary, we may consider a situation in which a plane wave incident on the medium is partly reflected and partly transmitted. The above matching conditions then yield that "the angle of incidence equals the angle of reflection" and that the transmitted wave obeys "Snell's law" (see problem 1 at the end of this chapter). The matching conditions also yield formulas for the amplitudes of the reflected and transmitted waves in terms of the amplitude of the incident wave and the values of ϵ and μ (see problem 2 for the case of normal incidence). The reflection and transmission amplitudes will depend on the polarization of the incident wave (see problem 3).

Finally, note that the same calculation that led to eq. (5.17) now yields[1]

$$\frac{\partial}{\partial t}\left[\frac{1}{2}\langle \boldsymbol{D}\rangle \cdot \langle \boldsymbol{E}\rangle + \frac{1}{2}\langle \boldsymbol{H}\rangle \cdot \langle \boldsymbol{B}\rangle\right] + \nabla \cdot [(\langle \boldsymbol{E}\rangle \times \langle \boldsymbol{H}\rangle)] = -\langle \boldsymbol{J}_f\rangle \cdot \langle \boldsymbol{E}\rangle\,. \tag{6.17}$$

It would be natural to interpret eq. (6.17) as representing conservation of energy and thereby interpret

$$\tilde{\mathcal{E}} = \frac{1}{2}\langle \boldsymbol{D}\rangle \cdot \langle \boldsymbol{E}\rangle + \frac{1}{2}\langle \boldsymbol{H}\rangle \cdot \langle \boldsymbol{B}\rangle \tag{6.18}$$

as the electromagnetic energy density in the medium and

$$\tilde{\boldsymbol{\mathcal{S}}} = \langle \boldsymbol{E}\rangle \times \langle \boldsymbol{H}\rangle \tag{6.19}$$

as the flux of electromagnetic energy in the medium. However, as discussed in section 3.2, the true electromagnetic energy density in the medium has contributions from the electromagnetic self-energy density, which is not calculable without a microscopic model of the medium. Furthermore, as previously noted, the value of $\langle \boldsymbol{D}\rangle$ depends on what charges are declared to be "free," so $\tilde{\mathcal{E}}$ depends on this choice. (And $\langle \boldsymbol{J}_f\rangle$ similarly depends on this choice, so eq. (6.17) holds independently of this choice.) Nevertheless, the conservation law eq. (6.17) is very useful, and little harm is done by referring to eq. (6.18) and eq. (6.19) as the electromagnetic "energy density" and "energy flux," respectively, as long as it is understood that one should not take this terminology too seriously.

We can similarly obtain an equation valid in the medium that corresponds to conservation of momentum by writing down eq. (5.20) and then making the substitutions eq. (6.12). However, in this case, there has historically been a controversy as to what quantity should be referred to as the "electromagnetic momentum density" $\tilde{\boldsymbol{\mathcal{P}}}$. If one accepts eq. (6.19) as representing the energy flux, then either of the following two alternative choices for $\tilde{\boldsymbol{\mathcal{P}}}$ are arguably natural:

$$\tilde{\boldsymbol{\mathcal{P}}} = \begin{cases} \frac{n^2}{c^2}\tilde{\boldsymbol{\mathcal{S}}} = \frac{n^2}{c^2}\langle \boldsymbol{E}\rangle \times \langle \boldsymbol{H}\rangle, & \text{Minkowski,} \\ \frac{1}{c^2}\tilde{\boldsymbol{\mathcal{S}}} = \frac{1}{c^2}\langle \boldsymbol{E}\rangle \times \langle \boldsymbol{H}\rangle, & \text{Abraham}\,. \end{cases} \tag{6.20}$$

The first was proposed by Minkowski in 1908, and the second was proposed by Abraham in 1909. The resulting Abraham-Minkowski controversy has continued for more than a century. Since neither formula represents the true electromagnetic momentum

[1]For an isotropic medium as we are considering here, $\langle \boldsymbol{D}\rangle$ and $\langle \boldsymbol{E}\rangle$ are proportional, so we could have written $\langle \boldsymbol{D}\rangle \cdot \langle \boldsymbol{E}\rangle$ as $\epsilon|\langle \boldsymbol{E}\rangle|^2$. We have chosen to write this term as $\langle \boldsymbol{D}\rangle \cdot \langle \boldsymbol{E}\rangle$, because this form would continue to hold for a linear, homogeneous, anisotropic medium. A similar remark applies to the term $\langle \boldsymbol{H}\rangle \cdot \langle \boldsymbol{B}\rangle$.

density, the reader should feel free to side with either Minkowski or Abraham—or to remain neutral.

6.2 Linear, Homogeneous, Isotropic Medium with a Delayed Response

A more realistic model of a linear, homogeneous, isotropic dielectric medium would assume that the polarization $\langle \boldsymbol{P}\rangle$ at time t depends not only on $\langle \boldsymbol{E}\rangle$ at time t but also on $\langle \boldsymbol{E}\rangle$ at earlier times $t' < t$ (i.e., that the response of the polarization to the electric field may be somewhat delayed). In other words, rather than assume that $\langle \boldsymbol{P}\rangle(t,\boldsymbol{x}) = \epsilon_0\chi\langle \boldsymbol{E}\rangle(t,\boldsymbol{x})$ with χ a constant (and thus $\langle \boldsymbol{D}\rangle(t,\boldsymbol{x}) = \epsilon\langle \boldsymbol{E}\rangle(t,\boldsymbol{x})$ with ϵ a constant), we assume that[2]

$$\langle \boldsymbol{P}\rangle(t,\boldsymbol{x}) = \int_{-\infty}^{\infty} f(t-t')\langle \boldsymbol{E}\rangle(t',\boldsymbol{x})dt', \tag{6.21}$$

where f is a smooth function that has the property $f(\eta)=0$ for all $\eta<0$; this condition on f implies that the polarization at a given time is influenced only by the electric field in the past, not in the future. We further assume that $f(\eta)\to 0$ rapidly as $\eta\to\infty$ (i.e., we assume that the behavior of $\langle \boldsymbol{E}\rangle$ in the distant past has very little influence on $\langle \boldsymbol{P}\rangle$).

The time Fourier transform of $\langle \boldsymbol{P}\rangle$ is given by

$$\langle \hat{\boldsymbol{P}}\rangle(\omega) = \frac{1}{\sqrt{2\pi}}\int_{-\infty}^{\infty} dt\, e^{i\omega t}\int_{-\infty}^{\infty} dt' f(t-t')\langle \boldsymbol{E}\rangle(t'), \tag{6.22}$$

where, for notational simplicity, we have omitted writing $\boldsymbol{x}$, but it is understood that this equation holds at each $\boldsymbol{x}$. Changing variables from t to $\eta = t-t'$, we obtain

$$\begin{aligned}\langle \hat{\boldsymbol{P}}\rangle(\omega) &= \frac{1}{\sqrt{2\pi}}\int_{-\infty}^{\infty}\int_{-\infty}^{\infty} d\eta dt' e^{i\omega(\eta+t')} f(\eta)\langle \boldsymbol{E}\rangle(t') \\ &= \sqrt{2\pi}\hat{f}(\omega)\langle \hat{\boldsymbol{E}}\rangle(\omega).\end{aligned} \tag{6.23}$$

In the case of instantaneous response considered in section 6.1, we have $\langle \boldsymbol{P}\rangle(t) = \epsilon_0\chi\langle \boldsymbol{E}\rangle(t)$, in which case the Fourier transforms satisfy $\langle \hat{\boldsymbol{P}}\rangle(\omega) = \epsilon_0\chi\langle \hat{\boldsymbol{E}}\rangle(\omega)$ with χ constant. For the delayed response, we have just shown that the relationship between $\langle \boldsymbol{P}\rangle$ and $\langle \boldsymbol{E}\rangle$ in Fourier transform space takes the form

$$\langle \hat{\boldsymbol{P}}\rangle(\omega) = \epsilon_0\chi(\omega)\langle \hat{\boldsymbol{E}}\rangle(\omega), \tag{6.24}$$

with

$$\chi(\omega) = \sqrt{2\pi}\hat{f}(\omega)/\epsilon_0\,. \tag{6.25}$$

In other words, in a delayed-response model, the susceptibility depends on frequency. We then have

$$\langle \hat{\boldsymbol{D}}\rangle(\omega) = \epsilon(\omega)\langle \hat{\boldsymbol{E}}\rangle(\omega), \tag{6.26}$$

where

$$\epsilon(\omega) = \epsilon_0(1+\chi(\omega))\,. \tag{6.27}$$

---

[2]The assumption that the function f depends on t and t' only in the combination $(t-t')$ ensures that the response properties of the medium do not change with time.

By similar arguments, in a delayed-response model, the relationship between $\langle \boldsymbol{H} \rangle$ and $\langle \boldsymbol{B} \rangle$ in Fourier transform space takes the form

$$\langle \hat{\boldsymbol{B}} \rangle(\omega) = \mu(\omega) \langle \hat{\boldsymbol{H}} \rangle(\omega) \,. \tag{6.28}$$

The dependence of $\chi(\omega)$ on the frequency ω is constrained by the fact that it is the Fourier transform of a smooth function f with the property that $f(\eta) = 0$, for all $\eta < 0$ and $f(\eta) \to 0$ suitably rapidly as $\eta \to \infty$. It follows that $\chi(\omega)$ can be extended to a holomorphic function in the upper half complex ω-plane (i.e., for real ω, $\chi(\omega)$ is the boundary value of a function that is holomorphic in the upper half-plane). Furthermore, $\chi(\omega)$ has suitable decay properties as $|\omega| \to \infty$ in this half-plane.[3] It follows that for real ω, the real and imaginary parts of $\chi(\omega)$ satisfy the *Kramers-Kronig relations* (see problem 4):

$$\mathrm{Re}\,[\chi(\omega)] = \frac{1}{\pi} \mathscr{P} \int_{-\infty}^{\infty} \frac{\mathrm{Im}\,[\chi(\omega')]}{\omega' - \omega} d\omega' , \tag{6.29}$$

$$\mathrm{Im}\,[\chi(\omega)] = -\frac{1}{\pi} \mathscr{P} \int_{-\infty}^{\infty} \frac{\mathrm{Re}\,[\chi(\omega')]}{\omega' - \omega} d\omega' , \tag{6.30}$$

where $\mathscr{P}$ denotes that the principal value prescription[4] is to be used to evaluate the integral through the singularity at $\omega' = \omega$. In addition, since the function f appearing in eq. (6.21) is real, we may also write the Kramers-Kronig relations in the following alternative form, where we integrate only over positive frequencies (see problem 4):

$$\mathrm{Re}\,[\chi(\omega)] = \frac{2}{\pi} \mathscr{P} \int_{0}^{\infty} \frac{\omega' \mathrm{Im}\,[\chi(\omega')]}{\omega'^2 - \omega^2} d\omega' , \tag{6.31}$$

$$\mathrm{Im}\,[\chi(\omega)] = -\frac{2\omega}{\pi} \mathscr{P} \int_{0}^{\infty} \frac{\mathrm{Re}\,[\chi(\omega')]}{\omega'^2 - \omega^2} d\omega' \,. \tag{6.32}$$

A very important immediate consequence of eq. (6.29) (or eq. (6.31)) is that we cannot have $\mathrm{Im}\,[\chi(\omega)] = 0$ at all real ω unless we also have $\mathrm{Re}\,[\chi(\omega)] = 0$ at all ω. An imaginary part of χ at frequency ω means that $\langle \hat{\boldsymbol{P}} \rangle$ and $\langle \hat{\boldsymbol{E}} \rangle$ oscillate out of phase at that frequency. This will occur if the medium absorbs/dissipates electromagnetic radiation at that frequency. If $\mathrm{Im}\,[\chi(\omega)] \neq 0$, then $\mathrm{Im}\,[\epsilon(\omega)] \neq 0$, and—unless exactly canceled by an imaginary part of $\mu(\omega)$—the index of refraction $n(\omega) = c\sqrt{\epsilon(\omega)\mu(\omega)}$ will have a nonvanishing imaginary part. In this case, the propagation vector $\boldsymbol{k}$ of plane wave solutions oscillating in time with frequency ω will have an imaginary part, corresponding to exponential decay in space. For example, we will see in section 6.3, that a plane wave propagating into a conductor of finite conductivity decays exponentially.[5]

Although in general, $n(\omega)$ must be complex, it is still possible for the imaginary part of $n(\omega)$ to be negligible over a wide range of frequencies. In such a frequency range,

---

[3]Note that if $f(\eta) \propto \delta(\eta)$, as in the instantaneous response model of section 6.1, then the Fourier transform of f will not have the required decay properties, and the Kramers-Kronig relations (6.29) and (6.30) will not apply.

[4]The *principal value prescription* for evaluating an integral where the integrand has a pole consists of removing an interval of size ϵ centered at the pole and then taking the limit $\epsilon \to 0$.

[5]The index of refraction also will be imaginary—and exponential decay will occur—if ϵ is real and negative. As we shall see in section 6.3, this is the case for a plasma at low frequencies. In this case, however, there is no absorption/dissipation (see problem 9).

plane wave solutions have spacetime dependence $\exp(-i\omega t + i\boldsymbol{k}\cdot\boldsymbol{x})$, with $\boldsymbol{k}$ real and

$$\omega = \frac{c}{n(\omega)} k, \tag{6.33}$$

where $k = |\boldsymbol{k}|$. The surfaces of constant phase of such plane waves in spacetime are the surfaces $-\omega t + \boldsymbol{k}\cdot\boldsymbol{x} = \text{const}$; that is, the *phase velocity* of such plane waves is

$$v_p = \frac{\omega}{k} = \frac{c}{n(\omega)}. \tag{6.34}$$

The variation of v_p with ω is referred to as *dispersion*.

Now consider a wave packet obtained by taking a superposition of such plane waves such that (i) all the plane waves in the superposition have a fixed propagation direction, which we take to be the z-direction, and (ii) the superposition consists of k very near k_0. Then the amplitude of any electromagnetic field quantity ψ (such as components of $\langle \boldsymbol{A}\rangle$, $\langle \boldsymbol{E}\rangle$, or $\langle \boldsymbol{B}\rangle$) will vary in spacetime as

$$\psi(t,z) = \int F(k) e^{-i\omega(k)t + ikz} dk, \tag{6.35}$$

where $\omega(k)$ is determined by eq. (6.33), and by assumption, the function F is sharply peaked around k_0. We Taylor expand $\omega(k)$ around $k = k_0$:

$$\omega(k) = \omega(k_0) + (k - k_0)\left.\frac{d\omega}{dk}\right|_{k_0} + \cdots. \tag{6.36}$$

Since F is sharply peaked around k_0, it should be a good approximation to keep only the first two terms shown. In that case, we have

$$\begin{aligned}
\psi(t,z) &= \exp\left[-i\left(\omega(k_0) - k_0\left.\frac{d\omega}{dk}\right|_{k_0}\right)t\right]\int F(k)\exp\left[-i\left(\left.\frac{d\omega}{dk}\right|_{k_0} t - z\right)k\right]dk \\
&= \exp\left[-i\left(\omega(k_0) - k_0\left.\frac{d\omega}{dk}\right|_{k_0}\right)t\right]\sqrt{2\pi}\hat{F}\left(\left.\frac{d\omega}{dk}\right|_{k_0} t - z\right),
\end{aligned} \tag{6.37}$$

where $\hat{F}$ denotes the Fourier transform of the function F. Thus, we have

$$|\psi(t,z)| \propto |\hat{F}(v_g t - z)|, \tag{6.38}$$

where we have defined the *group velocity* v_g by

$$v_g = \left.\frac{d\omega}{dk}\right|_{k_0}. \tag{6.39}$$

Equation (6.38) shows that the magnitude of the wave packet maintains its shape in time and moves in the z-direction with speed v_g. Differentiating eq. (6.33) with respect to k, we obtain

$$\frac{d\omega}{dk} = \frac{c}{n} - \frac{c}{n^2}\frac{dn}{d\omega}\frac{d\omega}{dk}k = \frac{c}{n} - \frac{\omega}{n}\frac{dn}{d\omega}\frac{d\omega}{dk}, \tag{6.40}$$

which yields

$$v_g = \frac{c}{n + \omega dn/d\omega}. \tag{6.41}$$

In special relativity, "nothing can travel faster than light." Thus it might be expected that we must have $v_p < c$ and $v_g < c$, which would place corresponding restrictions on n. However, the precise, correct version of the statement "nothing can travel faster than light" involves the initial value formulation: What happens to any matter and fields at event $(t, \boldsymbol{x})$ is determined using only the initial data for the matter and fields on and within the past light cone of $(t, \boldsymbol{x})$. As we saw in section 5.4, this property holds in electromagnetism, and it is believed to hold for all forms of physical matter. However, it does not preclude having patterns that move at velocity $> c$. For example, if one shines a laser at an extremely distant screen, there would be no difficulty in making the resulting spot on the screen move with velocity $> c$. A plane wave is spread out over all of space, and the phase of a plane wave in spacetime is a similar such pattern. There is no reason that we cannot have $v_p > c$, and indeed, it is common for materials to have $n(\omega) < 1$ over ranges of frequency ω. For example, we will see in section 6.3.2 that plasmas have $n(\omega) < 1$ at high frequencies. It is true that if we were to construct a wave packet that is initially of finite spatial extent, then special relativity would preclude the extent of the wave packet from spreading out faster than c. However, the construction of an initial wave packet of bounded spatial extent would require the use of all $\boldsymbol{k}$. Any wave packet constructed using only $\boldsymbol{k}$ near $k_0\hat{\boldsymbol{z}}$ will initially be spread out over all of space. Thus, the profile of such a wave will also have the status of a pattern. Consequently, we can have $v_g > c$, and indeed, such behavior in materials is not uncommon over narrow frequency ranges.

6.3 The Lorentz Model for $\epsilon(\omega)$

It is extremely useful to have a simple model for the type of medium discussed in section 6.2. The model presented below, known as the Lorentz model, is not intended to be realistic, but as we shall see, it gives considerable insight into the qualitative behavior of dielectrics, conductors, and plasmas in electrodynamics.

In this model, we take $\langle \rho_f \rangle = \langle \boldsymbol{J}_f \rangle = 0$. Although we set the "free charges and currents" to zero, the model allows unbound charges that are "free" to move around the medium. As explained in the paragraph below eq. (6.7), by setting $\langle \rho_f \rangle = \langle \boldsymbol{J}_f \rangle = 0$, we are simply treating the unbound charges on the same footing as the bound charges, and we take their effects into account via their contributions to polarization. In this model, we neglect all effects associated with magnetic dipoles, so we set $\mu(\omega) = \mu_0$. Thus, all macroscopic electromagnetic effects of the medium are determined by its dielectric constant $\epsilon(\omega)/\epsilon_0$. We neglect all effects associated with permanent electric dipole moments and treat the polarization as being entirely due to the motion of electrons. (Neglect of the motion of nuclei is not unreasonable, since the nuclei are much more massive and move less than the electrons.) We treat the electrons as moving independently in the macroscopically averaged field $\langle \boldsymbol{E} \rangle$, and we neglect magnetic forces. Finally, we treat each electron as a nonrelativistic point particle with charge $-e$ and mass m that is in a harmonic oscillator potential with frictional damping. Thus, we take the equation of motion of each electron to be of the form

$$m\ddot{\boldsymbol{x}}(t) + m\gamma\dot{\boldsymbol{x}}(t) + m\omega_0^2[\boldsymbol{x} - \boldsymbol{x}_0] = -e\langle \boldsymbol{E} \rangle, \tag{6.42}$$

where $\boldsymbol{x}_0$ is the minimum of the harmonic potential, and for convenience, we have written the damping coefficient as $m\gamma$ and the "spring constant" as $m\omega_0^2$, so that both γ and ω_0 have units of frequency. Of course, the electron is not connected to $\boldsymbol{x}_0$ by a spring, but it would not be unreasonable to treat it as such if it is bound to an atom at $\boldsymbol{x}_0$. Similarly, the electron does not experience "friction" in the usual macroscopic sense, but the damping term is not an unreasonable representation of the effects of collisions and radiation damping, which may dissipate the energy of the electron.

We will use this simple model to describe behavior in nonconducting media as well as in plasmas and conducting media. For a nonconducting medium, we treat the electrons as bound and take $\omega_0^2 > 0$. In the case of a plasma, we set $\omega_0^2 = 0$ but keep the damping term. In the case of a conducting material, we consider only the electrons that are not bound to atoms,[6] and we also set $\omega_0^2 = 0$ and keep the damping term.

We first focus on a single electron. Setting $\boldsymbol{x}_0 = 0$ for convenience and taking the time Fourier transform of eq. (6.42) (or, equivalently, restricting attention to solutions that oscillate as $e^{-i\omega t}$), we obtain

$$m\left[-\omega^2 - i\omega\gamma + \omega_0^2\right]\hat{\boldsymbol{x}}(\omega) = -e\langle \hat{\boldsymbol{E}} \rangle(\omega), \tag{6.43}$$

which is easily solved for $\hat{\boldsymbol{x}}(\omega)$. Since the electric dipole moment of the electron is

$$\boldsymbol{p} = -e\boldsymbol{x}, \tag{6.44}$$

we obtain

$$\hat{\boldsymbol{p}}(\omega) = \frac{e^2/m}{\omega_0^2 - \omega^2 - i\omega\gamma} \langle \hat{\boldsymbol{E}} \rangle(\omega)\,. \tag{6.45}$$

The coefficient of $\langle \hat{\boldsymbol{E}} \rangle$ on the right side of this equation represents the contribution of the electron to ϵ_0 times the susceptibility $\chi(\omega)$.

To obtain the total susceptibility, we sum over the contributions of all electrons. Since different electrons may be in different types of bound states labeled by j, we replace $\omega_0 \to \omega_j$ and $\gamma \to \gamma_j$ in the previous formula and sum over j. We thereby obtain

$$\chi(\omega) = \frac{e^2}{\epsilon_0 m} \sum_j \frac{N_j}{\omega_j^2 - \omega^2 - i\omega\gamma_j}, \tag{6.46}$$

where N_j denotes the number density of electrons in the jth bound state. Note that, by inspection, $\chi(\omega)$ can be extended to complex ω in such a way that it is holomorphic in the upper half of the complex ω-plane, in accord with the discussion in section 6.2. In particular, $\chi(\omega)$ satisfies the Kramers-Kronig relations.

---

[6]For a metal, the unbound electrons interact strongly with one another (as well with the periodic potential provided by the atoms), and the fermionic nature of the electrons plays an essential role in their description. Nevertheless, this simple picture can still be applied with "electrons" replaced by "quasiparticles" of some effective mass m^*.

The dielectric constant in this model is given by

$$\frac{\epsilon(\omega)}{\epsilon_0} = 1 + \chi(\omega) = 1 + \frac{e^2}{\epsilon_0 m} \sum_j \frac{N_j}{\omega_j^2 - \omega^2 - i\omega\gamma_j}. \tag{6.47}$$

Since in this model, we have assumed $\langle \boldsymbol{J}_f \rangle = 0 = \langle \boldsymbol{M} \rangle = 0$, eq. (6.6) yields

$$\langle \boldsymbol{J} \rangle = \frac{\partial \langle \boldsymbol{P} \rangle}{\partial t}. \tag{6.48}$$

Thus, we have

$$\langle \hat{\boldsymbol{J}} \rangle(\omega) = -i\omega \langle \hat{\boldsymbol{P}} \rangle(\omega) = -i\omega\epsilon_0 \chi(\omega) \langle \hat{\boldsymbol{E}} \rangle(\omega). \tag{6.49}$$

We see that, in this model, there is a linear relationship between $\langle \hat{\boldsymbol{J}} \rangle(\omega)$ and $\langle \hat{\boldsymbol{E}} \rangle(\omega)$ of the form

$$\langle \hat{\boldsymbol{J}} \rangle(\omega) = \sigma(\omega) \langle \hat{\boldsymbol{E}} \rangle(\omega), \tag{6.50}$$

where the *conductivity* $\sigma(\omega)$ is given by

$$\sigma(\omega) = -i\omega\epsilon_0 \chi(\omega) = -i\omega \frac{e^2}{m} \sum_j \frac{N_j}{\omega_j^2 - \omega^2 - i\omega\gamma_j}. \tag{6.51}$$

We are now ready to apply this model to various media.

6.3.1 NONCONDUCTING MEDIUM

We consider a medium where all electrons are bound to atoms, so we take $\omega_j^2 > 0$ for all j. The case of most relevance is that of small damping: $\gamma_j \ll \omega_j$ for all j. In that case, we see from eq. (6.47) that $\epsilon(\omega)$ is nearly real and does not vary rapidly with ω *except* near the frequencies ω_j, which are referred to as *resonant frequencies.*[7] Near a resonant frequency, it can be seen that $\epsilon(\omega)$ varies rapidly with ω, and the jth contribution to $\epsilon(\omega)$ becomes large and imaginary at the resonant frequency itself. This corresponds to having large absorption near the resonant frequencies. Note that the jth contribution to the real part of $\epsilon(\omega)$ is positive for $\omega < \omega_j$ and negative for $\omega > \omega_j$. The decrease of the jth contribution to the real part of $\epsilon(\omega)$ near $\omega = \omega_j$ can result in the real part of $n(\omega)$ being a decreasing function of frequency in this regime, which is referred to as *anomalous dispersion.*

In the limit as $\omega \to 0$, ϵ goes to a constant, whereas $\sigma(\omega) \to 0$. Thus, the electrostatic behavior becomes that of the dielectrics studied in chapter 3, and the medium becomes truly nonconducting.

[7]In a more realistic model, the resonant frequencies would correspond to the transition frequencies between bound states

6.3.2 PLASMA OR CONDUCTING MEDIUM

For the adaptation of our model to a plasma or conducting medium, we set $\omega_j = 0$. Our formulas for the dielectric constant and conductivity become

$$\frac{\epsilon(\omega)}{\epsilon_0} = 1 - \frac{1}{\epsilon_0} \frac{Ne^2/m}{\omega^2 + i\omega\gamma}, \tag{6.52}$$

$$\sigma(\omega) = i\omega \frac{Ne^2/m}{\omega^2 + i\omega\gamma}. \tag{6.53}$$

Here, in the case of a plasma, N denotes the number of electrons per unit volume, whereas in the case of a conducting medium, it represents the number of electrons[8] per unit volume that are not bound to atoms.

Low-frequency behavior: If $\omega \ll \gamma$, then we may neglect ω^2 compared with $i\gamma\omega$ in the above formulas, and our expressions for $\epsilon(\omega)/\epsilon_0$ and $\sigma(\omega)$ become

$$\frac{\epsilon(\omega)}{\epsilon_0} \approx 1 + i\frac{Ne^2}{\epsilon_0 m\omega\gamma}, \tag{6.54}$$

$$\sigma(\omega) \approx \frac{Ne^2}{m\gamma}. \tag{6.55}$$

Thus, $\sigma(\omega)$ is real and independent of ω. If we are considering phenomena involving only low-frequency behavior, then in position space, we have

$$\langle \boldsymbol{J} \rangle(t, \boldsymbol{x}) = \sigma \langle \boldsymbol{E} \rangle(t, \boldsymbol{x}), \tag{6.56}$$

with $\sigma = \sigma(0)$. This relationship is referred to as *Ohm's law*. Our model in this context corresponds to the *Drude model* of conductivity.

If we further assume that frequencies are sufficiently low so that $\epsilon_0\omega \ll \sigma$, then the dielectric constant eq. (6.54) becomes purely imaginary and is extremely large. Since $\mu = \mu_0$ in our model, the index of refraction is given by $n = \sqrt{\epsilon/\epsilon_0}$, so we have

$$n(\omega) = \sqrt{\frac{iNe^2}{\epsilon_0 m\omega\gamma}} = \sqrt{\frac{i\sigma}{\epsilon_0\omega}} = \frac{1+i}{\sqrt{2}}\sqrt{\frac{\sigma}{\epsilon_0\omega}}. \tag{6.57}$$

Thus, a plane wave of frequency ω propagating in the z-direction will have propagation vector $\boldsymbol{k} = k\hat{\boldsymbol{z}}$ with

$$k = \frac{\omega n}{c} = (1+i)\sqrt{\frac{\omega\sigma}{2c^2\epsilon_0}} = (1+i)\sqrt{\frac{\mu_0\omega\sigma}{2}}. \tag{6.58}$$

[8] More accurately, N represents the number density of quasiparticles (see footnote 6).

Thus, for such a plane wave solution, the amplitude of any electromagnetic field quantity ψ (such as components of $\langle\hat{\boldsymbol{A}}\rangle$, $\langle\hat{\boldsymbol{E}}\rangle$, or $\langle\hat{\boldsymbol{B}}\rangle$) will vary in spacetime as

$$\psi(t,z) = e^{-i\omega t + iz/\delta} e^{-z/\delta}, \tag{6.59}$$

where[9]

$$\delta \equiv \sqrt{\frac{2}{\mu_0 \omega \sigma}} \tag{6.60}$$

is called the *skin depth* of the plasma or conducting medium. Thus, a plane wave in such a medium will be exponentially damped on the scale of the skin depth. An electromagnetic wave in vacuum that is incident on a plasma or conducting medium will, in effect, penetrate into the medium only to depth δ. The reflected wave will have smaller amplitude than the incident wave on account of the dissipation in the medium (see problem 9 at the end of the chapter). For the above plane wave solution in a conducting medium, by eq. (6.4), we have

$$i\omega\langle\hat{\boldsymbol{B}}\rangle = -\frac{\partial\langle\hat{\boldsymbol{B}}\rangle}{\partial t} = \nabla \times \langle\hat{\boldsymbol{E}}\rangle = -\frac{1-i}{\delta}\hat{\boldsymbol{z}} \times \langle\hat{\boldsymbol{E}}\rangle \,. \tag{6.61}$$

Thus, $\langle\hat{\boldsymbol{E}}\rangle$ and $\langle\hat{\boldsymbol{B}}\rangle$ are orthogonal and oscillate 45° out of phase. Furthermore, we have

$$|\langle\hat{\boldsymbol{E}}\rangle| = \left|\frac{\omega\delta}{(1-i)}\right| |\langle\hat{\boldsymbol{B}}\rangle| = \sqrt{\frac{\omega}{\mu_0\sigma}}|\langle\hat{\boldsymbol{B}}\rangle| = c\sqrt{\frac{\epsilon_0\omega}{\sigma}}|\langle\hat{\boldsymbol{B}}\rangle| \,. \tag{6.62}$$

Thus, for high conductivity, $\sigma \gg \epsilon_0\omega$, we have $|\langle\hat{\boldsymbol{E}}\rangle| \ll c|\langle\hat{\boldsymbol{B}}\rangle|$.

High-frequency behavior: If $\omega \gg \gamma$, then we can neglect γ in eq. (6.52) and thereby obtain

$$\frac{\epsilon(\omega)}{\epsilon_0} \approx 1 - \frac{\omega_P^2}{\omega^2}, \tag{6.63}$$

$$\sigma(\omega) \approx i\frac{\epsilon_0\omega_P^2}{\omega} \,. \tag{6.64}$$

Here the *plasma frequency* ω_P is defined by

$$\omega_P^2 = \frac{Ne^2}{\epsilon_0 m} \,. \tag{6.65}$$

Thus, at high frequencies, $\epsilon(\omega)$ is real, and $\sigma(\omega)$ is purely imaginary. An imaginary $\sigma(\omega)$ means that $\langle\boldsymbol{J}\rangle$ and $\langle\boldsymbol{E}\rangle$ oscillate completely out of phase, in accord with the fact that we have eliminated dissipation in the model at high frequencies by neglecting γ. For $\omega > \omega_P$, we have $\epsilon(\omega) > 0$, so the index of refraction $n = \sqrt{\epsilon/\epsilon_0}$ also is real. Thus, at frequencies above the plasma frequency, plane waves can propagate without decay

[9] If we had allowed a nontrivial magnetic permeability μ in the model, the corresponding formula for δ would be $\delta = \sqrt{2/\mu\omega\sigma}$.

in the plasma or conducting medium. Note that $n(\omega) < 1$ for all $\omega > \omega_P$, so the phase velocity eq. (6.34) of all such plane waves satisfies $v_p > c$. However, the group velocity eq. (6.41) satisfies $v_g < c$ for all $\omega > \omega_P$ (see problem 7).

If $\omega \gg \gamma$ but $\omega < \omega_P$, then $\epsilon(\omega) < 0$, so $n(\omega)$ is purely imaginary. Thus plane wave solutions at frequency ω are purely exponentially damped in space. If a plasma has a planar boundary, and a plane wave in vacuum is incident on it, the wave will effectively penetrate into the plasma only to depth $c/\sqrt{\omega_P^2 - \omega^2}$. This explains why radio waves are reflected by the ionosphere. A wave reflected by a plasma will have the same amplitude as the incident wave (see problem 9(b)), corresponding to the fact that there is no dissipation.

In summary, the Lorentz model considered in this section gives a good qualitative description of phenomena that occur in different media in a wide range of circumstances.

6.4 Magnetohydrodynamics

Magnetohydrodynamics (MHD) is an approximate description of a plasma in which a large magnetic field may be present. The MHD approximation is expected to be valid under the following physical assumptions: (i) The plasma can be treated as a fluid. (ii) The fluid is highly conducting. (iii) The fluid motion is nonrelativistic. (iv) The macroscopically averaged electromagnetic field and the fluid variables vary slowly in time compared with all relevant timescales associated with the microscopic physics—in particular, the timescales set by collision times, the gyrofrequency of orbits (see eq. (8.89)), and the plasma frequency eq. (6.65).

When we considered the simple model of a medium in section 6.3, we took the "free charges and currents" to vanish and correspondingly treated each electron as contributing an electric dipole moment $\boldsymbol{p} = -e\boldsymbol{x}$. The time variations of these dipole moments then produced a macroscopic current $\langle \boldsymbol{J} \rangle = \partial \langle \boldsymbol{P} \rangle / \partial t$ (see eq. (6.48)). By proceeding in this manner, we were able to give a uniform treatment of both nonconducting media (where the electrons are bound) and plasmas (where the electrons are unbound). In the present context of MHD—where all charges of interest are unbound—it is more convenient to treat the charges and currents (of both the electrons and ions) as being free and thereby not contributing to dipole moments. In particular, each electron would then contribute $\boldsymbol{j} = -e\boldsymbol{v}$ to the free current density $\langle \boldsymbol{J}_f \rangle$. The macroscopic current would be given by the free current, $\langle \boldsymbol{J} \rangle = \langle \boldsymbol{J}_f \rangle$. It is not difficult to see that these viewpoints are equivalent (i.e., each electron would contribute $-e\boldsymbol{v}$ to $\langle \boldsymbol{J} \rangle$ under either viewpoint). By treating the charges and currents in the plasma as free, as we shall below, we have $\langle \boldsymbol{P} \rangle = \langle \boldsymbol{M} \rangle = 0$, and hence $\langle \boldsymbol{D} \rangle = \epsilon_0 \langle \boldsymbol{E} \rangle$ and $\langle \boldsymbol{B} \rangle = \mu_0 \langle \boldsymbol{H} \rangle$.

Thus, the macroscopic variables describing the electromagnetic field are $\langle \boldsymbol{E} \rangle$ and $\langle \boldsymbol{B} \rangle$, together with the (free) charge density $\langle \rho \rangle$ and (free) current density $\langle \boldsymbol{J} \rangle$. These quantities satisfy Maxwell's equations (6.3), (6.4), (6.5), and (6.7) with $\langle \boldsymbol{D} \rangle = \epsilon_0 \langle \boldsymbol{E} \rangle$ and $\langle \boldsymbol{B} \rangle = \mu_0 \langle \boldsymbol{H} \rangle$. In addition, the macroscopic quantities describing the fluid (composed of the electrons and ions) are its mass density ρ_M, fluid velocity $\boldsymbol{v}$, and pressure P. These quantities are also obtained by averaging corresponding microscopic quantities, so in principle, these quantities should have angle brackets around them to denote that they are macroscopic averages. However, there is no need to do so in this case, since there is no danger of them being confused with microscopic quantities. For notational consistency, instead of putting angle brackets around the fluid quantities, for the remainder of this chapter, I will simply drop the angle brackets on all macroscopic averages. *Thus, below,* **E**, **B**, ρ, *and* **J** *will denote the macroscopically averaged electromagnetic field variables.*

The equations governing the motion of the fluid are the mass conservation equation,

$$\frac{\partial \rho_M}{\partial t} + \nabla \cdot (\rho_M \boldsymbol{v}) = 0, \tag{6.66}$$

and the *Euler equation*

$$\rho_M \left(\frac{\partial}{\partial t} + \boldsymbol{v} \cdot \nabla \right) \boldsymbol{v} = \boldsymbol{f} - \nabla P, \tag{6.67}$$

where the force density $\boldsymbol{f}$ is given by the Lorentz force eq. (5.21) plus any other external forces[10] that act on the fluid (such as gravity). The operator appearing on the left side of eq. (6.67) is often denoted by

$$\frac{\partial}{\partial t} + \boldsymbol{v} \cdot \nabla \equiv \frac{d}{dt} \tag{6.68}$$

and is called the *convective derivative*. It corresponds to taking the time derivative in a frame moving instantaneously with the local velocity of the fluid.

To have deterministic motion of the fluid, one must give further information about the pressure P of the fluid. It is usually assumed that the fluid is locally in thermal equilibrium and can be assigned an entropy density S. The pressure is then assumed to be a function of ρ_M and S. If there is no dissipation or heat flow in the fluid, the entropy density will be conserved in the sense that

$$\frac{\partial S}{\partial t} + \nabla \cdot (S\boldsymbol{v}) = 0\,. \tag{6.69}$$

If we define $s \equiv S/\rho_M$, it then follows from eq. (6.66) and eq. (6.69) that s will be constant along the flow lines of the fluid. If we write the pressure as $P = P(\rho_M, s)$, then P will depend only on ρ_M and the initial values of s (i.e., it will follow an "adiabatic law" as ρ_M changes). For example, for an ideal gas, we obtain

$$\left[\frac{\partial}{\partial t} + \boldsymbol{v} \cdot \nabla \right] \left(\frac{P}{\rho_M^\gamma} \right) = 0, \tag{6.70}$$

where γ is the ratio of specific heats C_P/C_V ($= 5/3$ for a monatomic gas). The adiabatic law of the fluid gives the necessary equation for the evolution of P.

Thus, we have a coupled system of Maxwell's equations (6.3), (6.4), (6.5), and (6.7) and the fluid evolution equations (6.66), (6.67), and (6.70) (or a suitable replacement of eq. (6.70) for a different equation of state of the plasma fluid). This system is not yet deterministic, because we do not have equations determining the evolution of ρ and $\boldsymbol{J}$. We now state the assumptions and approximations of MHD, which will provide us with a deterministic system of equations.

---

[10]One could also allow consideration of viscous forces. The Euler equation with the addition of viscous force terms is called the *Navier-Stokes equation*.

The first assumption is that, since we consider slow time variations of all macroscopic quantities, it will not be possible for significant macroscopic "charge separation" to occur between the electrons and ions.[11] Thus, we set $\rho = 0$. The Lorentz force term $\boldsymbol{f}$ in eq. (6.67) is then

$$\boldsymbol{f} = \boldsymbol{J} \times \boldsymbol{B} . \tag{6.71}$$

The second assumption is that, since the plasma is assumed to be very highly conducting, there should be vanishing force on a charged particle that is co-moving with the plasma fluid. That is, we should have

$$\boldsymbol{E} + \boldsymbol{v} \times \boldsymbol{B} = 0 . \tag{6.72}$$

The argument for eq. (6.72) is essentially the same argument that the electric field should vanish inside a conductor in electrostatics. Indeed, as we shall see at the end of section 8.2, to leading order in v/c, the quantity $\boldsymbol{E}' = \boldsymbol{E} + \boldsymbol{v} \times \boldsymbol{B}$ is just the electric field as determined by an observer who is co-moving with the fluid, so eq. (6.72) states that $\boldsymbol{E}' = 0$. The curl of eq. (6.72) together with Maxwell's equation (6.4) yields the *induction equation*

$$-\frac{\partial \boldsymbol{B}}{\partial t} + \nabla \times (\boldsymbol{v} \times \boldsymbol{B}) = 0 . \tag{6.73}$$

The third assumption is that the time derivative of $\boldsymbol{E}$ can be neglected in Maxwell's equation (6.7). By eq. (6.72), $\boldsymbol{E}$ is already smaller than $c\boldsymbol{B}$ by $O(v/c)$. For slow time variations of quantities, the time derivative of $\boldsymbol{E}$ should be very much smaller than the space derivatives of $c^2\boldsymbol{B}$. Thus, we replace eq. (6.7) with

$$\nabla \times \boldsymbol{B} = \mu_0 \boldsymbol{J} . \tag{6.74}$$

Using this equation, we may eliminate $\boldsymbol{J}$ in favor of $\boldsymbol{B}$. In particular, the Euler equation (6.67) with no nonelectromagnetic external forces becomes

$$\rho_M \left(\frac{\partial}{\partial t} + \boldsymbol{v} \cdot \nabla \right) \boldsymbol{v} = -\frac{1}{\mu_0} \boldsymbol{B} \times (\nabla \times \boldsymbol{B}) - \nabla P . \tag{6.75}$$

The *equations of ideal MHD* are the Maxwell equation $\nabla \cdot \boldsymbol{B} = 0$ together with the mass conservation equation (6.66), the Euler equation (6.75), the induction equation (6.73), and the equation (6.70) for the evolution of the pressure (or the corresponding evolution equation for the pressure if the plasma cannot be treated as an ideal gas). These equations form a closed system for the evolution of $\boldsymbol{B}$, ρ_M, $\boldsymbol{v}$, and P. After these equations are solved, the electric field can then be obtained from eq. (6.72), and the current is obtained from eq. (6.74).

If we do not idealize the plasma as having infinite conductivity, then we should replace eq. (6.72) with Ohm's law eq. (6.56) holding in the local fluid rest frame:

$$\boldsymbol{J} = \sigma \left(\boldsymbol{E} + \boldsymbol{v} \times \boldsymbol{B} \right) , \tag{6.76}$$

[11] Such charge separation can occur for high-frequency oscillations of a plasma.

where σ is the conductivity of the plasma. The corresponding induction equation becomes

$$\nabla \times \boldsymbol{J} = \sigma \left[-\frac{\partial \boldsymbol{B}}{\partial t} + \nabla \times (\boldsymbol{v} \times \boldsymbol{B}) \right] . \tag{6.77}$$

Eliminating $\boldsymbol{J}$ using eq. (6.74), we obtain

$$\frac{1}{\mu_0 \sigma} \nabla \times (\nabla \times \boldsymbol{B}) = -\frac{1}{\mu_0 \sigma} \nabla^2 \boldsymbol{B} = -\frac{\partial \boldsymbol{B}}{\partial t} + \nabla \times (\boldsymbol{v} \times \boldsymbol{B}) . \tag{6.78}$$

Thus, for the equations of *nonideal MHD*, we would replace eq. (6.73) with the finite conductivity version, eq. (6.78). In nonideal MHD, we may also allow additional dissipation terms from viscosity in the Euler equation (6.75). Since entropy conservation no longer holds in the presence of dissipation, we must also replace eq. (6.69) with an equation for the entropy evolution to determine the evolution of P.

The ideal magnetic induction equation (6.73) has a very important consequence by itself. Consider an arbitrary 2-dimensional surface $\mathcal{S}_0$ in the fluid at time $t = t_0$. ($\mathcal{S}_0$ need not have any particular orientation with respect to either $\boldsymbol{v}$ or $\boldsymbol{B}$.) Now carry this surface element along with the fluid, so that it goes to the surface $\mathcal{S}_1$ at time t_1, as indicated in figure 6.1. Then we have[12]

$$\int_{\mathcal{S}_0} \boldsymbol{B}(t_0) \cdot \hat{\boldsymbol{n}} \, dA = \int_{\mathcal{S}_1} \boldsymbol{B}(t_1) \cdot \hat{\boldsymbol{n}} \, dA, \tag{6.79}$$

where $\hat{\boldsymbol{n}}$ denotes the unit normal to the surface over which the integration is done. Equation (6.79) can be interpreted as saying that the magnetic field is *frozen into* the fluid (i.e., it is carried along by the fluid in such a way that its flux through any comoving surface does not change with time).

To prove eq. (6.79), let $\mathcal{S}(t)$ denote the surface that $\mathcal{S}_0$ is carried into at time t, and consider

$$\frac{d}{dt} \int_{\mathcal{S}(t)} \boldsymbol{B}(t) \cdot \hat{\boldsymbol{n}} \, dA = \int_{\mathcal{S}} \frac{\partial \boldsymbol{B}}{\partial t} \cdot \hat{\boldsymbol{n}} \, dA + \frac{\delta}{\delta t} \int_{\mathcal{S}(t)} \boldsymbol{B} \cdot \hat{\boldsymbol{n}} \, dA . \tag{6.80}$$

Here, "d/dt" means the total rate of change of the integral—due both to the fact that $\boldsymbol{B}$ changes with time and to the fact that the surface over which we are integrating is also changing. On the right side of this equation, this total time derivative is broken up into a contribution coming purely from the time variation of $\boldsymbol{B}$ and a contribution coming purely from the variation of the integration surface—holding $\boldsymbol{B}$ fixed at its value at t—for which I have invented the notation "$\delta/\delta t$." To evaluate the latter contribution, note that, as illustrated in figure 6.1, the flow of $\mathcal{S}$ between two times sweeps out a volume $\mathcal{V}$, whose boundary, $\partial\mathcal{V}$, consists of the faces $\mathcal{S}_0$ and $\mathcal{S}_1$ and a surface, $\mathscr{S}$, generated by the flow lines of $\boldsymbol{v}$. By Gauss's law, for $\boldsymbol{B}$ at any fixed time, we have

[12] This result has a much more elegant formulation and proof in special relativistic notation. Let u^μ be the 4-velocity of the fluid flow lines (see eq. (8.16)), and let $F_{\mu\nu}$ be the electromagnetic field-strength tensor (see eq. (8.43)). The statement of the "frozen in" property is $\pounds_u F_{\mu\nu} = 0$, where $\pounds$ denotes the Lie derivative (which I will not attempt to define here). The proof follows immediately from the Cartan formula for the Lie derivative, which, in differential forms notation, yields $\pounds_u F = d(u \cdot F) + u \cdot dF$. The first term vanishes by eq. (6.72), and the second term vanishes by Maxwell's equations.

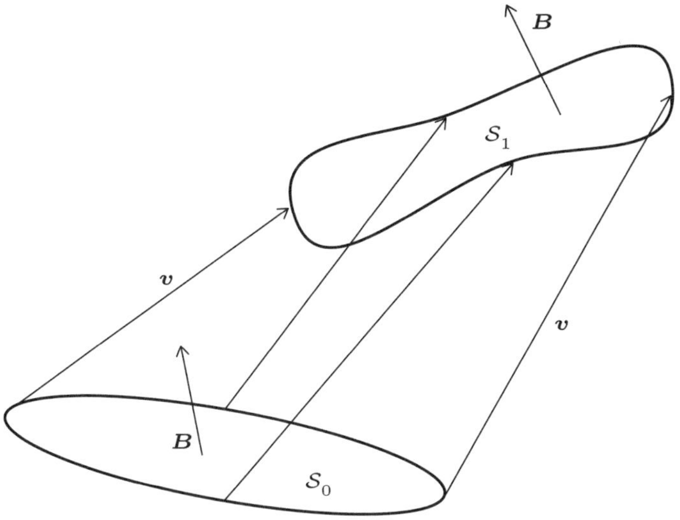

Figure 6.1. A 2-dimensional surface $\mathcal{S}_0$ at time t_0 carried along by the fluid flow to the surface $\mathcal{S}_1$ at time t_1.

$$\int_{\partial\mathcal{V}} \boldsymbol{B}\cdot\hat{\boldsymbol{N}}\,dA = 0, \tag{6.81}$$

where $\hat{\boldsymbol{N}}$ denotes the unit outward normal to $\partial\mathcal{V}$. We may assume that $\hat{\boldsymbol{N}} = \hat{\boldsymbol{n}}$ on $\mathcal{S}_1$, and hence, $\hat{\boldsymbol{N}} = -\hat{\boldsymbol{n}}$ on $\mathcal{S}_0$. It follows that the difference in the flux of $\boldsymbol{B}$ through the top and bottom faces is given by minus its flux through $\mathscr{S}$. However, on $\mathscr{S}$, we have

$$\hat{\boldsymbol{N}} dA = \boldsymbol{dl}\times\boldsymbol{v}\,dt, \tag{6.82}$$

where $\boldsymbol{dl}$ denotes the line element[13] along the 1-dimensional boundary $\mathcal{C}(t)$ of $\mathcal{S}(t)$. Putting all of this together, we see that

$$\begin{aligned}\frac{\delta}{\delta t}\int_{\mathcal{S}(t)} \boldsymbol{B}\cdot\hat{\boldsymbol{n}}\,dA &= -\int_{\mathcal{C}(t)} \boldsymbol{B}\cdot(\boldsymbol{dl}\times\boldsymbol{v}) = -\int_{\mathcal{C}(t)} (\boldsymbol{v}\times\boldsymbol{B})\cdot\boldsymbol{dl}\\ &= -\int_{\mathcal{S}(t)} [\boldsymbol{\nabla}\times(\boldsymbol{v}\times\boldsymbol{B})]\cdot\hat{\boldsymbol{n}}\,dA,\end{aligned} \tag{6.83}$$

where Stokes's theorem was used in the last line. Thus, we find

$$\frac{d}{dt}\int_{\mathcal{S}(t)} \boldsymbol{B}(t)\cdot\hat{\boldsymbol{n}}\,dA = \int_{\mathcal{S}(t)} \left[\frac{\partial\boldsymbol{B}}{\partial t} - \boldsymbol{\nabla}\times(\boldsymbol{v}\times\boldsymbol{B})\right]\cdot\hat{\boldsymbol{n}}\,dA = 0, \tag{6.84}$$

as we desired to show.

[13]The direction of $\boldsymbol{dl}$ is given by the right-hand rule with respect to $\hat{\boldsymbol{N}}$.

For finite conductivity, eq. (6.73) must be replaced by eq. (6.78), and the magnetic field is no longer frozen in. Indeed, in the special case $v = 0$, eq. (6.78) simplifies to

$$\frac{\partial \boldsymbol{B}}{\partial t} = \frac{1}{\mu_0 \sigma} \nabla^2 \boldsymbol{B}, \tag{6.85}$$

which is the same equation that arises in the study of diffusion and heat flow. The magnetic field will thereby decay by "magnetic diffusion" (see problem 10), although the decay time can be very long if the conductivity is high.

A simple solution to the equations of ideal MHD is $\boldsymbol{B} = \boldsymbol{B}_0$, $\boldsymbol{v} = 0$, $\rho_M = \rho_{M0}$, and $P = P_0$, where $\boldsymbol{B}_0$, ρ_{M0}, and P_0 are independent of $(t, \boldsymbol{x})$. The equations of ideal MHD for linearized perturbations $\delta \boldsymbol{B}$, $\delta \boldsymbol{v}$, $\delta \rho_M$, and δP about this solution are

$$\nabla \cdot \delta \boldsymbol{B} = 0, \tag{6.86}$$

$$\frac{\partial \delta \rho_M}{\partial t} + \rho_{M0} \nabla \cdot \delta \boldsymbol{v} = 0, \tag{6.87}$$

$$\rho_{M0} \frac{\partial \delta \boldsymbol{v}}{\partial t} + \frac{1}{\mu_0} \boldsymbol{B}_0 \times (\nabla \times \delta \boldsymbol{B}) + \nabla \delta P = 0, \tag{6.88}$$

$$-\frac{\partial \delta \boldsymbol{B}}{\partial t} + \nabla \times (\delta \boldsymbol{v} \times \boldsymbol{B}_0) = 0, \tag{6.89}$$

together with the relations that determine δP. If we assume that the perturbation is adiabatic, then $\delta s = 0$, and we have

$$\delta P = \left(\frac{\partial P}{\partial \rho} \right)_s \delta \rho_M \,. \tag{6.90}$$

Equations (6.86)–(6.90) describe perturbations of an ideal MHD fluid away from the above background solution.

If $\boldsymbol{B}_0 = 0$, then eqs. (6.87), (6.88), and (6.90) correspond to perturbations of an ordinary fluid. Taking the divergence of eq. (6.88) with $\boldsymbol{B}_0 = 0$ and using (6.87) and (6.90), we see that $\delta \rho_M$ satisfies a wave equation with propagation speed c_s, where

$$c_s^2 = (\partial P / \partial \rho)_s \,. \tag{6.91}$$

Such waves are called *sound waves*, and they have the essential feature that they are compressive (i.e., that $\nabla \cdot \delta \boldsymbol{v} \neq 0$). If $\boldsymbol{B}_0 \neq 0$, we must solve the coupled system (6.86)–(6.90). There exist propagating wave solutions of a similar, compressive character, called *magnetoacoustic waves*. Indeed, there exist two classes of such solutions, usually referred to as "fast" and "slow" magnetoacoustic waves (see problem 11). But there also exist waves propagating in the plasma of a qualitatively different character, for which $\delta \rho_M = \delta P = \nabla \cdot \delta \boldsymbol{v} = 0$. Such waves are called *Alfvén waves*.

A simple example of an Alfvén wave is as follows. Taking $\boldsymbol{B}_0 = B_0 \hat{\boldsymbol{z}}$, we seek solutions with $\delta \rho_M = \delta P = 0$ and with $\delta \boldsymbol{v}$ and $\delta \boldsymbol{B}$ of the form

$$\delta \boldsymbol{v} = \boldsymbol{v}_1(t, z) \,, \qquad \delta \boldsymbol{B} = \boldsymbol{B}_1(t, z), \tag{6.92}$$

where $\hat{z} \cdot \boldsymbol{v}_1 = \hat{z} \cdot \boldsymbol{B}_1 = 0$. This ansatz implies that $\nabla \cdot \delta \boldsymbol{v} = \nabla \cdot \delta \boldsymbol{B} = 0$. Thus, eqs.(6.86), (6.87), and (6.90) hold automatically, and we need only solve (6.88) and (6.89). These equations yield

$$\rho_{M0} \frac{\partial \boldsymbol{v}_1}{\partial t} = \frac{B_0}{\mu_0} \frac{\partial \boldsymbol{B}_1}{\partial z}, \tag{6.93}$$

$$\frac{\partial \boldsymbol{B}_1}{\partial t} = B_0 \frac{\partial \boldsymbol{v}_1}{\partial z}. \tag{6.94}$$

Thus, $\boldsymbol{v}_1$ satisfies[14]

$$\frac{\partial^2 \boldsymbol{v}_1}{\partial t^2} - \left(\frac{B_0^2}{\mu_0 \rho_{M0}} \right) \frac{\partial^2 \boldsymbol{v}_1}{\partial z^2} = 0, \tag{6.95}$$

with $\boldsymbol{B}_1$ then determined by eqs. (6.93) and (6.94). This equation describes waves propagating along the direction of $\boldsymbol{B}_0$ with transverse oscillations of $\delta \boldsymbol{v}$ and $\delta \boldsymbol{B}$. These waves propagate with speed

$$c_A = \frac{B_0}{\sqrt{\mu_0 \rho_{M0}}}, \tag{6.96}$$

which is referred to as the *Alfvén speed.*

Problems

1. The half-space $z < 0$ is filled with a medium of permittivity ϵ_1 and permeability μ_1, whereas the half-space $z > 0$ is filled by a medium of permittivity ϵ_2 and permeability μ_2. A plane wave of frequency ω and wave vector $\boldsymbol{k}_I$ is incident from the region $z < 0$; that is, the incident wave has $\phi_I = 0$, and

$$\boldsymbol{A}_I = \boldsymbol{C}_I e^{-i\omega t} e^{i\boldsymbol{k}_I \cdot \boldsymbol{x}}.$$

The reflected plane wave has frequency ω and wave vector $\boldsymbol{k}_R$; that is, $\phi_R = 0$ and

$$\boldsymbol{A}_R = \boldsymbol{C}_R e^{-i\omega t} e^{i\boldsymbol{k}_R \cdot \boldsymbol{x}}.$$

The plane wave transmitted through the second medium ($z > 0$) has $\phi_T = 0$ and

$$\boldsymbol{A}_T = \boldsymbol{C}_T e^{-i\omega t} e^{i\boldsymbol{k}_T \cdot \boldsymbol{x}}.$$

Let θ_I, θ_R, and θ_T denote the angles between the normal to the boundary and $\boldsymbol{k}_I$, $\boldsymbol{k}_R$, and $\boldsymbol{k}_T$, respectively, as illustrated in the diagram. Show that (a) the wave vectors $\boldsymbol{k}_R$, and $\boldsymbol{k}_T$ lie in the plane defined by $\boldsymbol{k}_I$ and the normal to the medium (so, in particular, $\boldsymbol{k}_I$, $\boldsymbol{k}_R$, and $\boldsymbol{k}_T$ are co-planar); (b) $\theta_I = \theta_R$ (i.e., the angle of incidence equals the angle of reflection); (c) $n_1 \sin \theta_I = n_2 \sin \theta_T$, where $n_i = c\sqrt{\mu_i \epsilon_i}$ for $i = 1, 2$ is the index of refraction. This relation is known as *Snell's law.* (d) If $n_1 > n_2$, then $\boldsymbol{A}_T = 0$ if the angle of incidence is such that $\sin \theta_I > n_2/n_1$. This phenomenon is known as *total internal reflection.* [Hint: The full matching conditions between the solutions at $z < 0$ and $z > 0$ are relatively complicated to write out and solve (see problem 2 for a simple case).

[14] See problem 11 for the corresponding equation satisfied by $\delta \boldsymbol{v}$ for arbitrary perturbations.

However, the above results can be obtained in a relatively simple manner by noting that the matching must hold at $z = 0$ for all (x, y).]

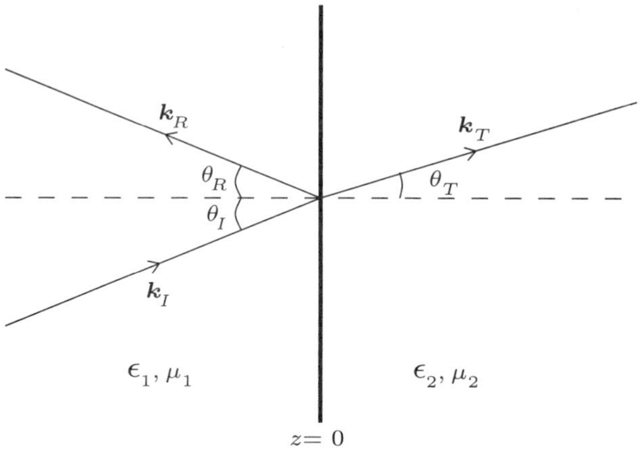

2. Consider the same setup as in problem 1, but consider the case where $\epsilon_1 = \epsilon_0$, $\mu_1 = \mu_0$ (i.e., the region $z < 0$ is vacuum), and the incoming plane wave is normally incident on the medium ($\theta_I = 0$), so we have

$$\boldsymbol{A}_I = \boldsymbol{C}_I e^{-i\omega t} e^{ikz}.$$

Solve for $\boldsymbol{A}_R$ and $\boldsymbol{A}_T$.

3. Consider, again, the same setup as in problem 1, but now suppose that $\mu_1 = \mu_2 = \mu_0$. Consider the case where $\boldsymbol{C}_I$ lies in the plane determined by $\boldsymbol{k}_I$ and the normal to the medium. ($\boldsymbol{C}_I$ is, of course, perpendicular to $\boldsymbol{k}_I$.) Define the *Brewster angle* to be the angle such that $\tan\theta_B = n_2/n_1$. Show that if $\theta_I = \theta_B$, then the matching conditions yield $\boldsymbol{C}_R = 0$. Thus, if an unpolarized electromagnetic wave (see footnote 6 in section 7.2) is incident at the Brewster angle, the reflected wave will be polarized in the direction perpendicular to the plane determined by $\boldsymbol{k}_I$ and the normal to the medium.

4. (a) Derive the Kramers-Kronig relations (6.29) and (6.30). [Hint: Use Cauchy's theorem to show that for any $\epsilon > 0$, we have

$$\int_{-\infty}^{\infty} \frac{\chi(\omega')}{\omega' - \omega + i\epsilon} d\omega' = 0.$$

Then use the Plemelj relation

$$\lim_{\epsilon \downarrow 0} \frac{1}{x + i\epsilon} = \mathscr{P}\left(\frac{1}{x}\right) - i\pi\,\delta(x).\Big]$$

(b) Starting with the Kramers-Kronig relations in the form of eqs. (6.29) and (6.30), derive the Kramers-Kronig relations in the form of eqs. (6.31) and (6.32). [Hint: Note that for real f, we have $\hat{f}^*(\omega) = \hat{f}(-\omega)$.]

5. Suppose a dielectric medium is such that $\chi(\omega)$ is real (i.e., there is no absorption) for $\omega_1 < \omega < \omega_2$, where $\omega_1 \ll \omega_2$. Show that for $\omega_1 \ll \omega \ll \omega_2$, $\chi(\omega)$ must take the functional form

$$\chi(\omega) = a + b\frac{1}{\omega^2},$$

where a and b are real constants. [Hint: The Kramers-Kronig relations will be useful for this problem.]

6. In section 6.2, we showed that the magnitude of a wave packet of the form eq. (6.35) will propagate undistorted with group velocity v_g in the approximation where we keep only the first term in the Taylor series eq. (6.36) for $\omega(k)$. However, distortions will occur if we keep additional terms in this Taylor series. Calculate $\psi(t, z)$ for the case where $F(k)$ is a Gaussian,

$$F(k) = e^{-(k-k_0)^2/\alpha^2},$$

in the approximation

$$\omega(k) = \omega(k_0) + v_g(k - k_0) + \frac{1}{2}\lambda(k - k_0)^2,$$

where $\lambda = (d^2\omega/dk^2)|_{k_0}$.

7. (a) As we found in section 5.6.2, for a TE or TM mode, the relationship between the wave number and frequency is given by $k = \sqrt{(\omega/c)^2 - \gamma}$, where $\gamma > 0$ is a constant (see the text below eq. (5.192)). Find the phase velocity v_p and group velocity v_g of such TE or TM modes at frequency $\omega > c\sqrt{\gamma}$. Show that $v_p > c$, but $v_g < c$.
   (b) Find the phase velocity and group velocity of electromagnetic waves in a plasma with $\epsilon(\omega)/\epsilon_0$ given by eq. (6.63) for $\omega > \omega_P$. Show that $v_p > c$, but $v_g < c$.

8. A plane wave with electric field given by

$$\boldsymbol{E} = \mathrm{Re}\left[e^{-i\omega t}e^{ikx}E_0\hat{\boldsymbol{z}}\right]$$

is incident on an electron of charge $-e$ and mass m in otherwise empty space. As a consequence of its resulting motion, the electron will radiate. Show that, when averaged over a cycle of oscillation, the radiated power per unit solid angle is given by

$$\frac{dP}{d\Omega} = \frac{\mu_0}{32\pi^2 c}\frac{e^4|E_0|^2}{m^2}\sin^2\theta.$$

Note: This phenomenon is known as *Thompson scattering*. Since, when averaged over a cycle of oscillation, the power per unit area of the incident plane wave is $|E_0|^2/(2\mu_0 c)$, it follows that the differential cross section (see eq. (7.61)) for Thompson scattering is

$$\frac{d\sigma}{d\Omega} = \frac{dP/d\Omega}{|E_0|^2/(2\mu_0 c)} = \frac{\mu_0^2 e^4}{16\pi^2 m^2}\sin^2\theta = \frac{e^4}{(4\pi\epsilon_0)^2 m^2 c^4}\sin^2\theta\,.$$

9. (a) The half-space $z \geq 0$ is filled by a conducting medium with $\mu = \mu_0$ that is well described by the Drude model, eqs. (6.54)–(6.55). A plane wave of frequency $\epsilon_0\omega \ll \sigma$ is normally incident on the medium from the vacuum region, $z < 0$. Thus, the incident wave is of the form $\boldsymbol{A}_I = \boldsymbol{C}_I e^{-i\omega t} e^{ikz}$, the reflected wave is of the form $\boldsymbol{A}_R = \boldsymbol{C}_R e^{-i\omega t} e^{-ikz}$, and the transmitted wave is of the form $\boldsymbol{A}_T = \boldsymbol{C}_T e^{-i\omega t} e^{-(1-i)z/\delta}$ (see eq. (6.59)). Solve for $\boldsymbol{C}_R$ and $\boldsymbol{C}_T$ in terms of $\boldsymbol{C}_I$, ω, and σ. Show that the Poynting flux of the reflected wave is smaller than the Poynting flux of the ingoing wave.
   (b) Consider the same setup as in part (a), but with the conducting medium described by eq. (6.54) now replaced by a plasma described by eq. (6.63). A plane wave $\boldsymbol{A}_I = \boldsymbol{C}_I e^{-i\omega t} e^{ikz}$ with $\omega < \omega_P$ is incident from $z < 0$. Solve for the reflected and transmitted waves. Show that the Poynting flux of the reflected wave is equal to the Poynting flux of the incident wave.

10. Consider a nonideal MHD fluid of conductivity σ that fills all of space and has $\boldsymbol{v} = 0$, so that $\boldsymbol{B}$ satisfies eq. (6.85).
   (a) Use arguments parallel to those of section 5.5 to show that the general solution for $\boldsymbol{B}$ takes the form

$$\boldsymbol{B}(t, \boldsymbol{x}) = \int \boldsymbol{\mathcal{F}}(\boldsymbol{k}) \exp\left(-\frac{k^2}{\mu_0 \sigma} t\right) e^{i\boldsymbol{k}\cdot\boldsymbol{x}} d^3k + \text{c.c.},$$

   where "c.c" denotes the complex conjugate of the preceding term, and $\boldsymbol{\mathcal{F}}(\boldsymbol{k})$ satisfies $\boldsymbol{k} \cdot \boldsymbol{\mathcal{F}} = 0$. Thus, each $\boldsymbol{k}$-mode decays exponentially in time.
   (b) Show that for $t > 0$, $\boldsymbol{B}(t, \boldsymbol{x})$ is given in terms of its initial value $\boldsymbol{B}(0, \boldsymbol{x})$ by

$$\boldsymbol{B}(t, \boldsymbol{x}) = \left(\frac{\mu_0 \sigma}{4\pi t}\right)^{3/2} \int d^3x' \boldsymbol{B}(0, \boldsymbol{x}') \exp\left(-\frac{\mu_0 \sigma |\boldsymbol{x} - \boldsymbol{x}'|^2}{4t}\right).$$

11. Consider perturbations of an ideal MHD fluid, as described by eqs. (6.86)–(6.90).
   (a) Show that $\delta\boldsymbol{v}$ satisfies

$$\frac{\partial^2(\delta\boldsymbol{v})}{\partial t^2} = (c_s^2 + c_A^2)\nabla(\nabla \cdot \delta\boldsymbol{v}) + \frac{1}{\mu_0 \rho_{M0}} \boldsymbol{B}_0 \cdot \nabla\, [(\boldsymbol{B}_0 \cdot \nabla)\delta\boldsymbol{v} - \nabla(\boldsymbol{B}_0 \cdot \delta\boldsymbol{v}) - \boldsymbol{B}_0 \nabla \cdot \delta\boldsymbol{v}]\,.$$

   where c_s and c_A are given by eq. (6.91) and eq. (6.96), respectively.
   (b) Consider solutions to this equation of the form $\delta\boldsymbol{v} = \boldsymbol{V} e^{-i\omega t + i\boldsymbol{k}\cdot\boldsymbol{x}}$ with $\boldsymbol{V}$ constant. (i) For $\boldsymbol{k}$ perpendicular to $\boldsymbol{B}_0$, show that this equation has a solution with $\boldsymbol{V}$ parallel to $\boldsymbol{k}$ and $\omega^2 = (c_s^2 + c_A^2)k^2$. This is an example of a "fast" magnetoacoustic wave. (ii) For $\boldsymbol{k}$ parallel to $\boldsymbol{B}_0$, show that—in addition to the Alfvén wave solution found in the text—this equation has a solution with $\boldsymbol{V}$ parallel to $\boldsymbol{B}_0$ (and $\boldsymbol{k}$) and $\omega^2 = c_s^2 k^2$. This is an example of a "slow" magnetoacoustic wave.

CHAPTER 7

Geometric Optics, Interference, and Diffraction

In this chapter, we discuss the geometric optics approximation, interference phenomena, and diffraction. As we shall see in section 7.1, geometric optics gives a very simple description of waves that are much like plane waves but have an amplitude that varies slowly over space compared with variations in phase. Propagation of electromagnetic waves in this approximation can be described in terms of "light rays." Section 7.2 shows that interference phenomena arise when one superposes such plane-wave-like solutions. When these individual waves are composed of different frequencies, notions of coherence must be introduced to analyze interference phenomena. The term "diffraction" is used to describe any phenomena where the geometric optics description is wholly inadequate. In section 7.3, we discuss scattering by a dielectric ball and the propagation of a wave through an aperture as examples of diffraction.

7.1 Geometric Optics

Plane wave solutions to the source-free Maxwell's equations were considered in section 5.5. These solutions have the form

$$\boldsymbol{A}(t,\boldsymbol{x}) = \boldsymbol{C}e^{-i\omega t}e^{i\boldsymbol{k}\cdot\boldsymbol{x}} \tag{7.1}$$

(see eq. (5.161)). The amplitude $\boldsymbol{C}$ of a plane wave is exactly constant over spacetime, whereas the phase $-i\omega t + i\boldsymbol{k}\cdot\boldsymbol{x}$ is variable. It is of interest to consider approximate solutions for which the amplitude is variable, but the variation of the amplitude is much slower than the variation of the phase. As we shall see, such solutions can be given a very simple description in terms of the behavior of rays perpendicular to the surfaces of constant phase.

As usual, we first consider the wave equation

$$\Box\psi = 0 \tag{7.2}$$

and then consider the (minor) modifications involved in treating Maxwell's equations. Consider solutions of the wave equation with time dependence[1] $e^{-i\omega t}$,

[1] It is not necessary to make this restriction to solutions oscillating and a definite frequency. Instead, we could write ψ in the form $\mathscr{A}e^{i\mathscr{S}}$ with $\mathscr{A}$ and $\mathscr{S}$ now functions of $(t,\boldsymbol{x})$ and $\mathscr{S}$ assumed to be rapidly varying compared with $\mathscr{A}$ (see problem 1 at the end of this chapter).

$$\psi(t, \boldsymbol{x}) = \chi(\boldsymbol{x}) e^{-i\omega t}, \tag{7.3}$$

where it is understood that we should take the real part of ψ at the end of the calculation. Then χ satisfies the Helmholtz equation (see eq. (5.82))

$$\nabla^2 \chi + \frac{\omega^2}{c^2} \chi = 0 . \tag{7.4}$$

We seek approximate solutions to this equation of the form

$$\chi(\boldsymbol{x}) = \mathcal{A}(\boldsymbol{x}) e^{iS(\boldsymbol{x})}, \tag{7.5}$$

where the *phase* S and the *amplitude* $\mathcal{A}$ are real, and S is very rapidly varying relative to $\mathcal{A}$. Substituting eq. (7.5) into the Helmholtz equation (7.4), we obtain

$$\left[-|\nabla S|^2 \mathcal{A} + \nabla^2 \mathcal{A} + \frac{\omega^2}{c^2} \mathcal{A} + i(\nabla^2 S)\mathcal{A} + 2i \nabla S \cdot \nabla \mathcal{A} \right] e^{iS} = 0 . \tag{7.6}$$

This equation is exact. The approximation we shall now make consists—to leading order[2]—of neglecting the term $\nabla^2 \mathcal{A}$ in this equation, since this term should be small compared with $|\nabla S|^2 \mathcal{A}$ if S is much more rapidly varying than $\mathcal{A}$. Dropping $\nabla^2 \mathcal{A}$ and then setting the real and imaginary parts of the expression in square brackets in eq. (7.6) to zero, we obtain

$$|\nabla S|^2 = \frac{\omega^2}{c^2}, \tag{7.7}$$

and

$$2 \nabla S \cdot \nabla \mathcal{A} + (\nabla^2 S) \mathcal{A} = 0 . \tag{7.8}$$

The approximation to solutions of eq. (7.4) obtained by taking χ to be of the form eq. (7.5) with S and $\mathcal{A}$ satisfying eq. (7.7) and eq. (7.8) is known as the *geometric optics* or *Wentzel-Kramers-Brillouin (WKB)* approximation.

It is useful to write

$$\boldsymbol{k} \equiv \nabla S . \tag{7.9}$$

Then $\boldsymbol{k}$ is normal to the surfaces of constant phase S, and we have

$$|\boldsymbol{k}|^2 = \frac{\omega^2}{c^2} . \tag{7.10}$$

Note that plane waves correspond to the (exact) solution $S = \boldsymbol{k} \cdot \boldsymbol{x}$ with $\boldsymbol{k}$ and $\mathcal{A}$ constant. However, in a general WKB (approximate) solution, the surfaces of constant S need not be planar, and the direction of $\boldsymbol{k}$ depends on $\boldsymbol{x}$, although its magnitude is constant by eq. (7.10).

The *integral curves* of the vector field $\boldsymbol{k}$ are defined to be curves $x_i(\alpha)$ whose tangent everywhere coincides with $\boldsymbol{k}$:

$$\frac{dx_i}{d\alpha} = k_i, \tag{7.11}$$

[2] See problem 2 for higher-order corrections.

where α denotes the curve parameter. The integral curves of $\boldsymbol{k}$ are referred to as *rays*—or as *light rays* in the electromagnetic case discussed below. By construction, these rays are everywhere orthogonal to the surfaces of constant S. An immediate consequence of eq. (7.7) and the definition of $\boldsymbol{k}$ is

$$[(\boldsymbol{k}\cdot\nabla)\boldsymbol{k}]_i=\sum_j k_j\partial_j k_i=\sum_j k_j\partial_j\partial_i S=\sum_j k_j\partial_i\partial_j S=\sum_j k_j\partial_i k_j=\frac{1}{2}\partial_i|\boldsymbol{k}|^2=0\,. \quad (7.12)$$

Thus, the rays satisfy

$$0=\left(\frac{d\boldsymbol{x}}{d\alpha}\cdot\nabla\right)\frac{dx_i}{d\alpha}=\frac{d^2x_i}{d\alpha^2}\,. \quad (7.13)$$

The solutions to this equation are straight lines. Thus, the rays (and, in the electromagnetic case, the light rays) are straight lines. Furthermore, note that eq. (7.8) can be written as

$$\nabla\cdot(\mathcal{A}^2\boldsymbol{k})=0\,. \quad (7.14)$$

Now, consider an ensemble of particles in a steady, space-filling flow, with number density proportional to $\mathcal{A}^2$ and with the particles moving in straight lines with velocity c in the direction $\boldsymbol{k}$. Then, eq. (7.14) corresponds to conservation of particles, $\nabla\cdot\boldsymbol{\mathcal{N}}=0$, where $\boldsymbol{\mathcal{N}}\propto\mathcal{A}^2\boldsymbol{k}$ denotes the particle number current. It follows that in the geometric optics approximation, the behavior of $|\psi|^2$—as well as the behavior of $(\mathrm{Re}[\psi])^2$ averaged over times $\gg 1/\omega$—is identical to the behavior of the number density of such an ensemble of particles. We shall see below that similar behavior holds for electromagnetic waves. This helps explain why, in the early history of the subject, the debate over whether light is described by waves or particles was not straightforward to resolve, since one needs to consider phenomena where the solution cannot be described by geometric optics to tell the difference.

It is worth pointing out that, as seen in problem 3, if the rays are converging, a caustic always will result, wherein $|\mathcal{A}|\to\infty$ in a finite time. The geometric optics approximation always breaks down in a neighborhood of a caustic, since the variations of $\mathcal{A}$ are not small compared with variations of S near the caustic. Thus, one must return to the exact wave equation (7.2) to properly describe a geometric optics solution in a neighborhood of a caustic.

The geometric optics approximation can also be applied to modifications of the wave equation (7.2). In particular, suppose ψ satisfies the equation

$$\nabla^2\psi-\frac{n^2(\boldsymbol{x})}{c^2}\frac{\partial^2\psi}{\partial t^2}=0, \quad (7.15)$$

which, as we will see below, corresponds closely to equations that arise in electromagnetism for a medium with a spatially variable index of refraction $n(\boldsymbol{x})$. Again we consider solutions with harmonic time dependence (eq. (7.3)) with χ of the form eq. (7.5). In this case, the geometric optics approximation yields

$$|\nabla S|^2=n^2(\boldsymbol{x})\frac{\omega^2}{c^2} \quad (7.16)$$

together with eq. (7.8). Again writing $\boldsymbol{k}=\nabla S$, we now find

$$(\boldsymbol{k}\cdot\nabla)\boldsymbol{k}=\frac{1}{2}\nabla|\boldsymbol{k}|^2=\frac{1}{2}\frac{\omega^2}{c^2}\nabla(n^2)=\frac{\omega^2}{c^2}n\nabla n\,. \tag{7.17}$$

Thus, the rays no longer move on straight lines but "bend" in proportion to ∇n. In more explicit form, the equation satisfied by rays in a medium with a variable index of refraction is

$$\frac{d^2\boldsymbol{x}}{d\alpha^2}=\frac{\omega^2}{c^2}n(\boldsymbol{x})\nabla n(\boldsymbol{x})\,. \tag{7.18}$$

In addition, by eq. (7.16), we have $|d\boldsymbol{x}/d\alpha|=n\omega/c$.

The paths traversed by rays satisfying eq. (7.18) are exactly the same paths that would be obtained by extremizing an action for particle motion given by

$$\mathcal{S}=\int n(\boldsymbol{x})\sqrt{\frac{d\boldsymbol{x}}{d\lambda}\cdot\frac{d\boldsymbol{x}}{d\lambda}}\,d\lambda, \tag{7.19}$$

that is, the solutions $\boldsymbol{x}(\lambda)$ to the Euler-Lagrange equations obtained from variation of this action agree (up to curve reparametrization) with the solutions to eq. (7.18). Now, since $ds=|d\boldsymbol{x}/d\lambda|d\lambda$ is the element of length along a curve, if we assign the velocity c/n to a particle moving along the curve, the action eq. (7.19) can be interpreted as being proportional to the elapsed time along the curve. In this sense, we can say that for the modified wave equation (7.15), the rays in the geometric optics approximation are extrema of elapsed time. This fact is known as *Fermat's principle*.

We now turn to the source-free Maxwell equations in vacuum. As discussed in section 5.5, we may work in a restricted Lorenz gauge such that $\phi=0$, in which case $\boldsymbol{A}$ satisfies

$$\Box\boldsymbol{A}=0\,,\qquad \nabla\cdot\boldsymbol{A}=0\,. \tag{7.20}$$

We seek approximate solutions of the form

$$\boldsymbol{A}(t,\boldsymbol{x})=\boldsymbol{C}(\boldsymbol{x})e^{iS(\boldsymbol{x})}e^{-i\omega t} \tag{7.21}$$

with S and $\boldsymbol{C}$ real and S rapidly varying compared to $\boldsymbol{C}$. Neglecting the term $\nabla^2\boldsymbol{C}$ in the analog of eq. (7.6) and again writing $\boldsymbol{k}=\nabla S$, we obtain

$$|\nabla S|^2=|\boldsymbol{k}|^2=\frac{\omega^2}{c^2}, \tag{7.22}$$

$$2(\boldsymbol{k}\cdot\nabla)\boldsymbol{C}+(\nabla\cdot\boldsymbol{k})\boldsymbol{C}=0, \tag{7.23}$$

$$\boldsymbol{k}\cdot\boldsymbol{C}=0\,, \tag{7.24}$$

where the first two relations arise from $\Box\boldsymbol{A}=0$, and the last condition arises from $\nabla\cdot\boldsymbol{A}=0$. We can again introduce the notion of rays—now called "light rays"—as integral curves of $\boldsymbol{k}$. By the same calculation as in eq. (7.12), we see that light rays in vacuum are straight lines. Equation (7.24) states that the direction of $\boldsymbol{C}$ is orthogonal to the light ray, and eq. (7.23) implies that the direction of $\boldsymbol{C}$ is constant along each light ray. Furthermore, dotting eq. (7.23) with $\boldsymbol{C}$, we find

$$\nabla\cdot(|\boldsymbol{C}|^2\boldsymbol{k})=0, \tag{7.25}$$

so the magnitude of $|\boldsymbol{C}|^2$ varies in the same manner as the number density of particles moving at constant speed along the rays.

The electric and magnetic fields are given by

$$\boldsymbol{E} = -\frac{\partial \boldsymbol{A}}{\partial t} = i\omega \boldsymbol{C}(\boldsymbol{x}) e^{iS(\boldsymbol{x})} e^{-i\omega t}, \tag{7.26}$$

$$\boldsymbol{B} = \nabla \times \boldsymbol{A} = i\boldsymbol{k} \times \boldsymbol{C}(\boldsymbol{x}) e^{iS(\boldsymbol{x})} e^{-i\omega t}, \tag{7.27}$$

where we have neglected the term involving $\nabla \times \boldsymbol{C}$ in eq. (7.27), since derivatives of $\boldsymbol{C}$ are much smaller than derivatives of S. Again, one should take the real parts of eqs. (7.21), (7.26), and (7.27) to get the corresponding real electromagnetic fields. The Poynting vector $\frac{1}{\mu_0}\text{Re}[\boldsymbol{E}] \times \text{Re}[\boldsymbol{B}]$ points in the direction of $\boldsymbol{k}$ and its magnitude—when averaged over times $\gg 1/\omega$—is proportional to $|\boldsymbol{C}|^2$ (see eq. (7.39) below). Thus, in the geometric optics approximation, the energy flux of the electromagnetic field is the same as that of an ensemble of particles of number density proportional to $|\boldsymbol{C}|^2$ moving at uniform speed in the direction $\boldsymbol{k}$.

Now consider Maxwell's equations in a medium that is linear and isotropic, as in section 6.1, but is *not* homogeneous, so that[3]

$$\langle \boldsymbol{D}\rangle(t,\boldsymbol{x}) = \epsilon(\boldsymbol{x})\langle \boldsymbol{E}\rangle(t,\boldsymbol{x}), \qquad \langle \boldsymbol{B}\rangle(t,\boldsymbol{x}) = \mu(\boldsymbol{x})\langle \boldsymbol{H}\rangle(t,\boldsymbol{x}), \tag{7.28}$$

that is, ϵ and μ (and, hence $n = c\sqrt{\epsilon\mu}$) are now allowed to depend on $\boldsymbol{x}$. Then in the modified Lorenz gauge

$$\frac{n^2(\boldsymbol{x})}{c^2}\frac{\partial\langle\phi\rangle}{\partial t} + \nabla\cdot\langle \boldsymbol{A}\rangle = 0, \tag{7.29}$$

(see eq. (6.13)) Maxwell's equations with $\langle\rho_f\rangle = \langle \boldsymbol{J}_f\rangle = 0$ take the form

$$\left[\nabla^2 - \frac{n^2(\boldsymbol{x})}{c^2}\frac{\partial^2}{\partial t^2}\right]\langle\phi\rangle = \text{lower derivative terms}, \tag{7.30}$$

$$\left[\nabla^2 - \frac{n^2(\boldsymbol{x})}{c^2}\frac{\partial^2}{\partial t^2}\right]\langle \boldsymbol{A}\rangle = \text{lower derivative terms}, \tag{7.31}$$

where the "lower derivative terms" involve spatial derivatives of ϵ or μ and no more than one space or time derivative of $\langle\phi\rangle$ or $\langle \boldsymbol{A}\rangle$. We seek solutions of the form

$$\langle\phi\rangle = \Phi(\boldsymbol{x}) e^{iS(\boldsymbol{x})} e^{-i\omega t}, \qquad \langle \boldsymbol{A}\rangle = \boldsymbol{C}(\boldsymbol{x}) e^{iS(\boldsymbol{x})} e^{-i\omega t}, \tag{7.32}$$

where S varies rapidly compared not only with Φ and $\boldsymbol{C}$ but also compared with ϵ and μ. The term involving $|\nabla S|^2$ is therefore much larger than the contributions arising from the "lower derivative terms" in eqs. (7.30) and (7.31), so in parallel with eq. (7.16), we find

$$|\nabla S|^2 = n^2(\boldsymbol{x})\frac{\omega^2}{c^2}. \tag{7.33}$$

[3] Our analysis would also apply to an inhomogeneous medium of the type considered in section 6.2 with a frequency-dependent ϵ and μ, provided we restrict consideration to frequencies at which ϵ and μ are real.

It follows immediately from our previous analysis that light rays in a medium with a variable index of refraction satisfy eq. (7.17) and thereby also obey Fermat's principle (i.e., they are the curves that extremize the action eq. (7.19)).

Finally, note that the ray tracing procedures used in elementary optics follow from the geometric optics approximation.

7.2 Interference and Coherence

The geometric optics approximation discussed in section 7.1 gives a good description of a wide range of phenomena involving the propagation of electromagnetic radiation. However, for many phenomena, the geometric optics approximation is not adequate. Such phenomena are usually referred to as "interference" and "diffraction." I use these terms as follows: By *interference*, I mean phenomena that cannot be described by a single geometric optics solution but can be described by a sum of geometric optics solutions. By *diffraction* (to be discussed in section 7.3), I mean phenomena for which the geometric optics approximation description is entirely inadequate. For example, the geometric optics approximation cannot describe the scattering of electromagnetic radiation by a dielectric body of size comparable to the wavelength of the radiation, so such scattering would be referred to as "diffraction." Non-geometric-optics phenomena arising from the propagation of electromagnetic radiation through an aperture would also be considered to be diffraction. However, it is not always possible to draw a sharp distinction between interference and diffraction. For example, non-geometric-optics phenomena arising from emission from an extended body could be viewed as diffraction or as interference arising from the emission from different parts of the body.

Before considering interference, it is important to recognize what aspects of electromagnetic radiation can be observed. The oscillation frequency of radio waves is low enough that a radio antenna can respond directly to the individual cycles of oscillation of the wave. Thus, the response of a radio antenna to the electromagnetic waves emitted by a radio station could be viewed as a direct measurement of the amplitude of the electromagnetic field of the radiation. However, at much higher frequencies—such as visible light—the oscillations are too rapid for the individual cycles of oscillation to be detected. What typically can be observed is the Poynting flux and/or other quantities quadratic in the electromagnetic field averaged over a time large compared with the inverse frequency of the wave. We will refer to the magnitude of the time average of the Poynting flux as the *intensity* of the radiation:

$$I(t) = \frac{1}{\mu_0} |\overline{\boldsymbol{E} \times \boldsymbol{B}}|(t), \tag{7.34}$$

where the overbar denotes the time average over many cycles of oscillation, that is,

$$|\overline{\boldsymbol{E} \times \boldsymbol{B}}|(t) \equiv \left| \frac{1}{2T} \int_{t-T}^{t+T} \boldsymbol{E}(t') \times \boldsymbol{B}(t') dt' \right|, \tag{7.35}$$

where $T \gg 1/\omega$. Note that the averaging is done before the magnitude of the vector is taken. Let us now focus our attention on the properties of the intensity.

For an electromagnetic wave of the geometric optics form

$$\boldsymbol{A}(t, \boldsymbol{x}) = \boldsymbol{C}(\boldsymbol{x}) e^{iS(\boldsymbol{x})} e^{-i\omega t} \tag{7.36}$$

(see eq. (7.21)), the intensity is

$$I = \frac{1}{\mu_0}\left|\overline{\mathrm{Re}[\boldsymbol{E}] \times \mathrm{Re}[\boldsymbol{B}]}\right|, \tag{7.37}$$

where $\boldsymbol{E}$ and $\boldsymbol{B}$ are given by eqs. (7.26) and (7.27). We substitute

$$\mathrm{Re}[\boldsymbol{E}] = \frac{1}{2}(\boldsymbol{E}+\boldsymbol{E}^*), \qquad \mathrm{Re}[\boldsymbol{B}] = \frac{1}{2}(\boldsymbol{B}+\boldsymbol{B}^*), \tag{7.38}$$

where $*$ denotes complex conjugation. It is easily seen that the $\boldsymbol{E}\times\boldsymbol{B}$ and $\boldsymbol{E}^*\times\boldsymbol{B}^*$ terms oscillate as $e^{-2i\omega t}$ and $e^{+2i\omega t}$, respectively, and so their time average over a time $T \gg 1/\omega$ is negligibly small. Thus, we obtain

$$I = \frac{1}{4\mu_0}\left|\overline{\boldsymbol{E}^*\times\boldsymbol{B}+\boldsymbol{E}\times\boldsymbol{B}^*}\right| = \frac{1}{4\mu_0}\left|\boldsymbol{E}^*\times\boldsymbol{B}+\boldsymbol{E}\times\boldsymbol{B}^*\right| = \frac{\omega|\boldsymbol{k}|}{2\mu_0}|\boldsymbol{C}|^2 = \frac{\omega^2}{2\mu_0 c}|\boldsymbol{C}|^2, \tag{7.39}$$

where the overline was dropped in the second equality, since the quantity being averaged is manifestly time independent. Thus, the intensity of a geometric optics solution depends only on the magnitude of the amplitude $\boldsymbol{C}$ of the wave—not its phase—and thus varies slowly in space.

Suppose now that we have two sources, each of which emits radiation at exactly frequency ω. Further suppose that the emission from each source can be well described by a geometric optics solution. Then the full solution is

$$\boldsymbol{A}(t,\boldsymbol{x}) = \left[\boldsymbol{C}_1(\boldsymbol{x})e^{iS_1(\boldsymbol{x})} + \boldsymbol{C}_2(\boldsymbol{x})e^{iS_2(\boldsymbol{x})}\right]e^{-i\omega t}. \tag{7.40}$$

By a similar calculation to eq. (7.39), the intensity is

$$\begin{aligned} I &= \frac{1}{4\mu_0}\left|(\boldsymbol{E}_1^*+\boldsymbol{E}_2^*)\times(\boldsymbol{B}_1+\boldsymbol{B}_2)+(\boldsymbol{E}_1+\boldsymbol{E}_2)\times(\boldsymbol{B}_1^*+\boldsymbol{B}_2^*)\right| \\ &= \left|\frac{\omega}{2\mu_0}\boldsymbol{k}_1|\boldsymbol{C}_1|^2+\frac{\omega}{2\mu_0}\boldsymbol{k}_2|\boldsymbol{C}_2|^2+\boldsymbol{I}_{\text{int}}\right| \end{aligned} \tag{7.41}$$

where the "interference term," $\boldsymbol{I}_{\text{int}}$, is given by

$$\begin{aligned} \boldsymbol{I}_{\text{int}} &= \frac{1}{4\mu_0}\left[\boldsymbol{E}_1^*\times\boldsymbol{B}_2+\boldsymbol{E}_2^*\times\boldsymbol{B}_1+\text{c.c.}\right] \\ &= \frac{\omega}{4\mu_0}\left[e^{-i[S_1(\boldsymbol{x})-S_2(\boldsymbol{x})]}\boldsymbol{C}_1\times(\boldsymbol{k}_2\times\boldsymbol{C}_2)+1\leftrightarrow 2+\text{c.c.}\right] \\ &= \frac{\omega}{2\mu_0}\left[\boldsymbol{C}_1\times(\boldsymbol{k}_2\times\boldsymbol{C}_2)\cos[S_1(\boldsymbol{x})-S_2(\boldsymbol{x})]+1\leftrightarrow 2\right], \end{aligned} \tag{7.42}$$

where "c.c." again denotes the complex conjugate of the preceding terms, and "$1\leftrightarrow 2$" denotes the same quantity as the previous term with 1 and 2 interchanged.

The nature of the interference term and the resulting interference pattern is best illustrated in the case where the individual geometric optics solutions are plane waves, so

$\mathcal{C}_1, \mathcal{C}_2, \boldsymbol{k}_1$, and $\boldsymbol{k}_2$ are constant, and $S_1 = \boldsymbol{k}_1 \cdot \boldsymbol{x} + \varphi_1$, $S_2 = \boldsymbol{k}_2 \cdot \boldsymbol{x} + \varphi_2$, where the phases[4] φ_1 and φ_2 are constant in space. We further restrict our discussion to the case where $\boldsymbol{k}_1$ and $\boldsymbol{k}_2$ are nearly parallel: $|\boldsymbol{k}_1 - \boldsymbol{k}_2| \ll |\boldsymbol{k}_1| = |\boldsymbol{k}_2| = \omega/c$. In that case, we have $\boldsymbol{k}_1 \cdot \mathcal{C}_2 \approx 0$ and $\boldsymbol{k}_2 \cdot \mathcal{C}_1 \approx 0$, so

$$\boldsymbol{I}_{\text{int}} = \frac{\omega}{2\mu_0} \mathcal{C}_1 \cdot \mathcal{C}_2 (\boldsymbol{k}_2 + \boldsymbol{k}_1) \cos[(\boldsymbol{k}_1 - \boldsymbol{k}_2) \cdot \boldsymbol{x} + \varphi_1 - \varphi_2] . \tag{7.43}$$

Note that the interference term vanishes when the polarizations of the two solutions are orthogonal (i.e., orthogonally polarized waves do not interfere). Neglecting all terms quadratic in $|\boldsymbol{k}_1 - \boldsymbol{k}_2|$, we obtain

$$I = \frac{\omega^2}{2\mu_0 c} \left[|\mathcal{C}_1|^2 + |\mathcal{C}_2|^2 + 2\mathcal{C}_1 \cdot \mathcal{C}_2 \cos[(\boldsymbol{k}_1 - \boldsymbol{k}_2) \cdot \boldsymbol{x} + \varphi_1 - \varphi_2] \right] . \tag{7.44}$$

The first two terms are just the intensities, I_1 and I_2, of the individual solutions, which are constant in space for plane waves. The interference term, however, is highly variable in space. The interference effect is largest when $\mathcal{C}_1 \approx \mathcal{C}_2 \equiv \mathcal{C}$, in which case, the intensity is given by

$$I = \frac{\omega^2}{\mu_0 c} |\mathcal{C}|^2 \left(1 + \cos[(\boldsymbol{k}_1 - \boldsymbol{k}_2) \cdot \boldsymbol{x} + \varphi_1 - \varphi_2]\right) . \tag{7.45}$$

Thus, in this case, the intensity oscillates in space between zero and twice the sum of the intensities of the individual plane waves. The interference effects are qualitatively similar (although not as pronounced) if the amplitudes are unequal and/or the waves are not nearly parallel. Qualitatively similar results also hold for the sum of any two geometric optics solutions at frequency ω.

It is worth noting that if a geometric optics solution arises from emission from a localized source, the surfaces of constant phase should become very nearly spherical at large distances from the source.[5] Thus, $\boldsymbol{k} = \boldsymbol{\nabla} S$ will point in the radial direction relative to the source. It follows that if one is observing two distant sources emitting at frequency ω, then $c(\boldsymbol{k}_1 - \boldsymbol{k}_2)/\omega$ gives the difference in the unit vectors pointing to the sources. By eq. (7.45), one can determine this quantity by measuring the spatial variation of the intensity pattern. Thus, the interference pattern produced by two distant sources oscillating at the same frequency ω can be used to determine the angular separation of the sources. Much more generally, the interference pattern resulting from emission by an extended body can be used to deduce properties of the body.

The above discussion applies to the case where the solutions are oscillating with exactly frequency ω. It would, of course, be much more realistic to consider the case where each of the solutions is a superposition of geometric optics solutions of different frequencies, that is, where each individual solution is a wave packet the form

$$\boldsymbol{A}(t, \boldsymbol{x}) = \int_0^\infty d\omega \, \mathcal{C}(\omega, \boldsymbol{x}) e^{iS(\omega, \boldsymbol{x})} e^{-i\omega t} , \tag{7.46}$$

---

[4]Since we take $\mathcal{C}_1$ and $\mathcal{C}_2$ to be real, we must allow such phases to obtain general (linearly polarized) plane waves.

[5]This follows because the shear decreases along expanding rays (see problem 3).

where, as usual, it is understood that the actual electromagnetic field is the real part of $\boldsymbol{A}$. We will assume that this superposition is peaked around some central frequency value $\omega_0 > 0$ with a frequency spread $\Delta\omega$ (i.e, $\boldsymbol{C}(\omega, \boldsymbol{x})$ is nearly zero except for frequencies within $\Delta\omega$ of ω_0). We require that $\Delta\omega < \omega_0$, but the case of most interest is $\Delta\omega \ll \omega_0$. The restriction that $\Delta\omega < \omega_0$ excludes frequencies in a neighborhood of zero but is otherwise general, since the negative frequencies are recovered when one takes the real part. We assume that the wave packet eq. (7.46) is such that for any fixed $\boldsymbol{x}$, the direction of $\boldsymbol{k}(\omega) = \boldsymbol{\nabla} S(\omega, \boldsymbol{x})$ is essentially constant in ω. (Of course, since $|\boldsymbol{k}| = \omega/c$, the magnitude of $\boldsymbol{k}$ varies with ω, and thus $S(\omega, \boldsymbol{x})$ necessarily varies significantly with ω at a fixed $\boldsymbol{x}$.) We further assume that the direction of $\boldsymbol{C}(\omega, \boldsymbol{x})$ is essentially constant[6] at fixed $\boldsymbol{x}$. For the remainder of this section, our aim is to analyze how a finite spread of frequencies for such wave packets affects interference phenomena.

Note that three important timescales are associated with the wave packet eq. (7.46). The first is the oscillation timescale $t_0 \equiv 1/\omega_0$. The second is $t_c \equiv 1/\Delta\omega$. For reasons that will be explained below, we refer to t_c as the *coherence time*. Since we have assumed that $\Delta\omega < \omega_0$, we have $t_c > t_0$ (and if $\Delta\omega \ll \omega_0$, then $t_c \gg t_0$). The third timescale, t_d, is the duration of the wave packet eq. (7.46), that is, the time span over which this wave packet is nonvanishing at any fixed $\boldsymbol{x}$. By the same mathematical argument as used to prove the uncertainty relations in quantum mechanics, we necessarily have $t_d \Delta\omega \gtrsim 1$ (i.e., $t_d \gtrsim t_c$). We will assume that $t_d \gg t_c$ (i.e., that the duration of the wave is very long compared to the minimum duration for its frequency range). This would naturally be the case if the wave arises from a source (e.g., a light bulb or a star) that consists of emission from many similar, independent events (e.g., emission from individual atoms) that occur over a very extended period of time.

To simplify our notation, instead of considering the wave packet (7.46), let us instead analyze the properties of a function $F(t)$ of the form

$$F(t) = \mathrm{Re} \int_0^\infty d\omega e^{-i\omega t} H(\omega), \tag{7.47}$$

where $H(\omega)$ is nonvanishing only for frequencies within $\Delta\omega$ of $\omega_0 > 0$, where $\Delta\omega < \omega_0$ (and thus, in particular, $H(\omega) = 0$ for $\omega < 0$). This corresponds to the behavior of the real part of the wave packet (7.46) at a fixed point of space $\boldsymbol{x}$ in the direction of $\boldsymbol{C}$. We will then consider the superposition of two functions of the form eq. (7.47). It is straightforward to translate the results we obtain for superpositions of functions F of the form (7.47) into corresponding results for superpositions of wave packets of the form (7.46). Note that the Fourier transform of F is given by

$$\hat{F}(\omega) = \frac{\sqrt{2\pi}}{2} [H(\omega) + H^*(-\omega)]. \tag{7.48}$$

The reason for calling $t_c = 1/\Delta\omega$ the "coherence time" above can be understood from the following analysis. The *time correlation function* $\xi(\tau)$ of F is defined by

[6]Such a wave packet is said to be *polarized*. If $\boldsymbol{C}$ varies in direction over the frequency spread $\Delta\omega$, the wave packet is said to be *partially polarized* (or, in the extreme case, *unpolarized*). The degree of polarization of the wave packet and the nature of its polarization can be usefully characterized via Stokes parameters (see problem 7).

$$\xi(\tau) = \frac{\int_{-\infty}^{\infty} F(t+\tau)F(t)dt}{\int_{-\infty}^{\infty} F^2(t)dt} . \tag{7.49}$$

Note that we have $\xi(0) = 1$, and by the Schwartz inequality, we have $|\xi(\tau)| \leq 1$ for all τ. If $|\xi(\tau)|$ is close to 1, then the values of F at time separation τ are closely correlated, whereas if $\xi(\tau) \approx 0$, then there is essentially no correlation. Note that if we define $G(t) = F(t+\tau)$, we have

$$\begin{aligned}\hat{G}(\omega) &= \frac{1}{\sqrt{2\pi}} \int_{-\infty}^{\infty} e^{i\omega t} G(t)dt = \frac{1}{\sqrt{2\pi}} \int_{-\infty}^{\infty} e^{i\omega t} F(t+\tau)dt \\ &= \frac{1}{\sqrt{2\pi}} \int_{-\infty}^{\infty} e^{i\omega(t'-\tau)} F(t')dt' = e^{-i\omega\tau}\hat{F}(\omega),\end{aligned} \tag{7.50}$$

where we made the variable substitution $t' = t + \tau$ in the second line. Using the Plancherel theorem[7] and the reality of F, we have

$$\begin{aligned}\int_{-\infty}^{\infty} F(t+\tau)F(t)dt &= \int_{-\infty}^{\infty} G(t)F(t)dt = \int_{-\infty}^{\infty} \hat{G}(\omega)\hat{F}^*(\omega)d\omega \\ &= \int_{-\infty}^{\infty} e^{-i\omega\tau}\hat{F}(\omega)\hat{F}^*(\omega)d\omega \\ &= \frac{\pi}{2}\int_{-\infty}^{\infty} e^{-i\omega\tau}[H(\omega)+H^*(-\omega)][H^*(\omega)+H(-\omega)]d\omega \\ &= \frac{\pi}{2}\int_{-\infty}^{\infty} e^{-i\omega\tau}\left[|H(\omega)|^2+|H(-\omega)|^2\right]d\omega \\ &= \pi \,\mathrm{Re}\int_{-\infty}^{\infty} e^{-i\omega\tau}|H(\omega)|^2 d\omega,\end{aligned} \tag{7.51}$$

where, in the fourth line, we used $H(\omega)H^*(-\omega) = 0$, since $H(\omega) = 0$ for $\omega < 0$. Thus, we obtain

$$\xi(\tau) = \frac{\mathrm{Re}\int_{-\infty}^{\infty} e^{-i\omega\tau}|H(\omega)|^2 d\omega}{\int_{-\infty}^{\infty} |H(\omega)|^2 d\omega} . \tag{7.52}$$

The quantity $|H(\omega)|^2$ is referred to as the *power spectrum* of F. Thus, we see that the time correlation function $\xi(\tau)$ is directly related to the Fourier transform of the power spectrum.

If $\tau \ll 1/\Delta\omega$, then

$$\int_{-\infty}^{\infty} e^{-i\omega\tau}|H(\omega)|^2 d\omega = e^{-i\omega_0\tau}\int_{-\infty}^{\infty} e^{-i(\omega-\omega_0)\tau}|H(\omega)|^2 d\omega \approx e^{-i\omega_0\tau}\int_{-\infty}^{\infty} |H(\omega)|^2 d\omega . \tag{7.53}$$

[7] The Plancherel theorem states that for any square integrable functions f and g, we have $\int_{-\infty}^{\infty} f^*(t)g(t)dt = \int_{-\infty}^{\infty} \hat{f}^*(\omega)\hat{g}(\omega)d\omega$.

Thus, $\xi(\tau) \approx \mathrm{Re}(e^{-i\omega_0\tau})$, so $|\xi(\tau)| \sim 1$. This means that $F(t+\tau)$ is highly correlated with $F(t)$ for $\tau \ll 1/\Delta\omega$. Indeed, if one were observing F during a time interval of duration $T \ll 1/\Delta\omega$, it would be difficult to distinguish F from a function that is oscillating at exactly frequency ω_0.

To analyze the case when $\tau \gg 1/\Delta\omega$, we must make further assumptions about F. We assume that $F(t)$ is of long duration, $t_d \gg 1/\Delta\omega$, and consists of a sum of a very large number of similar (possibly overlapping) pulses of much shorter duration. For simplicity, we will take these individual pulses to have an identical time profile. Then $H(\omega)$ will be of the form

$$H(\omega) = \sum_n e^{-i\omega t_n} h(\omega), \tag{7.54}$$

where $h(\omega)$ is the frequency distribution for an individual pulse centered at $t = 0$, and t_n denotes the central time of the nth pulse. Thus, at any fixed ω, $H(\omega)$ is a sum of a large number of phase factors. Since t_n ranges over a time interval of duration t_d, $H(\omega)$ will fluctuate significantly as ω is varied over scales $\delta\omega \sim 1/t_d \ll \Delta\omega$. If the pulse times t_n are randomly spaced, then the phase of $H(\omega)$ should vary randomly as ω is varied by more than $\delta\omega$. However, one would not expect large variations[8] in $|H(\omega)|^2$ when averaged over frequency scales large compared with $\delta\omega$ but small compared with $\Delta\omega$. Consequently, if $t_d \gg \tau \gg 1/\Delta\omega$, the numerator in eq.(7.52) would be expected to nearly vanish on account of the oscillations of $e^{-i\omega\tau}$ and, hence, $\xi(\tau) \approx 0$. So the coherence time $t_c = 1/\Delta\omega$ should give a good measure of the time interval over which the values of F remain correlated.

We can see the qualitative effects of a frequency superposition on interference phenomena by considering $F(t) = F_1(t) + F_2(t)$, with $F_1(t)$ and $F_2(t)$ each of the form eq. (7.47) with $H_1(\omega)$ and $H_2(\omega)$ each peaked around ω_0 with spread $\Delta\omega$. We first consider the integral of F^2 over all time:

$$\int_{-\infty}^{\infty} F^2(t)dt = \int_{-\infty}^{\infty} \left[F_1^2(t) + F_2^2(t) + 2F_1(t)F_2(t)\right] dt. \tag{7.55}$$

The last term on the right side of this equation corresponds to the interference term we would obtain for wave packets of the form (7.46), except that we are integrating over all time rather than averaging over a finite time interval. We can estimate its magnitude compared with the first two terms in a manner similar to the derivation of eq. (7.51). We have

$$\begin{aligned} \int_{-\infty}^{\infty} F_1(t)F_2(t)dt &= \int_{-\infty}^{\infty} \hat{F}_1(\omega)\hat{F}_2^*(\omega)d\omega \\ &= \frac{\pi}{2}\int_{-\infty}^{\infty} [H_1(\omega) + H_1^*(-\omega)][H_2^*(\omega) + H_2(-\omega)]d\omega \\ &= \pi\,\mathrm{Re}\int_{-\infty}^{\infty} H_1(\omega)H_2^*(\omega)d\omega. \end{aligned} \tag{7.56}$$

---

[8] By contrast, if the pulses were spaced exactly evenly rather than randomly over the time interval t_d, then $|H(\omega)|^2$ would exhibit a regular "spiky" behavior in ω. We would then have $|\xi(\tau)| \sim 1$ at times τ that are integer multiples of the time interval between pulses.

As discussed above, if F_1 and F_2 are wave packets of long duration, the phases of $H_1(\omega)$ and $H_2(\omega)$ should change rapidly with ω. If F_1 and F_2 are independent (e.g., if they are produced by independent sources), then the rapid phase oscillations of H_1 and H_2 with ω will be uncorrelated. Consequently, the integral in eq. (7.56) should very nearly vanish, and there will be no interference effects. However, if $F_1(t)$ and $F_2(t)$ are strongly correlated (e.g., if $F_1 = F_2$), then the interference term will be of comparable size to the other terms in eq. (7.55).

Now consider the time average of F^2 over a finite time interval $2T$:

$$\overline{F^2}(t) = \frac{1}{2T}\int_{t-T}^{t+T} F^2(t')dt' = \frac{1}{2T}\int_{t-T}^{t+T} \left[F_1^2(t') + F_2^2(t') + 2F_1(t')F_2(t')\right] dt', \quad (7.57)$$

with $t_0 \ll T \ll t_d$, where, as above, $t_0 = 1/\omega_0$, and t_d is the duration of the wave packet. This corresponds to the intensity that would be observed at a fixed $\boldsymbol{x}$ for the superposition of two wave packets of the form (7.46). Suppose F_1 and F_2 arise from independent sources. Then if $T \gg t_c = 1/\Delta\omega$, the situation will be similar to what we found in the previous paragraph for the case of integrating over all time, and the interference term will be negligible. Thus, no interference effects will be observed. For example, if one has emission from two light bulbs, the integration time T required by the human eye is much greater than the coherence time t_c of the radiation emitted by the light bulbs. The total observed intensity will therefore be given by the sum of the intensities of the emission from the individual light bulbs, and no interference effects will be observed.

In contrast, suppose F_1 and F_2 arise from independent sources, but we now take the averaging time to be small compared with the coherence time for each source: $T \ll t_c$. Since we require $T \gg t_0$, this will be possible only if $t_0 \ll t_c$ (i.e., $\Delta\omega \ll \omega_0$). As we have seen, $F_1(t')$ and $F_1(t'')$ will be strongly correlated for $t', t'' \in [t - T, t + T]$. Similarly, $F_2(t')$ and $F_2(t'')$ will be strongly correlated in this time interval. It follows that—even though they are independent—$F_1(t)$ and $F_2(t)$ will be strongly correlated over this time interval. Consequently, the interference term in eq. (7.57) will not be negligible, and interference effects will be important when the averaging time is smaller than the coherence time.

In the case where F_1 and F_2 are not independent, then important interference effects can occur even when $T \gg t_c$. An important instance of this occurs when one has a wave packet of the form eq. (7.46) that is split into two separate wave packets (e.g., by means of a beamsplitter device as used in interferometers). Suppose that the resulting wave packets are then recombined (i.e., suppose that they are directed back toward each other and superpose). At a fixed $\boldsymbol{x}$ in the region of recombination, the recombined wave packets will then have profiles of the form $F_1(t)$ and $F_2(t) = F_1(t + \tau)$, where $\tau = \Delta L/c$, and ΔL denotes the difference in path length for the paths followed by the two wave packets between splitting and recombination. If $\tau \ll t_c$, then $F_1(t)$ and $F_2(t)$ will be strongly correlated over all time, and even when $T \gg t_c$, one will observe a spatial interference pattern similar to that previously found for solutions oscillating exactly with frequency ω. However, if $\tau \gg t_c$, then $F_1(t)$ and $F_2(t)$ will be essentially independent, and no interference effects will be observed for $T \gg t_c$.

In summary, the basic effects of a frequency superposition, eq. (7.46), on interference phenomena are as follows. If one has two independent sources emitting at frequencies near ω_0, then the interference effects will be similar to that of sources emitting with frequency exactly equal to ω_0 if the observation/averaging time is smaller than the coherence time $t_c = 1/\Delta\omega$. However, for longer observation times, the interference pattern will wash out. Similarly, in an interferometer, if the path length difference is smaller than ct_c, then the beams will interfere much as they would if oscillating at exactly frequency ω_0, and one will have a stable interference pattern. However, if the path length difference is greater than ct_c, then the interference effects will wash out over times larger than the coherence time.

Finally, we previously noted in the paragraph below eq. (7.45) that in the case of emission at a single frequency ω_0 from two distant sources with small angular separation, the spatial interference pattern can be used to determine the angular separation of the sources. And more generally, for emission from an extended body, the interference pattern allows one to deduce properties of the body. If instead, we consider the physically realistic case of emission from two distant sources with a frequency spread $\Delta\omega$ about ω_0, then for times short compared with the coherence time t_c of the sources, there will be an interference pattern similar to what one would have if both sources emitted at exactly frequency ω_0. However, this interference pattern will shift in space on a time-scale of the order of t_c, and as just discussed, the interference pattern will therefore wash out at any fixed $\boldsymbol{x}$ over times longer than t_c. Nevertheless, the properties of the interference pattern, such as the spacing between the maxima, will remain fixed as the interference pattern shifts. Thus, if one can measure the intensity at different spatial locations $\boldsymbol{x}_1$ and $\boldsymbol{x}_2$ with averaging time $T \lesssim t_c$, then there will be persistent *correlations* between $I(\boldsymbol{x}_1)$ and $I(\boldsymbol{x}_2)$:

$$C(\boldsymbol{x}_1, \boldsymbol{x}_2) \equiv \langle I(\boldsymbol{x}_1) I(\boldsymbol{x}_2) \rangle - \langle I(\boldsymbol{x}_1) \rangle \langle I(\boldsymbol{x}_2) \rangle \neq 0, \tag{7.58}$$

where the angle brackets denote a time average over a long timescale. By seeing how $C(\boldsymbol{x}_1, \boldsymbol{x}_2)$ varies with the separation between $\boldsymbol{x}_1$ and $\boldsymbol{x}_2$, one can deduce properties of the interference pattern and, thereby, properties of the source. This technique was used by R. Hanbury Brown and R. Q. Twiss to measure the radius of the star Sirius A. This technique also underlies radio interferometry, although in that case, the radio dishes are sensitive to the amplitude of the wave, so one can directly correlate the amplitude rather than the intensity.

7.3 Diffraction

As previously mentioned, the term "diffraction" can be applied to describe any phenomenon where the geometric optics approximation is wholly inadequate. Thus, diffraction encompasses a very wide range of phenomena.

A good example of diffraction is the scattering of electromagnetic radiation by a body whose size is comparable to the wavelength of the radiation. In this section, we outline how to solve the problem of the scattering of a plane wave by a dielectric ball, where an exact solution can be obtained at all wavelengths.

Another important example of diffraction is the propagation of electromagnetic radiation through an aperture in a screen. The geometric optics approximation is manifestly inadequate to describe this if the wavelength of the radiation is on the order of or larger than the size of the aperture. But even if the wavelength is much smaller than the size of the aperture, there are aspects of the propagation that cannot be adequately described by geometric optics. In particular, in the geometric optics approximation, the propagation would be described by rays (see section 7.1). The rays that make it through the aperture would be unaffected by the screen, and the rays that hit the screen would be blocked, so geometric optics would always predict a sharp boundary at the shadow cast by the screen. The actual behavior is far more complex. Below, we show how to approximately solve the problem of propagation of radiation through an aperture in the case where the wavelength of radiation is much smaller than the aperture size. For simplicity, we restrict our analysis of propagation of radiation through an aperture to the case of a scalar field.

7.3.1 SCATTERING BY A DIELECTRIC BALL

An example of a scattering problem that can be solved exactly is that of scattering of a plane electromagnetic wave of frequency ω by a dielectric ball of radius R and (real, homogeneous) dielectric constant ϵ/ϵ_0. This problem was solved by independently by Lorenz (in 1890) and Mie (in 1908) and is usually referred to as *Mie scattering*. Our aim in this section is to explain how to obtain the solution, but we will not carry out the algebra needed to obtain explicit formulas for the solution. The somewhat simpler problem of scattering of a plane wave by a perfectly conducting sphere can be solved in a similar manner (see problem 9).

We assume that for $r > R$, the solution consists of the plane wave we have chosen to send in plus a scattered wave at frequency ω that is purely outgoing. This solution at $r > R$ has to be matched to a regular solution at frequency ω inside the dielectric ball $(r < R)$. The key strategy is to write the solution in both regions as a series expansion in vector spherical harmonics.

The first step is to expand the incoming plane wave in vector spherical harmonics. This can be achieved by means of the formula

$$(\hat{\boldsymbol{x}} \pm i\hat{\boldsymbol{y}})e^{ikz} = \sum_{\ell=1}^{\infty} i^{\ell-1}\sqrt{\frac{4\pi(2\ell+1)}{\ell(\ell+1)}}\left[j_\ell(kr)\boldsymbol{r} \times \nabla Y_{\ell,\pm 1} \pm \frac{1}{k}\nabla \times \left(j_\ell(kr)\boldsymbol{r} \times \nabla Y_{\ell,\pm 1}\right)\right]. \tag{7.59}$$

To derive this formula, we write the vector field $(\hat{\boldsymbol{x}} \pm i\hat{\boldsymbol{y}})e^{ikz}$ as a sum of vector spherical harmonics with unknown radial function coefficients, as in eq. (4.38). On the left side of this equation, we then substitute the expansion of e^{ikz} in ordinary spherical harmonics obtained in problem 8 of chapter 5. The coefficients appearing in the general vector spherical harmonic expansion appearing on the right side then can be extracted by taking the inner product over a sphere of both sides of this equation with an arbitrary vector spherical harmonic, using the orthonormality of the vector spherical harmonics. This yields the desired formula, eq. (7.59). Circularly polarized plane waves have $\phi = 0$ and $\boldsymbol{A}_{\text{plane}} = e^{-i\omega t}(\hat{\boldsymbol{x}} \pm i\hat{\boldsymbol{y}})e^{ikz}$, so eq. (7.59) immediately gives an expansion for the vector potential of a circularly polarized plane wave. The expansion of a linearly polarized plane wave can then be obtained writing it as a superposition of circularly polarized plane waves.

For the scattered wave at $r > R$, we may work in a gauge where $\phi_{\text{scat}} = 0$ and $\nabla \cdot \boldsymbol{A}_{\text{scat}} = 0$. A general outgoing wave solution satisfying these gauge conditions can be expanded in vector spherical harmonics as

$$\boldsymbol{A}_{\text{scat}} = e^{-i\omega t}\sum_{\ell,m}\left[a_{\ell m}h_\ell^{(1)}(kr)\boldsymbol{r} \times \nabla Y_{\ell m} + b_{\ell m}\nabla \times \left(h_\ell^{(1)}(kr)\boldsymbol{r} \times \nabla Y_{\ell m}\right)\right], \tag{7.60}$$

with $k=\omega/c$, where the radial function is taken to be $h_\ell^{(1)}(kr)$ on account of the outgoing wave condition. Thus, the determination of $\boldsymbol{A}_{\text{scat}}$ reduces to finding the unknown coefficients $a_{\ell m}$ and $b_{\ell m}$. For $r<R$, a similar expansion holds for $\boldsymbol{A}_{\text{int}}$ with $h_\ell^{(1)}$ replaced by j_ℓ and with $k=n\omega/c$, where $n=\sqrt{\epsilon/\epsilon_0}$ is the index of refraction.

To solve for the unknown coefficients in the expansions of $\boldsymbol{A}_{\text{scat}}$ and $\boldsymbol{A}_{\text{int}}$, we match $\boldsymbol{E}_\parallel$, $\boldsymbol{H}_\parallel$, $\hat{\boldsymbol{r}}\cdot\boldsymbol{D}$, and $\hat{\boldsymbol{r}}\cdot\boldsymbol{B}$ across the boundary at $r=R$. This matching must occur independently for each ℓ, m and vector spherical harmonic type. We thereby obtain a system of algebraic equations for each ℓ, m that can be solved for all the unknown expansion coefficients. As already mentioned, we shall not carry out the fairly complicated algebra needed to obtain the explicit formulas for the coefficients, $a_{\ell m}$ and $b_{\ell m}$, appearing in the expansion of the scattered wave, since our aim here is merely to explain the method by which the solution to this problem can be obtained. However, we note that $a_{\ell m}$ and $b_{\ell m}$ are nonvanishing only for $m=\pm1$—a direct consequence of the plane wave expansion containing contributions only from $m=\pm1$.

The solution found in this manner is valid at all frequencies ω. If $\omega R \ll c$ (i.e., if the wavelength of the radiation is much larger than the size of the dielectric ball) then the scattering is dominated by the $l=1$ electric parity contribution. In this regime, the scattering is referred to as *Rayleigh scattering*. At $O(1/|\boldsymbol{x}|)$, the solution for the electric and magnetic fields of the scattered wave is then given by eqs. (5.75) and (5.78), where $\boldsymbol{p}$ is the induced electric dipole moment of the dielectric ball (see problem 1 of chapter 3).

For a plane wave of frequency ω and polarization vector $\hat{C}$ incident on a body of finite size, the *differential cross section* of the body is defined by[9]

$$\frac{d\sigma}{d\Omega}(\omega,\hat{C}) \equiv \frac{(dP/d\Omega)_{\text{scat}}}{|\boldsymbol{S}_{\text{plane}}|}. \tag{7.61}$$

Here, the numerator is the power of the scattered wave per unit solid angle radiated to infinity (see eq. (5.80)), whereas the denominator is the magnitude of the incoming Poynting flux of the plane wave (i.e., it represents the incoming power per unit area). (Averaging of these quantities over a cycle of oscillation is understood.) For a linearly polarized plane wave propagating in the z-direction, the differential cross section of a dielectric ball in the Rayleigh scattering limit, $\omega R \ll c$, is given by

$$\frac{d\sigma}{d\Omega}(\omega,\hat{C}) = \frac{\omega^4 R^6}{c^4}\left(\frac{\epsilon-\epsilon_0}{\epsilon+2\epsilon_0}\right)^2 \sin^2\alpha, \tag{7.62}$$

where α is the angle between $\hat{C}$ and the direction of the observation point. Averaging over the polarization directions, we obtain

$$\left(\frac{d\sigma}{d\Omega}\right)_{\text{avg}} = \frac{\omega^4 R^6}{c^4}\left(\frac{\epsilon-\epsilon_0}{\epsilon+2\epsilon_0}\right)^2 \frac{1+\cos^2\theta}{2}. \tag{7.63}$$

Although this result has been obtained for a perfect dielectric ball, the scattering of electromagnetic waves by molecules and particulate matter at wavelengths large compared with the size of the molecules or particles will similarly be dominated by the

[9]One also may define a more refined notion of cross section by considering the radiated power for a given polarization of the scattered wave.

induced dipole moment, and the scattering cross section will have a similar ω^4 dependence on frequency. This frequency dependence explains why the scattering of sunlight in the Earth's atmosphere is much more efficient in the blue part of the spectrum than in the red.

7.3.2 PROPAGATION OF WAVES THROUGH AN APERTURE

We turn now to the diffraction effects associated with an aperture. The propagation of a wave through an aperture in a conducting screen can be analyzed using the same basic ideas as we used to obtain the results on the initial value formulation in section 5.4. For simplicity, we consider here only the case of a scalar field ψ satisfying the homogeneous wave equation eq. (5.128). We model the screen by imposing the boundary conditions $\psi = 0$ on the screen, which is an analog of boundary conditions for a perfect conductor in electromagnetism.

Suppose we have a planar screen with an aperture (which may be of arbitrary shape). We choose our coordinates so that the screen lies at $z = 0$. A given wave (which need not be a plane wave and need not be propagating normally to the screen) originating from $z < 0$ at early times is incident on the screen, as illustrated in figure 7.1. We wish to obtain $\psi(t, \boldsymbol{x})$ at points $\boldsymbol{x} = (x, y, z)$ with $z > 0$ (i.e., we wish to find how ψ propagates through the aperture).

We can get a useful formula for $\psi(t, \boldsymbol{x})$ by using the spacetime version of Green's identity as in section 5.4 but choosing the spacetime rectangle $\mathscr{R}$ of figure 5.2 differently. In a spacetime diagram, the screen (including the aperture) will trace out the 3-dimensional timelike hypersurface defined by $z = 0$. Let us take this hypersurface to be one of the vertical faces of the rectangle $\mathscr{R}$ of figure 5.2, with the interior of the rectangle at $z > 0$. Let us push the bottom face of the rectangle (shown in figure 5.2 as being at $ct = x^0 = 0$) toward $t \to -\infty$. Let us also push all of the vertical faces[10] toward $\pm\infty$ except for the one at $z = 0$.

As in section 5.4, we integrate the identity eq. (5.125) over the new rectangle $\mathscr{R}'$, again choosing $\psi_1 = \psi$ and $\psi_2(x^\mu) = G_{\text{adv}}(x^\mu, x'^\mu)$, with x'^μ lying in $\mathscr{R}'$. The properties of $G_{\text{adv}}(x^\mu, x'^\mu)$ imply that the boundary terms arising from integration of the left side of eq. (5.125) are nonvanishing only on the intersection of the past light cone of x'^μ with the boundary (i.e., the faces) of $\mathscr{R}'$. However, with the exception of the face of $\mathscr{R}'$ representing the screen at $z = 0$, the intersection of the past light cone of x'^μ with the faces of $\mathscr{R}'$ occurs at $z > 0$ at early times. Since ψ is assumed to vanish for $z > 0$ at early times (because the wave originates from $z < 0$ at early times), the boundary term contribution from all faces except the face at $z = 0$ vanishes. Thus, in complete analogy with the derivation of eq. (5.133), we find that at any $(t, \boldsymbol{x})$ with $z > 0$, we have[11]

$$\psi(t, \boldsymbol{x}) = \int_{z'=0} \left[\psi(t', \boldsymbol{x}')\partial_{z'} G_{\text{ret}}(t, \boldsymbol{x}; t', \boldsymbol{x}') - G_{\text{ret}}(t, \boldsymbol{x}; t', \boldsymbol{x}')\partial_{z'}\psi(t', \boldsymbol{x}')\right] dt' dx' dy', \tag{7.64}$$

where G_{ret} is given by eq. (5.50).

---

[10] Only 4 timelike faces of $\mathscr{R}$ are shown in figure 5.2 since one spatial dimension has been suppressed. There are, in fact, 6 timelike faces.

[11] Aside from the fact that we are now integrating over $z = 0$ rather than $t = 0$, eq. (7.64) differs from eq. (5.133) by a minus sign (arising from $\eta^{00} = -\eta^{11}$) and a factor of c (arising from integrating over t' rather than $x'^0 = ct'$). We have also changed notation from x^μ to $(t, \boldsymbol{x})$.

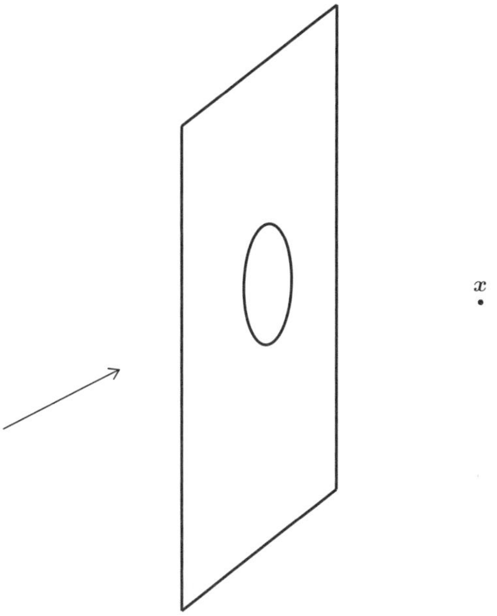

Figure 7.1. Radiation is incident on a planar screen with an aperture.

In the case of the initial value formulation, eq. (5.133) gave us precisely the result we were seeking, since it expressed the solution $\psi(t, \boldsymbol{x})$ in terms of the free initial data, $\psi(0, \boldsymbol{x})$ and $\partial_t \psi(0, \boldsymbol{x})$. However, in the present case, we are given an incoming wave at early times. Although the boundary conditions tell us that $\psi = 0$ on the screen, we do not a priori know ψ or $\partial_z \psi$ on the aperture, nor do we know $\partial_z \psi$ on the screen. However, if the wavelength of the radiation is much smaller than the size of the aperture, then over most of the aperture (i.e., more than a few wavelengths away from the edge of the aperture), the solution should be largely unaffected by the screen. Similarly, on the screen itself, we should have $\partial_z \psi \approx 0$ (i.e., the screen should block the radiation). Thus, for radiation of wavelength much smaller than the size of the aperture, it should be a good approximation to replace eq. (7.64) with

$$\psi(t, \boldsymbol{x}) = \int_{-\infty}^{\infty} dt' \int_{\mathscr{A}} dx' dy' \left[\psi_0(t', \boldsymbol{x}') \partial_{z'} G_{\text{ret}}(t, \boldsymbol{x}; t', \boldsymbol{x}') - G_{\text{ret}}(t, \boldsymbol{x}; t', \boldsymbol{x}') \partial_{z'} \psi_0(t', \boldsymbol{x}')\right]. \tag{7.65}$$

Here ψ_0 is the solution that would have been present if the screen were entirely absent, and the spatial integral is taken only over the aperture $\mathscr{A}$ (i.e., $z' = 0$, and x' and y' range only over the values corresponding to the aperture). Equation (7.65) is referred to as the *Kirchhoff approximation.*[12]

Now consider a solution oscillating with frequency ω:

$$\psi(t, \boldsymbol{x}) = \chi(\boldsymbol{x}) e^{-i\omega t}. \tag{7.66}$$

[12]The word "approximation" here is perhaps somewhat inappropriate, because this word is normally used in the context of a controlled scheme in which the size of the error terms could be estimated. The "Kirchhoff ansatz" might be better terminology.

Then the time integrals in eqs. (7.64) and (7.65) can be carried out, and they have the effect of converting the retarded Green's function G_{ret} to the Helmholtz Green's function G^{H} given by eq. (5.86). The Kirchhoff approximation eq. (7.65) then takes the explicit form

$$\chi(\boldsymbol{x}) = -\frac{1}{4\pi}\int_{\mathscr{A}} \frac{e^{i\omega|\boldsymbol{x}-\boldsymbol{x}'|/c}}{|\boldsymbol{x}-\boldsymbol{x}'|}\left[\partial_{z'}\chi_0(\boldsymbol{x}') + i\frac{\omega}{c}\frac{z}{|\boldsymbol{x}-\boldsymbol{x}'|}\chi_0(\boldsymbol{x}') - \frac{z}{|\boldsymbol{x}-\boldsymbol{x}'|^2}\chi_0(\boldsymbol{x}')\right]dx'dy', \tag{7.67}$$

where $\chi_0(\boldsymbol{x})e^{-i\omega t}$ is the incoming wave solution in the absence of the screen. The last term in brackets in the integrand normally can be dropped, since it is negligible if the observation point is many wavelengths from the aperture, $|\boldsymbol{x}-\boldsymbol{x}'| \gg c/\omega$ for all $\boldsymbol{x}'$ in $\mathscr{A}$, as will always be the case when $z \gg c/\omega$.

Now suppose that the incoming wave is a plane wave of wave vector $\boldsymbol{k}$ (i.e., suppose that $\chi_0 = Ce^{i\boldsymbol{k}\cdot\boldsymbol{x}}$ with $k = |\boldsymbol{k}| = \omega/c$). Suppose further that the observation point is much farther from the screen than the size of the aperture (i.e., $|\boldsymbol{x}| \gg L$, where L the maximum linear dimension of the aperture).[13] Then to leading order in $L/|\boldsymbol{x}|$, we obtain

$$\chi(\boldsymbol{x}) = -\frac{ik}{4\pi|\boldsymbol{x}|}(\cos\tilde{\theta} + \cos\theta)C\int_{\mathscr{A}} e^{ik|\boldsymbol{x}-\boldsymbol{x}'|}e^{i(k_x x' + k_y y')}dx'dy', \tag{7.68}$$

where $\cos\theta = z/|\boldsymbol{x}|$, and $\cos\tilde{\theta} = k_z/k$. Since we have

$$|\boldsymbol{x}-\boldsymbol{x}'| = |\boldsymbol{x}| - \hat{\boldsymbol{x}}\cdot\boldsymbol{x}' + \frac{1}{2|\boldsymbol{x}|}[|\boldsymbol{x}'|^2 - (\hat{\boldsymbol{x}}\cdot\boldsymbol{x}')^2] + \cdots, \tag{7.69}$$

we see that if[14]

$$\frac{kL^2}{|\boldsymbol{x}|} \ll 1, \tag{7.70}$$

then we can replace $e^{ik|\boldsymbol{x}-\boldsymbol{x}'|}$ by $e^{ik|\boldsymbol{x}|}e^{-ik\hat{\boldsymbol{x}}\cdot\boldsymbol{x}'}$ in the integrand of eq. (7.68). If this is the case, then eq. (7.68) reduces to

$$\chi(\boldsymbol{x}) = -\frac{ik}{4\pi|\boldsymbol{x}|}(\cos\tilde{\theta} + \cos\theta)Ce^{ik|\boldsymbol{x}|}\int_{\mathscr{A}} e^{-ik\hat{\boldsymbol{x}}\cdot\boldsymbol{x}'}e^{i(k_x x' + k_y y')}dx'dy'. \tag{7.71}$$

The diffractive phenomena described by eq. (7.71) are referred to as *Fraunhofer diffraction*. However, if eq. (7.70) does not hold, then we cannot make the approximation leading to eq. (7.71) and must use eq. (7.68). Diffractive phenomena in this regime are referred to as *Fresnel diffraction*.

An analysis of the propagation of electromagnetic waves through an aperture can be given in close parallel to the scalar case, but now we would have to impose appropriate conducting or other boundary conditions on the electromagnetic field at the screen. The results are very similar to the scalar case. However, we shall not discuss the details here.

---

[13]We assume here that the origin of the (x, y) coordinates has been chosen to lie in the aperture, so that $|\boldsymbol{x}|$ is the distance between the observation point and a point in the aperture.

[14]Note that the validity of the Kirchhoff approximation requires $kL \gg 1$, and we have assumed that $L/|\boldsymbol{x}| \ll 1$, but $kL^2/|\boldsymbol{x}|$ is not restricted by our assumptions.

Problems

1. Obtain a spacetime version of the geometric optics approximation for a scalar field[15] ψ as follows. As mentioned in footnote 1, instead of restricting the solutions ψ to oscillate with a definite frequency ω, we could write ψ in the form

$$\psi(t, \boldsymbol{x}) = \mathscr{A}(t, \boldsymbol{x}) e^{i\mathscr{S}(t, \boldsymbol{x})}.$$

   (a) Write the exact wave equation for ψ, eq. (7.2), in terms of $\mathscr{A}$ and $\mathscr{S}$. Then make the approximation that second derivatives of $\mathscr{A}$ can be neglected compared with squares of first derivatives of $\mathscr{S}$ to obtain the analogs of eq. (7.7) and eq. (7.8).
   (b) Define $k_0 = \frac{1}{c}\frac{\partial \mathscr{S}}{\partial t}$, and as before, define $\boldsymbol{k} = \nabla \mathscr{S}$. Show that $k_0^2 = |\boldsymbol{k}|^2$. As previously done in section 5.4 (and as will be fully elucidated in chapter 8), define $x^0 = ct$, and define ∂_μ by eq. (5.120). Define $k_\mu = (k_0, k_1, k_2, k_3)$, and define $k^\mu = \sum_\nu \eta^{\mu\nu} k_\nu$, where $\eta^{\mu\nu}$ is defined by eq. (5.121). Parallel to the derivation of eq. (7.12), show that

$$\sum_\nu k^\nu \partial_\nu k^\mu = 0\,.$$

   This result shows that in the geometric optics approximation, light rays move on null straight lines ("geodesics") in spacetime.

2. Systematically obtain higher-order corrections to the WKB approximation for solutions to the wave equation as follows. (Note that these corrections take account of the neglected term $\nabla^2 A$ in eq. (7.6) by leaving S unchanged but making complex corrections to $\mathcal{A}$.) Introduce a real parameter α, replace $\omega \to \omega/\alpha$, and write

$$\chi(\boldsymbol{x}) = \mathcal{A}(\boldsymbol{x}, \alpha) \exp(iS(\boldsymbol{x})/\alpha)$$

   with S real. Expand $\mathcal{A}$ as[16]

$$\mathcal{A}(\boldsymbol{x}, \alpha) = \sum_{n=0}^{\infty} \mathcal{A}^{(n)}(\boldsymbol{x}) \alpha^n$$

   with $\mathcal{A}^{(0)}$ real. Then substitute this expansion in eq. (7.6), and set the coefficient of each power of α to zero. Show that the coefficient of α^{-2} yields eq. (7.7), and the coefficient of α^{-1} yields eq. (7.8) for $\mathcal{A}^{(0)}$. Obtain the remaining equations for the additional corrections $\mathcal{A}^{(n)}$ for $n \geq 1$. Show that these corrections can be chosen to be real for even n and imaginary for odd n.

3. Consider a family of light rays in vacuum with tangents $\boldsymbol{k} = \nabla S$ satisfying eq. (7.10) and eq. (7.12). Define the *expansion* Θ and *shear* σ_{ij} of the family of light rays by

---

[15]This analysis also applies straightforwardly to Maxwell's equations with ψ replaced by A_μ and $\mathscr{A}$ replaced by a spacetime dual vector field $\mathscr{A}_\mu$, where the notation is that of chapter 8.

[16]This series provides an asymptotic expansion for solutions at large ω, but the series is not convergent in general.

$$\Theta = \nabla \cdot \boldsymbol{k}, \qquad \sigma_{ij} = \partial_i k_j - \frac{1}{2}\Theta[\delta_{ij} - n_i n_j],$$

where $n_i = k_i/(\omega/c)$ are the components of the unit vector in the $\boldsymbol{k}$ direction. Note that σ_{ij} is trace-free, orthogonal to $\boldsymbol{k}$, and also is symmetric, $\sigma_{ij} = \sigma_{ji}$, since $\partial_i k_j = \partial_i \partial_j S$.

(a) Show that Θ and σ satisfy

$$\boldsymbol{k} \cdot \nabla\Theta = -\frac{1}{2}\Theta^2 - \sum_{ij}(\sigma_{ij})^2, \qquad \boldsymbol{k} \cdot \nabla\sigma_{ij} = -\Theta\sigma_{ij}.$$

These may be viewed as evolution equations for the expansion and shear along the rays. Note that these equations imply that if $\Theta > 0$ at some point along a ray, then the expansion and the magnitude of the shear *decrease* as one moves along the ray. [Hint: To obtain the evolution equation for the shear, you will need to use the fact that σ_{ij} is effectively a 2×2 matrix.]

(b) Show that if the light rays are converging (i.e., $\Theta < 0$) at some point along a ray, then there will be a caustic (i.e., $\Theta \to -\infty$) at some finite distance along the ray. [Hint: Use the first of the equations of part (a) to obtain an inequality for the evolution of $1/\Theta$ along the ray.]

4. The half-space $z \geq 0$ is filled by a medium with index of refraction n. Consider a point $\boldsymbol{x}_1 = (x_1, y_1, z_1)$ in the vacuum region $z_1 < 0$ and a point $\boldsymbol{x}_2 = (x_2, y_2, z_2)$ in the medium ($z_2 > 0$). Find the path between $\boldsymbol{x}_1$ and $\boldsymbol{x}_2$ that minimizes the elapsed time in the sense of Fermat's principle. Show that the result agrees with Snell's law (see problem 1 of chapter 6).

5. An *optical fiber* is a cylindrical dielectric material that is used to transport light signals. The optical fiber is referred to as *step-index* if the index of refraction is nearly constant near the axis of the cylinder and then drops sharply, whereas it is referred to as *graded-index* if the index of refraction decreases gradually away from the axis. If the fiber diameter is sufficiently small compared to the inverse frequency of the light, the analysis of light propagation must be given in a manner similar to that of waveguides. (Such fibers are referred to as *single-mode.*) However, for a sufficiently large diameter fiber, the light propagation can be analyzed using the geometric optics approximation. (Such fibers are referred to as *multi-mode.*) Consider a graded-index, multi-mode optical fiber with $\mu = \mu_0$ and with dielectric constant ϵ/ϵ_0 varying for $x^2 + y^2 \leq R$ as

$$\epsilon(\boldsymbol{x})/\epsilon_0 = a - b(x^2 + y^2),$$

where a and b are positive constants, with $a \geq 1 + bR^2$. Write down and solve the ray propagation equation eq. (7.18). Show that rays that initially are sufficiently close to the axis and initially make a sufficiently small angle relative to the axis remain close to the axis forever.

6. In Young's interference experiment, a light source of frequency ω is located at $x = y = 0, z = -a < 0$. A screen is placed at $z = 0$, which is opaque but has

two "pinholes" (i.e., tiny apertures)[17] cut into it at $x = 0, y = \pm d/2$. Another screen is placed at $z = L > 0$, and the intensity of light on this screen at (x, y) is observed (see the sketch below). Treat the light emitted by the source in the geometric optics approximation, and assume that the phase S is spherically symmetric about the source. Treat the secondary waves (arising from diffraction by the pinholes) in the geometric optics approximation with phases S_1 and S_2 that are spherically symmetric about the pinholes and match the phase S at the pinholes.

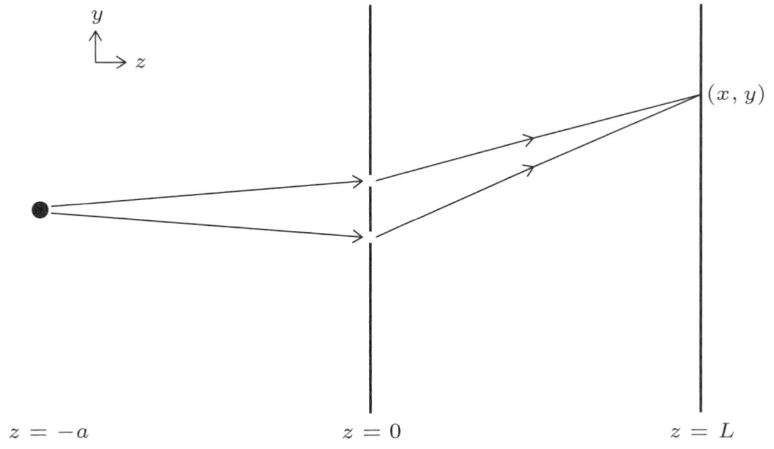

(a) Find the path length of the light rays reaching (x, y) from the source, assuming that $|x|, |y| \ll L$.
(b) Use the results of (a) to find the loci of points (x, y) on the screen with $|x|, |y| \ll L$ at which the intensity is minimum and the loci of points at which the intensity is maximum.
(c) Consider, independently, the following three modifications to the above setup: (i) The source is displaced in the y-direction by Δy. (ii) The source is treated as having finite size R rather than being treated as a point. (You may treat the emission from the different parts of the source as "incoherent"—as would be justified if the different parts of the source are emitting independently, there is some frequency spread, and the observation time is longer than the coherence time.) (iii) The source has frequency spread $\Delta\omega$ about ω rather than being purely monochromatic. In each of these cases, describe qualitatively the effect that these modifications would have on the interference pattern found on the screen at $z = L$. Estimate the magnitudes of Δy, R, and $\Delta\omega$ that would be needed to significantly affect the interference pattern.

7. Consider a superposition of plane waves of the form

$$\boldsymbol{A}(t, \boldsymbol{x}) = \int_0^\infty d\omega\, \boldsymbol{C}(\omega) e^{ikz} e^{-i\omega t},$$

[17]One could also do the experiment with parallel slits rather than pinholes.

where this superposition is peaked around frequency $\omega_0 > 0$ with spread $\Delta\omega < \omega_0$, (i.e., $\boldsymbol{C}(\omega)$ is nonvanishing only in this frequency range). Thus, the wave packet is as in eq. (7.46), except that the propagation direction is fixed, $\boldsymbol{k} = \frac{\omega}{c}\hat{\boldsymbol{z}}$, and $\boldsymbol{C}$ depends only on ω, not $\boldsymbol{x}$. However, now do *not* assume that the direction of $\boldsymbol{C}$ varies negligibly with ω. Fix a point $\boldsymbol{x}$ in space, and expand the electric field $\boldsymbol{E}(t)$ at $\boldsymbol{x}$ as

$$\boldsymbol{E}(t) = E_1(t)\hat{\boldsymbol{x}} + E_2(t)\hat{\boldsymbol{y}}.$$

Define the 2×2 Hermitian *coherency matrix* $\rho(t)$ by

$$\rho_{ij}(t) = \frac{1}{2T}\int_{t-T}^{t+T} E_i(t')E_j^*(t')dt' .$$

where $T \gg t_c = 1/\Delta\omega$. Note that $\mathcal{I} \equiv \mathrm{tr}\rho = 2\mu_0 c I$, where I is the intensity of the wave as defined by eq. (7.37). (Note, however, that in this context, $\mathcal{I}$ itself is commonly referred to as the intensity of the radiation.)

(a) Define the *degree of polarization*, $p \geq 0$, of the wave packet by

$$p^2 = 2\frac{\mathrm{tr}(\rho^2)}{[\mathrm{tr}(\rho)]^2} - 1 .$$

Show that p can take any value in $[0, 1]$, with $p = 1$ if and only if the direction of $\boldsymbol{E}(t)$ does not vary with t. (The case $p = 0$ is referred to as *unpolarized* radiation.) [Hint: It will be helpful to use the fact that any Hermitian matrix can be diagonalized.]

(b) Expand ρ in the basis of 2×2 Hermitian matrices consisting of the identity matrix $\mathbb{I}$ and the Pauli matrices $\sigma_x, \sigma_y, \sigma_z$:

$$\rho = \frac{1}{2}\mathcal{I}\,\mathbb{I} + \frac{1}{2}U\sigma_x + \frac{1}{2}V\sigma_y + \frac{1}{2}Q\sigma_z .$$

The four real numbers $\mathcal{I}$, U, V, and Q that appear in this basis expansion are known as the *Stokes parameters*. Show that

$$p = \frac{\sqrt{U^2 + V^2 + Q^2}}{\mathcal{I}} .$$

(c) For $p = 1$, show that (i) the case $U = V = 0$ corresponds to radiation that is linearly polarized in either the x- or y-direction; (ii) the case $V = Q = 0$ corresponds to radiation that is linearly polarized at 45° to the x- and y-axes; (iii) the case $U = Q = 0$ corresponds to radiation that is circularly polarized.

8. Let $H(\omega)$ be the Gaussian function

$$H(\omega) = \exp\left[-\frac{(\omega - \omega_0)^2}{\alpha^2}\right],$$

where $\omega_0 \gg \alpha > 0$, so that $H(\omega) \approx 0$ for $\omega < 0$. Explicitly obtain $F(t)$ (eq. (7.47)), and directly compute its time correlation function $\xi(\tau)$ from eq. (7.49). Then verify that eq. (7.52) holds (up to terms that are negligible when $\omega_0 \gg \alpha$).

9. Obtain the exact solution for the scattering of a circularly polarized plane wave of frequency ω off of a perfectly conducting ball of radius R. [Hint: Follow the same strategy as outlined in the text for the case of a dielectric ball. Start with the expansion of a plane wave corresponding to eq. (7.59)—consider only the case $m = +1$—and take the scattered wave to have the form eq. (7.60). Then apply the perfect conductor boundary condition $\boldsymbol{E}_{\parallel} = 0$ at $r = R$.]

10. A planar screen at $z = 0$ has a rectangular aperture, with the edges located at $x = \pm a$ and $y = \pm b$. A scalar plane wave of frequency ω is normally incident on this screen from $z < 0$, where $\omega/c \gg 1/a, 1/b$. Find the scalar field $\psi = \chi(\boldsymbol{x})e^{-i\omega t}$ at all points $\boldsymbol{x}$ with $z > 0$ and $|\boldsymbol{x}| \gg \omega(a^2 + b^2)/c$.

CHAPTER 8

Special Relativity

Special relativity is the theory of spacetime structure formulated by Einstein in 1905. Properties of the electromagnetic field played a central role in motivating special relativity. Specifically, electromagnetism is not compatible with pre-relativity notions of spacetime structure unless there is a preferred rest frame (the "aether"), since, as we have seen, Maxwell's equations predict that electromagnetic waves propagate with a particular velocity c, which can only be true in some preferred rest frame if pre-relativity notions of spacetime structure are valid. The Michelson-Morley experiment failed to find such a preferred rest frame. Furthermore, as Einstein realized, some physical phenomena in electromagnetism appear to obey an invariance with respect to moving observers even if the description of these phenomena in terms of a preferred rest frame does not have such an invariance.

The theory of electromagnetism is far more elegant and simple when formulated in the framework of special relativity. It therefore is somewhat of a travesty that, well into the twenty-first century, special relativity is discussed here—as in other texts on electromagnetism—as a separate chapter toward the end of the book. The reason, of course, is that even though special relativity has been a well-established theory for much more than a century, its basic concepts are still so unfamiliar to most physicists that it is not feasible to begin the treatment of electromagnetism by giving its formulation in the framework of special relativity. It is my hope that this situation will be rectified by the twenty-second century.

Einstein's original formulation of special relativity relied heavily on the transformations between the labeling of events by different inertial observers and the invariance of the laws of physics under such transformations. The theory was reformulated in a much more geometrical manner by Minkowski in 1908, wherein it was recognized that the underlying structure of spacetime in special relativity is that of a spacetime metric.[1] Our treatment of special relativity will emphasize the role of the spacetime metric. Although Einstein was initially unimpressed by Minkowski's reformulation, he soon incorporated it into his thinking about gravitation. This led him to the theory of general relativity, wherein the spacetime metric becomes a dynamical variable that describes

[1]Minkowski introduced an imaginary time coordinate so as to obtain a Euclidean spacetime metric. However, although this approach remains in use in many treatments of special relativity, it does not generalize to curved spacetime and cannot be used in general relativity. We shall use a real time coordinate in our treatment, and our spacetime metric will therefore be of Lorentzian signature.

not only spacetime structure but also all the effects of gravity. However, we shall not discuss general relativity here.

The framework of special relativity is presented in section 8.1. The formulation of electromagnetism in the framework of special relativity is then given in section 8.2. In section 8.3.1, we analyze the motion of a (relativistic) charged particle in an external electromagnetic field, including the solutions for motion in a uniform electric field and in a uniform magnetic field. The Lienard-Wiechert solution describing the retarded field of a point charge in arbitrary motion is given in section 8.3.2, and properties of the radiated power for this solution are analyzed there as well.

8.1 The Framework of Special Relativity

It is useful to think of space and time as composed of "events"—where each event corresponds to a point of space at an instant of time. The collection of all events comprises a 4-dimensional continuum, which I refer to as "spacetime."

I take as a starting point that there exist global families of inertial observers who "fill" all of spacetime (i.e., within each family, one and only one of these observers passes through each event in spacetime). I further assume that the observers in each family are all "at rest" with respect to one another, that they can consistently synchronize their clocks by some physical procedure, and that the spatial relationships between these observers are described by Euclidean geometry. Finally, I also assume that different families of such inertial observers all move at uniform velocity with respect to one another, so that the different families may be labeled by their velocity $\boldsymbol{v}$ with respect to some reference family. These assumptions are true in both pre-relativity physics and in special relativity, so they make a good starting point for describing the differences between these theories of spacetime structure. However, these assumptions are *not* true in general relativity, so they would make a very poor starting point from a fundamental viewpoint.

By the above assumptions, the inertial observers in a given family can uniquely label events by $(t, \boldsymbol{x})$, where t denotes the time of the event on the synchronized clock of the observer who passes through the event, and $\boldsymbol{x} = (x, y, z)$ are the Cartesian coordinates of that observer. I refer to the labeling of events in this way as *inertial coordinates*. The assumption that events can be labeled in this way is implicit in every physics text or other reference where a "t" or "$\boldsymbol{x}$" appears in an equation. However, this labeling depends on (i) a choice of origin of time (i.e., what time is labeled as $t = 0$); (ii) a choice of origin of space (i.e., what observer in the family is at $\boldsymbol{x} = 0$); (iii) a choice of orientation of axes (to define the x-, y-, and z-directions); and, most importantly for our present purposes, (iv) a choice of which family of inertial observers to use (i.e., a choice of the velocity $\boldsymbol{v}$ of the family). Different choices of origin of t and $\boldsymbol{x}$, orientations of axes, and $\boldsymbol{v}$ will give rise to different labelings $(t', \boldsymbol{x}')$ of events.

Usually, treatments of special relativity focus entirely on the difference in the labeling of events between families of inertial observers who are moving with velocity $\boldsymbol{v}$ relative to one another. This is is given by a Galilean transformation in pre-relativity physics and by a Lorentz transformation in special relativity. Although it certainly is useful to know the explicit form of this transformation, a nearly exclusive focus on this obscures the geometrical content of the theory. It is analogous to studying ordinary Euclidean geometry by focusing on how Cartesian coordinates transform under rotations.

I therefore focus on the "invariant structure" of spacetime. The four numbers $(t, \boldsymbol{x})$ associated with an event do not, by themselves, convey meaningful information about the event, since they depend as much, for example, on the choice of origin in t and $\boldsymbol{x}$ as they do on the event itself. Even if we consider the differences $(\Delta t, \Delta\boldsymbol{x})$ in the labeling of two events by a given family of inertial observers so as to eliminate the origin dependence, the values of $(\Delta t, \Delta\boldsymbol{x})$ will depend on the choice of orientation of axes as well as on the choice of family of inertial observers. It is of great interest to determine what quantities constructed out of $(\Delta t, \Delta\boldsymbol{x})$ are invariant (i.e., independent of these choices). Such quantities are well defined without making any arbitrary choices and hence can be considered as attributable to the structure of spacetime itself.

In pre-relativity physics, there are two such invariant quantities: (i) the time interval Δt between events, and (ii) the space interval $|\Delta\boldsymbol{x}|^2$ between simultaneous events (i.e., events with $\Delta t = 0$). The space interval between nonsimultaneous events is not invariant, because if the family O' of inertial observers moves with velocity $\boldsymbol{v}$ with respect to the family O, then we have

$$\Delta\boldsymbol{x}' = \Delta\boldsymbol{x} - \boldsymbol{v}\Delta t, \tag{8.1}$$

so $|\Delta\boldsymbol{x}'|^2 \neq |\Delta\boldsymbol{x}|^2$ if $\Delta t \neq 0$. In addition, the collection of "worldlines" of inertial observers (i.e., the possible paths in spacetime of inertial observers) also can be viewed as an additional aspect of spacetime structure. The worldlines of inertial observers contain additional information independent of (i) and (ii) in that they cannot be constructed from knowing only Δt for all pairs of events and knowing $|\Delta\boldsymbol{x}|^2$ for simultaneous events.

The situation in special relativity is much simpler. In special relativity, there is a single invariant quantity, the spacetime interval I between any pair of events, given by

$$I = -c^2(\Delta t)^2 + |\Delta\boldsymbol{x}|^2 . \tag{8.2}$$

Furthermore, it can be shown that the worldlines of inertial observers can be constructed from a knowledge of I between all pairs of events. Thus, I provides the complete description of spacetime structure in special relativity.

To tie the previous paragraph to what people are usually taught in special relativity, note that, in special relativity, the labeling $(t, \boldsymbol{x})$ of events by a family O of observers is related to the labeling $(t', \boldsymbol{x}')$ by a family O' of observers moving with velocity v in the x-direction relative to O (and with the same origin event and the same orientation of axes as O) by a *Lorentz transformation*:

$$\begin{aligned} t' &= \gamma(t - vx/c^2), \\ x' &= \gamma(x - vt), \\ y' = y, &\quad z' = z, \end{aligned} \tag{8.3}$$

where

$$\gamma \equiv \frac{1}{\sqrt{1 - v^2/c^2}} . \tag{8.4}$$

It is easily checked that I is invariant under Lorentz transformations. Furthermore, it can be shown that the most general transformation that preserves I is a *Poincare transformation* (i.e., a composition of Lorentz transformations, rotations, and translations, as well as parity and time reversal transformations). Thus, Lorentz transformations naturally arise as (part of) the symmetry group of I.

The spacetime interval I has the same mathematical form as the squared distance in Euclidean geometry except for the minus sign in front of the contribution coming from

$(\Delta t)^2$. To pursue this further, we switch notation from $(t, \boldsymbol{x})$ to x^μ with $\mu = 0, 1, 2, 3$, where

$$x^0 = ct\,, \quad x^1 = x\,, \quad x^2 = y\,, \quad x^3 = z\,. \tag{8.5}$$

Note the superscript position of μ in x^μ, which will be important in order to align with notational conventions explained further below. We view x^μ as representing a spacetime displacement vector (relative to some origin in spacetime) in much the same way as we normally view $\boldsymbol{x}$ as representing a spatial displacement vector (relative to some origin in space). We view I, eq. (8.2), as arising from an "inner product" on spacetime displacement vectors, where the inner product of x_1^μ and x_2^μ is given in any inertial coordinates by

$$I(x_1^\mu, x_2^\mu) = \sum_{\mu,\nu=0}^{3} \eta_{\mu\nu} x_1^\mu x_2^\nu, \tag{8.6}$$

where[2]

$$\eta_{\mu\nu} \equiv \begin{pmatrix} -1 & 0 & 0 & 0 \\ 0 & 1 & 0 & 0 \\ 0 & 0 & 1 & 0 \\ 0 & 0 & 0 & 1 \end{pmatrix}. \tag{8.7}$$

I put "inner product" in quotes, because although I is linear in each variable, symmetric, and nondegenerate (i.e., $I(x_1^\mu, x_2^\mu) = 0$ for all x_2^μ if and only if $x_1^\mu = 0$), it fails to be positive definite. Nevertheless, it is closely analogous to the inner product on vectors in ordinary Euclidean geometry,

$$(\boldsymbol{x}_1, \boldsymbol{x}_2) = \boldsymbol{x}_1 \cdot \boldsymbol{x}_2 = \sum_{i,j=1}^{3} e_{ij} x^i x^j, \tag{8.8}$$

where

$$e_{ij} \equiv \begin{pmatrix} 1 & 0 & 0 \\ 0 & 1 & 0 \\ 0 & 0 & 1 \end{pmatrix}. \tag{8.9}$$

We refer e_{ij} as the *metric of space* in Euclidean geometry. Similarly, we refer to $\eta_{\mu\nu}$ as the *metric of spacetime* in special relativity.

It should be mentioned that the ability to give finite spatial or spacetime displacements a vector space structure, as implicitly assumed in the discussion above, is very special to a *flat* geometry. In a curved geometry—such as the 2-dimensional surface of a potato—there is no natural notion of adding two finite displacements about a point. Nevertheless, in a curved geometry, a notion of "infinitesimal displacements" about any point p can be defined, and these infinitesimal displacements have a vector space structure. The vector space of infinitesimal displacements about p is referred to as the *tangent space* at p. In differential geometry, a metric would be defined as a (not necessarily positive-definite) inner product defined on the tangent space at p for all p. However,

[2] Many authors define $\eta_{\mu\nu}$ with an opposite sign convention, which results in sign changes in some formulas. The reader is advised to check the sign convention used for $\eta_{\mu\nu}$ when comparing formulas in different references. As mentioned in footnote 1 in this chapter, some authors use an imaginary time coordinate, in which case $\eta_{\mu\nu}$ would take a Euclidean form and normally would not be written down explicitly at all.

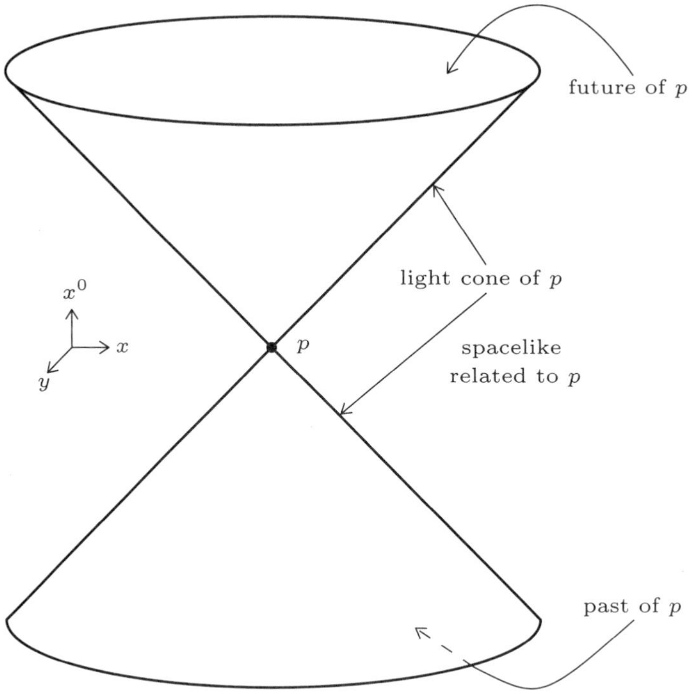

Figure 8.1. The light cone of an event p.

the spacetime geometry of special relativity is flat, so we may treat *finite* spacetime displacements $\Delta x^\mu \equiv x^\mu - x^\mu(p)$ about a point p as "vectors"—and we can treat the metric as an inner product on these finite displacement vectors—as we have done above.

A striking feature of the spacetime metric eq. (8.7) is that there are nonzero spacetime displacement vectors Δx^μ about any event p that are *null*, in other words, such that

$$\sum_{\mu,\nu} \eta_{\mu\nu} \Delta x^\mu \Delta x^\nu = 0 \,. \tag{8.10}$$

The collection of all null spacetime displacement vectors about p comprise a cone with vertex at p, as illustrated in figure 8.1. The portion of the cone with $\Delta x^0 \geq 0$ is referred to as the *future light cone* of p, whereas the portion with $\Delta x^0 \leq 0$ is referred to as its *past light cone*. In the precise sense discussed in chapter 5, electromagnetic radiation emitted at p propagates along the future light cone of p (see section 5.2), whereas electromagnetic radiation observed at p propagated to p along its past light cone (see section 5.4). The interior of the future light cone of p is referred to as the *future* of p. It is composed of events that, in principle, can be reached by an observer initially present at p. Similarly, the interior of the past light cone of p is referred to as the *past* of p. It is composed of events with the property that an observer starting at that event can, in principle, arrive at p. The events lying outside the light cone of p are said to be *spacelike related* to p. No observer can be present both at p and at an event spacelike related to p. In other words, in special relativity, nothing can travel faster than light.

If q is an event that lies in the future of p, then there is a unique inertial observer who is present at both p and q. The proper time $\Delta\tau$ that elapses on a clock carried by this observer between p and q is given in any inertial coordinates by[3]

$$\Delta\tau = \frac{1}{c}\sqrt{-\sum_{\mu,\nu}\eta_{\mu\nu}\Delta x^{\mu}\Delta x^{\nu}}\,, \tag{8.11}$$

where $\Delta x^{\mu} = x^{\mu}(q) - x^{\mu}(p)$. A general, noninertial observer will trace out a curve in spacetime, called the *worldline* of the observer. Any curve in spacetime may be specified by giving $x^{\mu}(\lambda)$, where λ is an arbitrary parametrization of the curve. The tangent T^{μ} to the curve in this parametrization is defined by

$$T^{\mu} = \frac{dx^{\mu}}{d\lambda}\,. \tag{8.12}$$

The tangent T^{μ} to the worldline of any observer must be timelike (i.e., $\sum_{\mu,\nu}\eta_{\mu\nu}T^{\mu}T^{\nu} < 0$), since the observer must stay within the light cone of any event that he/she passes through. The proper time elapsed on the clock of an arbitrary noninertial observer going between events p and q is given by

$$\Delta\tau = \frac{1}{c}\int_{\lambda(p)}^{\lambda(q)} d\lambda\,\sqrt{-\sum_{\mu,\nu}\eta_{\mu\nu}T^{\mu}T^{\nu}}\,. \tag{8.13}$$

It is not difficult to show that the inertial observer who passes through events p and q maximizes the elapsed proper time relative to all observers who pass between p and q. This fact is often referred to as the *twin paradox* (see problem 1 at the end of the chapter).

It is convenient to parametrize the worldline of an arbitrary observer by the proper time τ elapsed along the worldline. The tangent to the worldline in this parametrization is called the 4-*velocity* of the observer:

$$u^{\mu} = \frac{dx^{\mu}}{d\tau}\,. \tag{8.14}$$

It follows from eq. (8.13) with $\lambda = \tau$ that $\sum_{\mu,\nu}\eta_{\mu\nu}u^{\mu}u^{\nu} = -c^2$. In any inertial coordinates, the velocity $\boldsymbol{v}$ of this observer is given by

$$\boldsymbol{v} = \frac{d\boldsymbol{x}}{dt} = c\frac{d\boldsymbol{x}}{dx^0} = c\frac{d\boldsymbol{x}/d\tau}{dx^0/d\tau} = c\frac{\boldsymbol{u}}{u^0}\,. \tag{8.15}$$

It follows from this relation and the normalization condition on u^{μ} that, in any inertial coordinates, the components of u^{μ} are given in terms of $\boldsymbol{v}$ by

$$u^{\mu} = (c\gamma, \gamma\boldsymbol{v}), \tag{8.16}$$

where γ is given by eq. (8.4).

[3]This can be seen by noting that the right side of eq. (8.11) is the spacetime interval between p and q and does not depend on the choice of inertial coordinates. In the frame of the inertial observer who goes from p to q, we have $\Delta\boldsymbol{x} = 0$, so $\sum\eta_{\mu\nu}\Delta x^{\mu}\Delta x^{\nu} = -c^2(\Delta t)^2$.

As discussed above, because of the flat spacetime geometry, the finite displacements $\Delta x^\mu = x^\mu - x^\mu(p)$ about p comprise a 4-dimensional vector space, which we denote by V_p. The tangent T^μ to a curve passing through p can naturally be viewed as a vector in V_p. For an arbitrary spacetime vector, we will strictly adhere to the notational convention that an *upper* Greek letter index will be attached to the letter representing the vector (e.g., when a symbol such as W^μ appears below, one can immediately tell from the notation that W^μ is a spacetime vector).

Spacetime vectors in special relativity have a geometrical status in exactly the same sense that ordinary vectors have a geometrical status in ordinary, 3-dimensional space. Although the representation of a spacetime vector in terms of its components depends on a choice of inertial coordinate system, one may view the spacetime vector as having a geometrical meaning that does not depend on a choice of coordinates—in exactly the same manner as an ordinary vector can be viewed as having a geometrical meaning, independent of the representation of its components in a particular Cartesian coordinate system.

Another class of objects that have a similar geometrical status are linear maps taking vectors to numbers. A linear map taking V_p into $\mathbb{R}$ is called a *dual vector*. The collection of all dual vectors at p comprises a vector space of the same dimension as V_p, known as the *dual space* to V_p and denoted by V_p^*. Given any vector W^μ in V_p, we can use the spacetime metric $\eta_{\mu\nu}$ to construct a dual vector L_W taking V_p into $\mathbb{R}$ by

$$S^\mu \rightarrow \sum_{\mu,\nu} \eta_{\mu\nu} S^\mu W^\nu, \tag{8.17}$$

that is, L_W maps the arbitrary vector S^μ into the number given by the right side of this equation. It is not difficult to show that, in the presence of a metric, all dual vectors arise in this manner. It is natural to denote the dual vector L_W by

$$W_\mu \equiv \sum_{\nu=0}^{3} \eta_{\mu\nu} W^\nu, \tag{8.18}$$

so that the linear map L_W is given by $S^\mu \rightarrow \sum_\nu S^\nu W_\nu$. In this notation, the lowered index on W_μ on the left side of eq. (8.18) indicates that this quantity is the dual vector obtained from W^μ rather than W^μ itself. For an arbitrary dual vector, we will strictly adhere to the notational convention that a *lower* Greek letter index will be attached to the letter representing the dual vector (e.g., when a symbol such as U_μ appears below, one can immediately tell from the notation that U_μ is a spacetime dual vector).

Since the spacetime metric is nondegenerate, eq. (8.18) can be inverted to give

$$W^\mu = \sum_{\nu=0}^{3} \eta^{\mu\nu} W_\nu \tag{8.19}$$

where $\eta^{\mu\nu}$ denotes the inverse spacetime metric, whose matrix of components is given by the inverse matrix[4] of $\eta_{\mu\nu}$. Following standard conventions, if U_μ is a dual vector,

[4] In an inertial coordinate system, the components of $\eta_{\mu\nu}$ are given by eq. (8.7), and the inverse of eq. (8.7) then has exactly the same components as eq. (8.7) (see eq. (5.121)). However, $\eta^{\mu\nu}$ and $\eta_{\mu\nu}$ are fundamentally very different objects, and this equality of their components would not hold in a general, noninertial coordinate system.

we denote the corresponding vector given by eq. (8.19) as U^μ. Thus, we use $\eta^{\mu\nu}$ and $\eta_{\mu\nu}$ to raise and lower indices in the manner given by eq. (8.18) and eq. (8.19).

Vectors and dual vectors are fundamentally very different objects. Nevertheless, when a metric is present, vectors and dual vectors can be identified via the correspondence eq. (8.18) or equivalently, eq. (8.19). When dealing with a vector $\boldsymbol{v}$ in ordinary Euclidean space, the components (v^1, v^2, v^3) of the vector in Cartesian coordinates are equal to the components (v_1, v_2, v_3) of the corresponding dual vector on account of the trivial form, eq. (8.9), of the Euclidean metric. Thus, in this case, if one treats a vector as a 3-tuple of numbers, one can get away with not making a distinction between a vector and its corresponding dual vector.[5] This accounts for why many students of physics have never explicitly encountered the notion of a dual vector. However, in special relativity, in any inertial coordinates, the dual vector W_μ corresponding to W^μ has $W_0 = -W^0$, and we cannot get away with ignoring the distinction between vectors and dual vectors.

A prime example of a dual vector at p is the spacetime gradient $\partial_\mu f = \partial f/\partial x^\mu$ of a function f evaluated at p. This is seen to be a dual vector as follows. For any curve with tangent T^μ at p, the quantity $\sum_\mu T^\mu \partial_\mu f$ represents the derivative $df/d\lambda$ of f along the curve and thus is well defined, independently of any choice of coordinates. Thus, $\partial_\mu f$ at p naturally can be viewed as a linear map taking tangent vectors T^μ at p into numbers (i.e., a dual vector). Note that in ordinary Euclidean space, the gradient of f would normally be denoted by $\boldsymbol{\nabla} f$ in vector calculus notation (i.e., it would be treated as a vector rather than a dual vector). This is because the correspondence between vectors and dual vectors that is provided by the Euclidean metric is being used. Fundamentally, however, the gradient of f is a dual vector.

Another important example of a dual vector arises when we consider plane wave solutions or, more generally, solutions in the geometric optics approxmation. Let ψ be a plane wave solution to the scalar wave equation

$$\psi(t, \boldsymbol{x}) = Ce^{-i\omega t + i\boldsymbol{k}\cdot\boldsymbol{x}} . \tag{8.20}$$

The phase of ψ,

$$\mathscr{S} \equiv -\omega t + \boldsymbol{k}\cdot\boldsymbol{x}, \tag{8.21}$$

may be viewed as a linear map taking spacetime displacement vectors $x^\mu = (ct, \boldsymbol{x})$ into numbers. Thus,

$$k_\mu = (-\omega/c, k_1, k_2, k_3) \tag{8.22}$$

defines a spacetime dual vector. More generally, for any ψ of the spacetime geometric optics form (see problem 1 of chapter 7)

$$\psi(t, \boldsymbol{x}) = \mathscr{A}(t, \boldsymbol{x}) e^{i\mathscr{S}(t,\boldsymbol{x})}, \tag{8.23}$$

the quantity

[5] In this regard, it should be noted that in the previous chapters of this book, we denoted the Cartesian components of vectors such as the electric field $\boldsymbol{E}$ as E_i rather than E^i. This is not because we were working with the corresponding dual vector but instead because there was no need to make a distinction between vectors and dual vectors and, correspondingly, no need to adhere to the conventions on the index positions that we have just introduced. We wrote "E_i" simply because it was typographically more convenient to write E_i rather than E^i. However, from this point forward in this book, we will strictly adhere to our conventions on index positions.

$$k_\mu = \partial_\mu \mathscr{S} \tag{8.24}$$

defines a dual vector field on spacetime. These results also apply to plane wave and geometric optics solutions to Maxwell's equations. It is important to note that, as shown in problem 1 of chapter 7, the corresponding vector field

$$k^\mu = \sum_\nu \eta^{\mu\nu} k_\nu \tag{8.25}$$

is null, $\sum_{\mu,\nu} \eta_{\mu\nu} k^\mu k^\nu = 0$, and it is tangent to straight lines ("light rays") in spacetime. The upshot is that for plane waves—or, more generally, geometric optics solutions—the *wave 4-vector*

$$k^\mu = (\omega/c, \boldsymbol{k}) \tag{8.26}$$

plays a role for light rays that is very similar to the role played by the 4-velocity u^μ for particle motion. In particular, the change in frequency and direction of a light ray as seen by a moving observer follows from the fact that k^μ is a spacetime vector (see problem 4).

It should be noted that vectors and dual vectors transform differently under Lorentz transformations. Under the Lorentz transformation eq. (8.3) with origin at p, a vector W^μ at p transforms as

$$W^\mu \to W'^\mu = \sum_\nu \Lambda^\mu{}_\nu W^\nu, \tag{8.27}$$

where the matrix of components of the Lorentz transformation $\Lambda^\mu{}_\nu$ is given by[6]

$$\Lambda^\mu{}_\nu = \begin{pmatrix} \gamma & -\gamma v/c & 0 & 0 \\ -\gamma v/c & \gamma & 0 & 0 \\ 0 & 0 & 1 & 0 \\ 0 & 0 & 0 & 1 \end{pmatrix}. \tag{8.28}$$

In contrast, a dual vector U_μ transforms as[7]

$$U_\mu \to U'_\mu = \sum_\nu (\Lambda^{-1})^\nu{}_\mu U_\nu, \tag{8.29}$$

where $(\Lambda^{-1})^\mu{}_\nu$ is the inverse of $\Lambda^\mu{}_\nu$. It is easily checked that $(\Lambda^{-1})^\mu{}_\nu$ is given by the same formula as eq. (8.28) with $v \to -v$. The relations eq. (8.18) and eq. (8.19) are preserved under Lorentz transformations on account of the invariance of the metric under Lorentz transformations in the sense that (see problem 2)

$$\eta_{\alpha\beta} = \sum_{\mu,\nu} (\Lambda^{-1})^\mu{}_\alpha (\Lambda^{-1})^\nu{}_\beta \eta_{\mu\nu}\,. \tag{8.30}$$

---

[6]The "1 up" and "1 down" index positions in $\Lambda^\mu{}_\nu$ correspond to the fact that $\Lambda^\mu{}_\nu$ is a linear map on vectors and hence is a tensor of type $(1,1)$ (see the discussion later in this section).

[7]Note the index position of the summed index ν in eq. (8.29), which would correspond to taking the transpose of the matrix Λ^{-1} in usual matrix multiplication rules. Note also that the Cartesian coordinate components of a rotation matrix $R^\mu{}_\nu$ in Euclidean geometry satisfy $(R^{-1})^\nu{}_\mu = R^\mu{}_\nu$, so the transformations eq. (8.27) and eq. (8.29) take the same form for rotations in Euclidean geometry, consistent with being able to treat vectors and dual vectors as "the same" in this case.

The important thing about vectors and dual vectors is not the specific formulas describing how they transform under Lorentz transformations but rather that they represent geometrical quantities that are well defined, independently of any choice of inertial coordinates. The worldline of a particle in spacetime is a well-defined curve that has physical meaning independently of how the events in spacetime are labeled with inertial or other coordinates. The 4-velocity u^μ of the particle (i.e., the normalized tangent to its worldline) similarly has a well-defined physical meaning. It is therefore reasonable that it could enter the physical laws of motion of the particle. The fact that u^μ transforms according to eq. (8.27) under a change of inertial coordinates merely reflects that it has well-defined geometrical meaning as a vector. Similarly, the gradient $\partial_\mu f$ of a function f has a well-defined meaning as a dual vector, and the fact that it transforms as eq. (8.29) is merely a reflection of this. If f is a physical quantity, it is reasonable that $\partial_\mu f$ could enter laws of physics involving f.

A tensor is a more general geometrical quantity than vectors and dual vectors that can be defined in a mathematically precise manner as follows: A *tensor of type* (k, l) at p is a multilinear (i.e., linear in each variable separately) map taking k dual vectors and l vectors at p into $\mathbb{R}$. Thus, a tensor of type $(0, 1)$ is a linear map from vectors into numbers, i.e., it is a dual vector. A tensor of type $(1, 0)$ is a linear map from dual vectors into numbers. This might sound like yet another new object, but it is not difficult to see that such a "double dual vector" can be naturally identified with an ordinary vector,[8] so a tensor of type $(1, 0)$ is just a vector. A tensor of type $(0, 2)$ is a map that takes a pair of vectors into $\mathbb{R}$ and is linear in each variable. The spacetime metric $\eta_{\mu\nu}$ is an example of a tensor of type $(0, 2)$. A tensor of type $(1, 1)$ is a map that takes a dual vector and a vector into $\mathbb{R}$ and is linear in each variable. A tensor T of type $(1, 1)$ can be naturally identified[9] with a linear map taking V_p into V_p (and also can be identified with a linear map taking V_p^* into V_p^*).

Following the index conventions given above for vectors and dual vectors, we shall denote a tensor T of type (k, l) with k upper and l lower Greek indices, that is, as $T^{\mu_1 \ldots \mu_k}{}_{\nu_1 \ldots \nu_l}$, where all the indices range from 0 to 3. We can use $\eta^{\mu\nu}$ and $\eta_{\mu\nu}$ to raise and lower any of the indices in the manner given by eq. (8.18) and eq. (8.19), thereby enabling us to convert a tensor of type (k, l) into a tensor of any other type with the same rank, where the rank r of a tensor is defined by $r = k + l$. Note that a general tensor of rank r on a 4-dimensional space has 4^r independent components.

Under a Lorentz transformation, $T^{\mu_1 \ldots \mu_k}{}_{\nu_1 \ldots \nu_l}$ transforms as

$$\begin{aligned} T^{\mu_1 \ldots \mu_k}{}_{\nu_1 \ldots \nu_l} &\to T'^{\mu_1 \ldots \mu_k}{}_{\nu_1 \ldots \nu_l} \\ &= \sum_{\alpha_1 \ldots \beta_l} \Lambda^{\mu_1}{}_{\alpha_1} \cdots \Lambda^{\mu_k}{}_{\alpha_k} (\Lambda^{-1})^{\beta_1}{}_{\nu_1} \cdots (\Lambda^{-1})^{\beta_l}{}_{\nu_l} T^{\alpha_1 \ldots \alpha_k}{}_{\beta_1 \ldots \beta_l} . \end{aligned} \tag{8.31}$$

In other words, if a tensor of type (k, l) has components $T^{\mu_1 \ldots \mu_k}{}_{\nu_1 \ldots \nu_l}$ in some inertial coordinates x^μ, then the components of the tensor in the inertial coordinate system

[8] This can be seen as follows. The action of a dual vector U_μ on a vector S^μ is given by $S^\mu \to \sum_\mu U_\mu S^\mu$. However, $\sum_\mu U_\mu S^\mu$ could equally well be viewed as defining an action of S^μ on dual vectors given by $U_\mu \to \sum_\mu U_\mu S^\mu$. In this manner, we may view any vector S^μ as corresponding to a double dual vector.

[9] To see this, note that for any $v \in V_p$, the quantity $T(\cdot, v)$ is a linear map from V_p^* into $\mathbb{R}$ and thus is a double dual vector, which can be identified with a vector. Thus, T can be uniquely associated with a linear map from V_p into V_p.

$x'^{\mu} = \sum_{\nu} \Lambda^{\mu}{}_{\nu} x^{\nu}$ are given by eq. (8.31).[10] Again, the important thing about a tensor is not the explicit formula eq. (8.31) for how its components transform under a change of inertial coordinates but the fact that it represents a well-defined object, independent of any choice of coordinates.

A *tensor field* of type (k, l) is the specification of tensor of type (k, l) at p for all p. It is convenient to define a tensor of type $(0, 0)$ to be a number, in which case, a function may be viewed as a tensor field of type $(0, 0)$. We define the derivative of a tensor field $T^{\mu_1 \ldots \mu_k}{}_{\nu_1 \ldots \nu_l}$ of type (k, l) to be the tensor field $\partial_\alpha T^{\mu_1 \ldots \mu_k}{}_{\nu_1 \ldots \nu_l}$ of type $(k, l+1)$ given in any inertial coordinates by

$$\partial_\alpha T^{\mu_1 \ldots \mu_k}{}_{\nu_1 \ldots \nu_l} = \frac{\partial T^{\mu_1 \ldots \mu_k}{}_{\nu_1 \ldots \nu_l}}{\partial x^\alpha} . \tag{8.32}$$

Although this formula may look obvious/trivial, this simple notion of differentiation of a tensor field relies heavily on the flatness of the geometry of spacetime.[11] In a curved geometry, there is a still a unique, well-defined notion of differentiation of tensor fields that is determined by the metric, but this notion (usually referred to as "covariant differentiation") is not obtained by merely taking partial derivatives in a coordinate system. However, we are not concerned with curved geometries here, and the more general notion of differentiation in a curved geometry reduces to eq. (8.32) in inertial coordinates in a flat geometry. Thus, eq. (8.32) is entirely satisfactory for our purposes.

We now turn to the implications of what we have said above for the laws of physics in special relativity. We want the mathematical objects representing physical quantities to be well defined, using only the structure of spacetime in special relativity. Similarly, we want the equations satisfied by these objects to be well defined, using only the special relativistic structure of spacetime. This suggests the following two criteria for the formulation of the laws of physics in special relativity: (i) Physical quantities should be represented by spacetime tensors/tensor fields. (ii) The equations satisfied by these tensors/tensor fields should involve only these tensor fields, the spacetime metric, and operations that map spacetime tensors into spacetime tensors,[12] such as the notion of differentiation given by eq. (8.32). These criteria correspond to the more common statement that in special relativity, the laws of physics should be invariant under Lorentz transformations. Any relation equating two tensors will be well defined independently of any choice of inertial coordinates and thus automatically will remain valid under a change of inertial coordinates. We refer to a theory satisfying properties (i) and (ii) as being *special relativistically covariant*.

---

[10] Here we are taking a "passive view" of Lorentz transformations, wherein the tensor does not change, but its description in terms of coordinate components changes. Alternatively, we could take an "active view" of Lorentz transformations, wherein the inertial coordinate system is fixed but the Lorentz transformation is viewed as acting on the tensor, mapping the tensor T at x^μ to a new tensor T' at x'^μ via the formula (8.31). These views are equivalent (i.e., we obtain the same results by making a passive Lorentz transformation as we would by making an active transformation by the inverse of this Lorentz transformation).

[11] The notion of the gradient of a function (i.e., the derivative of a tensor field of type $(0, 0)$) does not require a flat geometry for the definition eq. (8.32) to yield a dual vector, but a flat geometry is needed to use eq. (8.32) to define the derivative of a higher rank tensor in a coordinate invariant way.

[12] An important example of such an operation is the trace $T^{\mu_1 \ldots \mu_k}{}_{\nu_1 \ldots \nu_l} \to \sum_\alpha T^{\mu_1 \ldots \alpha \ldots \mu_k}{}_{\nu_1 \ldots \alpha \ldots \nu_l}$ over one upper index and one lower index, which is a well-defined operation taking a tensor of type (k, l) into a tensor of type $(k-1, l-1)$. The fact that this operation is independent of the choice of inertial coordinates can be seen from eq. (8.31).

The implications of the previous paragraph are well illustrated by considering the nature of the laws of particle mechanics in special relativity. In pre-relativity physics, we define the momentum $\boldsymbol{p}$ of a particle by

$$\boldsymbol{p} = m\boldsymbol{v} = m\frac{d\boldsymbol{x}}{dt}, \tag{8.33}$$

where m denotes the mass of the particle, which is usually assumed to be constant. Newton's second law then takes the form

$$\frac{d\boldsymbol{p}}{dt} = \boldsymbol{f}, \tag{8.34}$$

where the form of the force $\boldsymbol{f}$ depends on the particular situation considered. However, spatial vectors have no well-defined status in special relativity, nor does t, so eq. (8.34) as it stands is not an acceptable law of physics in special relativity. We have two possibilities for formulating an acceptable version of the laws of particle mechanics: (a) It could be that eq. (8.34) is actually already special relativistically covariant, and all that needs to be done is to formulate/rewrite this equation in such a way that it involves only spacetime tensors. (b) Equation (8.34) does not correspond to a special relativistically covariant equation, in which case it must be discarded and replaced by a new law that involves only spacetime tensors. In the present case, it is possibility (b) that holds. Thus, we must find a new law of particle mechanics—reducing to eq. (8.34) for nonrelativistic motion—that involves only spacetime tensors.

The only reasonable candidate for a spacetime tensor that could represent the velocity of a particle is its 4-velocity, u^μ. The only reasonable candidate for a spacetime tensor that could represent the momentum of the particle is

$$p^\mu = mu^\mu . \tag{8.35}$$

We refer to p^μ as the 4-momentum of the particle and, in special relativity, the parameter m is usually referred to as the "rest mass" of the particle (to distinguish it from what some authors call the "relativistic mass," $p^0/c = \gamma m$). The only reasonable candidate for a modified version of Newton's second law that involves only tensor quantities is

$$\sum_\nu u^\nu \partial_\nu p^\mu = f^\mu, \tag{8.36}$$

where the 4-force f^μ depends on the particular situation considered. Thus, the principles of the previous paragraph lead, in an essentially unique way, to the modification eq. (8.36) of Newton's second law that is special relativistically covariant.

Note that $u^\mu \partial_\mu = d/d\tau$ (i.e., $u^\mu \partial_\mu$ is the derivative along the worldline of the particle using the proper time parametrization). Thus, we may rewrite eq. (8.36) in the form

$$\frac{d}{d\tau}\left(m\frac{dx^\mu}{d\tau}\right) = f^\mu . \tag{8.37}$$

It is also worth noting that if we assume that the 4-force f^μ is orthogonal to the 4-velocity u^μ, then

$$
\begin{aligned}
0 &= \sum_{\mu,\nu} \eta_{\mu\nu} u^\mu f^\nu = \sum_{\mu,\nu} \eta_{\mu\nu} u^\mu \frac{dp^\nu}{d\tau} \\
&= \frac{1}{2} m \frac{d}{d\tau}\left(\sum_{\mu,\nu} \eta_{\mu\nu} u^\mu u^\nu\right) + \frac{dm}{d\tau} \sum_{\mu,\nu} \eta_{\mu\nu} u^\mu u^\nu \\
&= -c^2 \frac{dm}{d\tau}\,. \qquad (8.38)
\end{aligned}
$$

Thus, if $\sum_{\mu,\nu} \eta_{\mu\nu} u^\mu f^\nu = 0$, then the rest mass of the particle does not change as the particle moves along its worldline.

The example of special relativistic particle mechanics also provides a good illustration of the manner in which spacetime tensors combine together—into a single quantity—quantities that would have been viewed as entirely distinct in pre-relativity physics. (This occurs whether or not the dynamical laws themselves need to be modified to be manifestly special relativistically covariant.) In pre-relativity physics, the energy of a particle, $E = \frac{1}{2} m v^2$, is a scalar quantity, and one has the freedom to add an arbitrary constant to it. It is entirely distinct from the momentum of the particle, $\boldsymbol{p} = m\boldsymbol{v}$. However, in special relativity, these quantities are identified as the components of the 4-momentum p^μ of the particle:

$$
(E/c, \boldsymbol{p}) = p^\mu = (m u^0, m\boldsymbol{u}) = (\gamma m c, \gamma m \boldsymbol{v}). \qquad (8.39)
$$

This gives rise to new formulas for momentum and energy, namely, $\boldsymbol{p} = \gamma m \boldsymbol{v}$, and $E = \gamma m c^2$. Most importantly, since E is now the component of a spacetime vector rather than a scalar quantity, there is no longer any freedom to modify its definition by the addition of a constant. Thus, a particle at rest must be assigned the energy $E = mc^2$.

We conclude this section by stating two standard notational conventions that will be used throughout the rest of this book: (i) Up until this point, we have explicitly written the metric $\eta_{\mu\nu}$ wherever it has been used to convert vectors to dual vectors via the correspondence eq. (8.18). We will no longer continue to do so (e.g., having introduced the 4-velocity u^μ, we will simply write u_μ for $\sum_\nu \eta_{\mu\nu} u^\nu$). Similarly, we will not write the inverse metric explicitly when converting a dual vector to a vector via eq. (8.19) (e.g., we will write $\partial^\mu f$ for $\sum_\nu \eta^{\mu\nu} \partial_\nu f$). (ii) We will omit the summation sign $\sum$ when summing over a repeated upper and lower index (e.g., we will simply write $u^\mu \partial_\mu f$ for $\sum_\mu u^\mu \partial_\mu f$). The omission of the summation sign in such sums was introduced by Einstein and is referred to as the *Einstein summation convention*. As a simple example of the notational conventions (i) and (ii) that will be used freely below, the first equality in (8.38) will now be written as $0 = u^\mu f_\mu$ (or, equivalently, as $0 = u_\mu f^\mu$).

8.2 The Formulation of Electromagnetic Theory in the Framework of Special Relativity

We wish to formulate the theory of electromagnetism in a manifestly special relativistically covariant manner. The discussion of electromagnetism thus far has involved spatial vector fields, such as $\boldsymbol{A}$, $\boldsymbol{E}$, and $\boldsymbol{B}$. Such spatial vectors have no well-defined status in special relativity. In accord with the discussion at the end of the previous section, we must either rewrite electromagnetic theory so that it involves only spacetime tensor fields, or we must discard the theory and replace it with a new one that does. In the case of Newtonian particle mechanics, we had to do the latter (i.e., we discarded the pre-relativity version, eq. (8.34), of Newton's second law and replaced it with the modified version, eq. (8.36), that is special relativistically covariant). Fortunately, in the case of electromagnetism, we need only do the former; we may simply rewrite the theory in a manner wherein it can be explicitly seen to involve spacetime tensors. Thus, in particular, all the equations and results regarding electromagnetism obtained in the previous chapters of this book remain valid in special relativity.

Our task is to reformulate all the equations of section 5.1 in spacetime tensor form. The first step is to make sense of $\boldsymbol{A}$. The most straightforward way of incorporating a spatial vector, such as $\boldsymbol{A}$, into the framework of special relativity would be to make it part of a spacetime vector A^μ. In fact, this can be done quite easily by combining the scalar potential ϕ with $\boldsymbol{A}$ to define a 4-vector:

$$A^\mu \equiv (\phi/c, \boldsymbol{A}) . \tag{8.40}$$

In other words, instead of viewing ϕ as a scalar function on spacetime—as would have been natural in pre-relativity physics—we now interpret ϕ/c as being the time component of a 4-vector field whose spatial components are $\boldsymbol{A}$.

In fact, it is more natural and convenient to work with the corresponding dual vector field (see eq. (8.18)):

$$A_\mu = (-\phi/c, A_1, A_2, A_3), \tag{8.41}$$

where the spatial components (A_1, A_2, A_3) of A_μ are numerically equal to $(A^1, A^2, A^3) = \boldsymbol{A}$. A general gauge transformation eq. (5.1) then takes the form[13]

$$A_\mu \to A'_\mu = A_\mu + \partial_\mu \chi . \tag{8.42}$$

Thus, we have written the potentials and their gauge transformations in a special relativistically covariant form.

Next, we define the *electromagnetic field-strength tensor* by

$$F_{\mu\nu} = \partial_\mu A_\nu - \partial_\nu A_\mu, \tag{8.43}$$

so that $F_{\mu\nu}$ is a spacetime tensor field of type (0, 2). Since $F_{\mu\nu}$ is antisymmetric, $F_{\mu\nu} = -F_{\nu\mu}$, it is easily seen that $F_{\mu\nu}$ has only 6 independent components. These independent components of $F_{\mu\nu}$ can be taken to be

$$E_i/c = F_{i0} = -\frac{1}{c}\frac{\partial \phi}{\partial x^i} - \frac{\partial A_i}{\partial x^0}, \qquad B_i = \frac{1}{2}\sum_{j,k=1}^{3} \epsilon_{ijk} F_{jk} = \sum_{j,k=1}^{3} \epsilon_{ijk} \frac{\partial A_k}{\partial x^j}, \tag{8.44}$$

where ϵ_{ijk} is defined by eq. (4.6). In more explicit terms, the components of the tensor $F_{\mu\nu}$ defined by eq. (8.43) are

[13]Note that the zeroth component of this equation is $-\phi/c \to -\phi/c + \partial\chi/\partial x^0 = -\phi/c + (1/c)\partial\chi/\partial t$, that is, $\phi \to \phi - \partial\chi/\partial t$.

$$F_{\mu\nu} = \begin{pmatrix} 0 & -E_1/c & -E_2/c & -E_3/c \\ E_1/c & 0 & B_3 & -B_2 \\ E_2/c & -B_3 & 0 & B_1 \\ E_3/c & B_2 & -B_1 & 0 \end{pmatrix}. \tag{8.45}$$

Thus, $F_{\mu\nu}$ combines $\boldsymbol{E}$ and $\boldsymbol{B}$ into a single spacetime tensor field. In this way, we have reformulated the definitions (5.2) and (5.3) of $\boldsymbol{E}$ and $\boldsymbol{B}$ in a special relativistically covariant manner.

Our next task is to write Maxwell's equations in a manifestly special relativistically covariant form. The definition of $F_{\mu\nu}$ together with the equality of mixed partials implies that

$$\partial_\alpha F_{\mu\nu} + \partial_\mu F_{\nu\alpha} + \partial_\nu F_{\alpha\mu} = 0\,. \tag{8.46}$$

This equation is equivalent to eqs. (5.6) and (5.7), and thus reformulates these equations in a special relativistically covariant manner. To express the remaining Maxwell equations (5.4) and (5.5) in a manifestly covariant form, we first combine the charge density ρ and the current density $\boldsymbol{J}$ into a single spacetime vector field J^μ, known as the *charge-current* 4-*vector*:

$$J^\mu = (c\rho, \boldsymbol{J})\,. \tag{8.47}$$

This makes the source terms in Maxwell's equations into a special relativistically covariant quantity. Charge-current conservation eq. (5.8) then takes the manifestly covariant form

$$\partial_\mu J^\mu = 0\,. \tag{8.48}$$

It is then straightforward to check that the remaining Maxwell equations (5.4) and (5.5) can be written as the single spacetime tensor equation[14]

$$\partial^\alpha F_{\alpha\mu} = -\mu_0 J_\mu\,, \tag{8.49}$$

where we are using the notational conventions described at the end of section 8.1.

Our remaining task is to write the energy density, momentum density, and stresses of the electromagnetic field in a special relativistically covariant form. This can be done by defining the following tensor field of type $(0, 2)$:

$$T^{\mathrm{EM}}_{\mu\nu} = \frac{1}{\mu_0}\left[F_\mu{}^\alpha F_{\nu\alpha} - \frac{1}{4}\eta_{\mu\nu} F_{\alpha\beta} F^{\alpha\beta}\right], \tag{8.50}$$

which is called the *stress-energy-momentum tensor* (or stress-energy tensor, for short) of the electromagnetic field. It is straightforward to check that

$$T^{\mathrm{EM}}_{00} = \frac{1}{2\mu_0}\sum_i\left(\frac{1}{c^2}E_i^2 + B_i^2\right) = \frac{1}{2}\left(\epsilon_0|\boldsymbol{E}|^2 + \frac{1}{\mu_0}|\boldsymbol{B}|^2\right), \tag{8.51}$$

$$T^{\mathrm{EM}}_{0i} = T^{\mathrm{EM}}_{i0} = \frac{1}{\mu_0}\sum_j F_{0j}F_{ij} = -\frac{1}{\mu_0 c}(\boldsymbol{E}\times\boldsymbol{B})_i\,, \tag{8.52}$$

[14] In particular, the $\mu = 0$ component of this equation is $\partial^\alpha F_{\alpha\mu} = (1/c)\nabla\cdot\boldsymbol{E} = -\mu_0(-c\rho)$, that is, $\nabla\cdot\boldsymbol{E} = \mu_0 c^2\rho = \rho/\epsilon_0$.

$$T_{ij}^{\rm EM} = -\frac{1}{\mu_0}\left[\frac{1}{c^2}E_iE_j + B_iB_j - \frac{1}{2}\delta_{ij}\left(\frac{1}{c^2}|\boldsymbol{E}|^2 + |\boldsymbol{B}|^2\right)\right]. \tag{8.53}$$

Thus, $T_{00}^{\rm EM}$ is the energy density of the electromagnetic field, eq. (5.11). It can be seen that $-T_{0i}^{\rm EM}/c$ is just the momentum density, eq. (5.12), or equivalently, $-cT_{0i}^{\rm EM}$ is the Poynting flux, eq. (5.14). Finally, $-T_{ij}^{\rm EM}$ is just the stress tensor Θ_{ij} of the electromagnetic field, eq. (5.13). Thus, $T_{\mu\nu}^{\rm EM}$ combines into a single spacetime tensor the stress, energy, and momentum properties of the electromagnetic field—quantities that would be viewed as distinct in pre-relativity physics. The energy and momentum conservation relations (5.17) and (5.20) then can be rewritten as the single spacetime tensor equation:

$$\partial^\alpha T_{\alpha\mu}^{\rm EM} = F_{\nu\mu}J^\nu . \tag{8.54}$$

Note that $T_{\mu\nu}^{\rm EM}$ has vanishing trace: $\eta^{\mu\nu}T_{\mu\nu}^{\rm EM} = 0$.

It is a fundamental requirement of general relativity that all other forms of matter also can be assigned a stress-energy-momentum tensor, $T_{\mu\nu}^{\rm M}$, whose components have the same interpretation as given above for the electromagnetic stress-energy-momentum tensor. It also is a fundamental requirement of general relativity that the total stress-energy-momentum tensor $T_{\mu\nu}^{\rm EM} + T_{\mu\nu}^{\rm M}$ must satisfy an equation that expresses local conservation of energy and momentum. In the flat spacetime of special relativity, this conservation equation takes the form

$$\partial^\alpha \left[T_{\alpha\mu}^{\rm EM} + T_{\alpha\mu}^{\rm M}\right] = 0, \tag{8.55}$$

and thus

$$\partial^\alpha T_{\alpha\mu}^{\rm M} = -F_{\nu\mu}J^\nu . \tag{8.56}$$

In any inertial coordinates, the time component of eq. (8.56) is equivalent to the formula (5.19) for the rate at which energy per unit volume is transferred from the electromagnetic field to matter. The space components of eq. (8.56) are equivalent to the formula (5.21) for the Lorentz force density.

Finally, it is worth noting that—when expressed in terms of A_μ rather than ϕ and $\boldsymbol{A}$—the Lorenz gauge condition eq. (5.24) takes the simple form

$$\partial^\mu A_\mu = 0 . \tag{8.57}$$

Thus, the Lorenz gauge condition is a special relativistically covariant condition. Maxwell's equations in Lorenz gauge take the form

$$\Box A_\mu \equiv \partial^\alpha \partial_\alpha A_\mu = -\mu_0 J_\mu, \tag{8.58}$$

which is manifestly special relativistically covariant. The retarded solution to eq. (8.58) is

$$A_\mu = \frac{\mu_0}{4\pi}\int \frac{[J_\mu(t',\boldsymbol{x}')]_{\rm ret}}{|\boldsymbol{x}-\boldsymbol{x}'|}d^3x', \tag{8.59}$$

which is equivalent to eqs. (5.56) and (5.57).

Thus we have succeeded in writing all the equations and results of section 5.1 in special relativistically covariant form. What has been gained by doing this? After all, as we have seen, the equations of electrodynamics as reformulated above are identical to the equations of electrodynamics as given in section 5.1. However, this fact alone is very significant, because it shows that—unlike Newton's second law of particle mechanics—the theory of electromagnetism does not have to be modified to be compatible with the spacetime structure of special relativity. Furthermore, as I shall now explain, the formulation of the theory of electromagnetism in special relativistically covariant form gives rise to a major change in viewpoint on electromagnetic phenomena.

From the perspective of pre-relativity physics, the equations of section 5.1 make sense only in a preferred rest frame, presumably provided by an "aether." Electromagnetic waves would then naturally be thought of as propagating disturbances of the aether (much as sound waves are propagating disturbances of the air), and it would be natural to try to give a purely mechanical explanation of all electromagnetic phenomena—as Maxwell himself attempted to do. However, with the need for a preferred rest frame eliminated, the notion of an aether becomes entirely unnatural, and it becomes natural to view the electromagnetic field as a physical entity in its own right.

In the pre-relativity viewpoint, it would not be any more useful to ask how an observer moving with respect to the aether would interpret electromagnetic phenomena than it would be to ask how an observer moving through air would interpret sound waves. A simple description would be available only in the rest frame of the medium, and it would be most sensible to do all calculations in this rest frame. If one wanted to give a description of electromagnetic phenomena from the viewpoint of an observer moving through the aether, it would not be obvious how to define the electric and magnetic fields "seen" by such a moving observer. Whatever definition of electric and magnetic fields that one gave, with pre-relativity notions of spacetime structure, they could not satisfy an unmodified form of Maxwell's equations in the frame of the moving observer.

However, we have now seen that in special relativity, electromagnetic theory in the form given in section 5.1 holds in all inertial coordinates. All inertial observers would give exactly the same description of electromagnetic phenomena in terms of spacetime tensors. But if they define electromagnetic quantities in terms of particular components of these tensors, different observers will assign different values to these electromagnetic quantities. This difference in values is given by the tensor transformation law eq. (8.31). It is of interest to see more explicitly how this works in the case of charge and current densities ρ and $\boldsymbol{J}$, and in the case of the electromagnetic fields $\boldsymbol{E}$ and $\boldsymbol{B}$.

First consider the charge current J^{μ}, eq. (8.47). Since this quantity is a spacetime vector, by eq. (8.27), an observer O' at event p who is moving in the x-direction with velocity v with respect to observer O would assign to the 4-current at p the components

$$J'^{\mu} = \Lambda^{\mu}{}_{\nu} J^{\nu} . \tag{8.60}$$

Writing this out, we see that the charge density ρ and current density $\boldsymbol{J}$ transform as

$$\rho' = \gamma\rho - \gamma\frac{v}{c^2}J^1 , \tag{8.61}$$

$$J'^1 = \gamma J^1 - \gamma v \rho . \tag{8.62}$$

It should not be surprising that a charge density ρ would contribute to the current density seen by a moving observer, as found in eq. (8.62). However, is perhaps surprising that a current density $\boldsymbol{J}$ contributes to the charge density seen by a moving observer.

As we have seen, the electromagnetic field strengths $\boldsymbol{E}$ and $\boldsymbol{B}$ are, in fact, components of the spacetime tensor $F_{\mu\nu}$, eq. (8.43). Since $F_{\mu\nu}$ is a tensor of type $(0,2)$, by eq. (8.31), its components transform as

$$F_{\mu\nu} \to F'_{\mu\nu} = \sum_{\alpha,\beta} (\Lambda^{-1})^{\alpha}{}_{\mu} (\Lambda^{-1})^{\beta}{}_{\nu} F_{\alpha\beta} \,. \tag{8.63}$$

Thus, if observer O sees fields $\boldsymbol{E}$ and $\boldsymbol{B}$ present at event p, we find from this equation that an observer O' moving in the x-direction with velocity v with respect to observer O would determine that the fields at p are

$$E'_1 = E_1, B'_1 = B_1, \tag{8.64}$$

$$E'_2 = \gamma\,(E_2 - vB_3)\,, B'_2 = \gamma\left(B_2 + \frac{v}{c^2}E_3\right), \tag{8.65}$$

$$E'_3 = \gamma\,(E_3 + vB_2)\,, B'_3 = \gamma\left(B_3 - \frac{v}{c^2}E_2\right). \tag{8.66}$$

In other words, we have

$$\boldsymbol{E}'_{\parallel} = \boldsymbol{E}_{\parallel}, \boldsymbol{B}'_{\parallel} = \boldsymbol{B}_{\parallel}, \tag{8.67}$$

$$\boldsymbol{E}'_{\perp} = \gamma\,(\boldsymbol{E}_{\perp} + \boldsymbol{v}\times\boldsymbol{B})\,, \boldsymbol{B}'_{\perp} = \gamma\left(\boldsymbol{B}_{\perp} - \frac{1}{c^2}\boldsymbol{v}\times\boldsymbol{E}\right), \tag{8.68}$$

where "$\parallel$" and "$\perp$" respectively denote components parallel and perpendicular to $\boldsymbol{v}$. That $\boldsymbol{E}$ and $\boldsymbol{B}$ transform into each other under Lorentz transformations is a reflection of their being different components of a single spacetime tensor $F_{\mu\nu}$.

8.3 Charged Particle Motion and Radiation

Thus far in this book, apart from our treatment of electrostatics, we have largely avoided discussing point charges. The main reason is that, as discussed in section 1.4, at a fundamental level, charged matter must be viewed as continuously distributed rather than point-like. Nevertheless, as already stated in section 1.4, the idealization of a point charge is very useful for considering (i) the motion of a small charged body whose self-fields can be neglected[15] and (ii) the radiation emitted by a small charged body undergoing arbitrary motion. We have delayed discussing these topics until now, so that we can give a fully relativistic treatment of both charged particle motion and radiation from a charged particle.

8.3.1 CHARGED PARTICLE MOTION

As discussed at the end of section 8.1, Newton's second law in special relativity takes the form

$$u^{\nu}\partial_{\nu}p^{\mu} = \frac{dp^{\mu}}{d\tau} = f^{\mu}, \tag{8.69}$$

where $p^{\mu} = mu^{\mu}$. The electromagnetic 4-force f^{μ} exerted on a particle of charge q is obtained by taking the limit of eq. (8.56) as the matter and charge distribution shrinks

[15] Self-field effects will be considered in depth in chapter 10.

down to a worldline. This is a highly nontrivial limit in that, if taken at fixed charge, the electromagnetic self-energy of the body would diverge. We will deal with the issue of how to properly take the point charge limit in chapter 10. For now, we merely ignore the effects of the "self-field" of the charged body and replace $F_{\nu\mu}$ on the right side of eq. (8.56) by the external field $F^{\text{ext}}_{\nu\mu}$, into which the body is placed. The integral of the left side of eq. (8.56) over space at time t yields

$$\int \partial^\alpha T^{\text{M}}_{\alpha\mu} d^3x = \int \left[-\frac{1}{c}\frac{\partial T^{\text{M}}_{0\mu}}{\partial t} + \sum_{i=1}^{3} \partial_i T^{\text{M}}_{i\mu} \right] d^3x = -\frac{1}{c}\frac{d}{dt}\int T^{\text{M}}_{0\mu} d^3x = \frac{dp_\mu}{dt}, \quad (8.70)$$

where p^μ is the total matter 4-momentum of the body,[16] and we used Gauss's law and the boundedness of the matter distribution in the second equality to set the integral of the spatial divergence of the matter stress-energy tensor to zero.

To evaluate the spatial integral of the right side of eq. (8.56), we need an expression for the charge-current 4-vector J^μ of the body in the limit that it has shrunk down to a worldline. Since J^μ is a spacetime vector, it is clear that it must point along the direction of the 4-velocity u^μ of the worldline in this limit—there simply is no other candidate for the direction of J^μ. Clearly, J^μ must be localized on the worldline of the particle. Thus if the particle motion is described in some inertial coordinate system as $\boldsymbol{x} = \boldsymbol{X}(t)$ (where $\boldsymbol{X}(t)$ is a vector function of t with $|d\boldsymbol{X}/dt| < c$), we must have

$$J^\mu(t, \boldsymbol{x}) \propto \delta(\boldsymbol{x} - \boldsymbol{X}(t)) u^\mu . \quad (8.71)$$

Since $J^0 = c\rho$ (see eq. (8.47)), the constant of proportionality in eq. (8.71) can be fixed by requiring that the spatial integral of J^0 be equal to cq. Since $u^0 = c\gamma$, we thereby obtain

$$J^\mu(t, \boldsymbol{x}) = \frac{q}{\gamma}\delta(\boldsymbol{x} - \boldsymbol{X}(t)) u^\mu . \quad (8.72)$$

In terms of ρ and $\boldsymbol{J}$, we obtain

$$\rho(t, \boldsymbol{x}) = q\delta(\boldsymbol{x} - \boldsymbol{X}(t)), \qquad \boldsymbol{J}(t, \boldsymbol{x}) = q\frac{d\boldsymbol{X}}{dt}\delta(\boldsymbol{x} - \boldsymbol{X}(t)) . \quad (8.73)$$

The spatial integral of the right side of eq. (8.56) with $F_{\nu\mu}$ replaced by $F^{\text{ext}}_{\nu\mu}$ then yields

$$-\int F^{\text{ext}}_{\nu\mu} J^\nu d^3x = -\frac{q}{\gamma} F^{\text{ext}}_{\nu\mu} u^\nu . \quad (8.74)$$

Thus, equating the spatial integrals of the left and right sides of eq. (8.56) in the point charge limit, we obtain

$$\frac{dp_\mu}{dt} = -\frac{q}{\gamma} F^{\text{ext}}_{\nu\mu} u^\nu . \quad (8.75)$$

Using the fact that along the worldline of the particle, we have

[16] We will see in chapter 10 that the electromagnetic self-energy of the body—which is being neglected here—will contribute to its rest mass and 4-momentum.

$$\frac{dt}{d\tau} = \frac{1}{c}\frac{dX^0}{d\tau} = \frac{u^0}{c} = \gamma\,, \tag{8.76}$$

we see that eq. (8.75) can be put in the manifestly covariant form (8.69) with

$$f_\mu = -qF^{\text{ext}}_{\nu\mu}u^\nu\,. \tag{8.77}$$

Note that $u^\mu f_\mu = -qF^{\text{ext}}_{\nu\mu}u^\mu u^\nu = 0$, since $F^{\text{ext}}_{\mu\nu} = -F^{\text{ext}}_{\nu\mu}$. Thus, by eq. (8.38), the rest mass m of a point charge does not change when it moves under the influence of an external electromagnetic field.

In any inertial coordinates, the components of u^μ are $(c\gamma, \gamma\boldsymbol{v})$, and the components of $F^{\text{ext}}_{\mu\nu}$ are given by eq. (8.45). Thus, the spatial components of eq. (8.69) with f^μ given by eq. (8.77) take the form

$$m\frac{d(\gamma\boldsymbol{v})}{d\tau} = q\gamma\,(\boldsymbol{E} + \boldsymbol{v}\times\boldsymbol{B})\,, \tag{8.78}$$

where here and in the following, it is understood that $\boldsymbol{E}$ and $\boldsymbol{B}$ are the external fields. Using eq. (8.76), we see that eq. (8.78) is equivalent to

$$m\frac{d(\gamma\boldsymbol{v})}{dt} = m\frac{d(\gamma\boldsymbol{v})}{d\tau}\frac{d\tau}{dt} = q\,(\boldsymbol{E} + \boldsymbol{v}\times\boldsymbol{B})\,. \tag{8.79}$$

This differs from the usual nonrelativistic equation of motion for a charged particle found in elementary texts only in that on the left side of this equation, $\boldsymbol{v}$ is replaced by $\gamma\boldsymbol{v}$.

We now consider two important, simple examples of charged particle motion. The first example is motion in a uniform electric field $\boldsymbol{E}$, that is, in some inertial coordinates $(t, \boldsymbol{x})$, we have that $\boldsymbol{E}$ is independent of $(t, \boldsymbol{x})$ and $\boldsymbol{B} = 0$. Without loss of generality, we may assume that $\boldsymbol{E}$ points in the x-direction: $\boldsymbol{E} = E\hat{\boldsymbol{x}}$. We may choose the origin of coordinates so that $\boldsymbol{X}(0) = 0$ (i.e., the particle is at $\boldsymbol{x} = 0$ at $t = 0$). For simplicity, consider the case where the particle is initially at rest: $d\boldsymbol{X}/dt = 0$ at $t = 0$. It is clear from eq. (8.79) that we will have $Y(t) = Z(t) = 0$ for all t, so we need only consider the x-motion. We have

$$\frac{d(\gamma v)}{dt} = \frac{q}{m}E, \tag{8.80}$$

where $v = dX/dt$. Integration of this equation using $v = 0$ at $t = 0$ yields

$$\gamma v = \frac{q}{m}Et, \tag{8.81}$$

and hence,

$$v = \frac{dX}{dt} = \frac{qEt/m}{\left[1 + (qEt/(mc))^2\right]^{1/2}}\,. \tag{8.82}$$

Integration of this equation with the initial condition $X = 0$ at $t = 0$ yields the solution

$$X(t) = \frac{mc^2}{qE}\left\{\left[1 + (qEt/(mc))^2\right]^{1/2} - 1\right\}. \tag{8.83}$$

For $t \ll mc/(qE)$, we have $v \ll c$, and the solution reduces to

$$X(t) \approx \frac{1}{2}\frac{qE}{m}t^2 . \tag{8.84}$$

The second example is motion in a uniform magnetic field $\boldsymbol{B}$; that is, in some inertial coordinates $(t, \boldsymbol{x})$, we have that $\boldsymbol{B}$ is independent of $(t, \boldsymbol{x})$ and $\boldsymbol{E} = 0$. The equation of motion eq. (8.79) becomes

$$\frac{d(\gamma \boldsymbol{v})}{dt} = \frac{q}{m}\boldsymbol{v} \times \boldsymbol{B} . \tag{8.85}$$

Dotting this equation with $\boldsymbol{v}$, we find that $v = |\boldsymbol{v}|$ is constant. Since γ is constant, we may rewrite this equation as

$$\frac{d\boldsymbol{v}}{dt} = \frac{q}{\gamma m}\boldsymbol{v} \times \boldsymbol{B} . \tag{8.86}$$

The general solution to this equation is helical motion (i.e., linear motion parallel to $\boldsymbol{B}$ and circular motion perpendicular to $\boldsymbol{B}$). More explicitly, taking $\boldsymbol{B}$ to point along the z-axis so that $\boldsymbol{B} = B\hat{\boldsymbol{z}}$, we find that the general solution to eq. (8.86) is

$$\begin{aligned} X(t) &= X_0 + R_g \cos(\omega_g t + \varphi_0), \\ Y(t) &= Y_0 - R_g \sin(\omega_g t + \varphi_0), \\ Z(t) &= v_{z0} t + z_0 . \end{aligned} \tag{8.87}$$

Here X_0, Y_0, φ_0, v_{z0}, and z_0 are constants, and

$$\omega_g \equiv \frac{qB}{\gamma m}, \qquad R_g \equiv \frac{v_\perp}{|\omega_g|}, \tag{8.88}$$

where $v_\perp$ is the (constant) magnitude of the velocity[17] in the x-y plane. We refer to ω_g as the *gyrofrequency* (or *cyclotron frequency*), and we refer to R_g as the *gyroradius* (or *Larmor radius*). For nonrelativistic motion, $\gamma \approx 1$, the gyrofrequency is approximately

$$\omega_g \approx \frac{qB}{m}, \tag{8.89}$$

which is independent of the velocity.

8.3.2 RADIATION FROM A POINT CHARGE IN ARBITRARY MOTION

As discussed in section 8.1, the worldline of a particle in special relativity must be a timelike curve, that is, a curve $\boldsymbol{x} = \boldsymbol{X}(t)$ with $|d\boldsymbol{X}/dt| < c$. In principle, any timelike curve represents a possible particle motion. We wish to find the retarded solution for the electromagnetic field associated with a particle of charge q whose worldline is an arbitrary timelike curve.[18]

---

[17] We must, of course, have $v_{z0}^2 + v_\perp^2 < c^2$.

[18] As we saw in section 6.1, Maxwell's equations in a medium are the same as Maxwell's equations in vacuum with $\epsilon_0 \to \epsilon$, $\mu_0 \to \mu$, and $c \to c/n$. If $n > 1$, it is possible to have a charged particle in the medium move with velocity $v > c/n$, so that effectively, the charge has a spacelike worldline. This gives rise to the phenomenon of Cherenkov radiation (see problem 10).

The retarded solution with a point charge source is obtained by plugging eq. (8.72) into

$$A^\mu(t,\boldsymbol{x}) = \frac{\mu_0}{4\pi}\int \frac{[J^\mu(t',\boldsymbol{x}')]_{\rm ret}}{|\boldsymbol{x}-\boldsymbol{x}'|}d^3x', \tag{8.90}$$

or equivalently, by plugging eq. (8.73) into eqs. (5.56) and (5.57). We obtain

$$\phi(t,\boldsymbol{x}) = q\frac{\mu_0 c^2}{4\pi}\int \frac{\delta(\boldsymbol{x}'-\boldsymbol{X}(t_{\rm ret}))}{|\boldsymbol{x}-\boldsymbol{x}'|}d^3x', \tag{8.91}$$

$$\boldsymbol{A}(t,\boldsymbol{x}) = q\frac{\mu_0}{4\pi}\int \frac{d\boldsymbol{X}}{dt}(t_{\rm ret})\frac{\delta(\boldsymbol{x}'-\boldsymbol{X}(t_{\rm ret}))}{|\boldsymbol{x}-\boldsymbol{x}'|}d^3x', \tag{8.92}$$

where

$$t_{\rm ret} = t - \frac{1}{c}|\boldsymbol{x}-\boldsymbol{x}'|, \tag{8.93}$$

and we used $\epsilon_0\mu_0 c^2 = 1$ to eliminate ϵ_0 in favor of μ_0 and c in all expressions here and below. It is important to notice that the argument of the δ-function in the above integrals has a nontrivial dependence on $\boldsymbol{x}'$ because of the dependence of $t_{\rm ret}$ on $\boldsymbol{x}'$. Therefore, to evaluate the integrals in eqs. (8.91) and (8.92), we must use the fact that if $\boldsymbol{f}(\boldsymbol{x})$ is any vector function of $\boldsymbol{x}$ and $g(\boldsymbol{x})$ is any function of $\boldsymbol{x}$, then[19]

$$\int g(\boldsymbol{x})\delta(\boldsymbol{f}(\boldsymbol{x}))d^3x = \left.\frac{g}{|\mathcal{J}|}\right|_{\boldsymbol{f}(\boldsymbol{x})=0}, \tag{8.94}$$

where $\mathcal{J}$ denotes the determinant of the Jacobian matrix:

$$\mathcal{J}^i{}_j = \frac{\partial f^i}{\partial x^j}. \tag{8.95}$$

The Jacobian matrix relevant to the evaluation of the integrals in eqs. (8.91) and (8.92) is

$$\mathcal{J}^i{}_j = \frac{\partial\left[x'^i - X^i(t-|\boldsymbol{x}-\boldsymbol{x}'|/c)\right]}{\partial x'^j} = \delta^i{}_j - \frac{x^j - x'^j}{c|\boldsymbol{x}-\boldsymbol{x}'|}\frac{dX^i}{dt}(t_{\rm ret}). \tag{8.96}$$

The determinant of $\mathcal{J}^i{}_j$ is

$$\mathcal{J} = 1 - \left(\frac{\boldsymbol{x}-\boldsymbol{x}'}{c|\boldsymbol{x}-\boldsymbol{x}'|}\right)\cdot\frac{d\boldsymbol{X}}{dt}(t_{\rm ret}). \tag{8.97}$$

We now may perform the integrals in eqs. (8.91) and (8.92) to obtain our final result for the potentials of the retarded solution for a point charge in arbitrary motion:

$$\phi(t,\boldsymbol{x}) = \frac{\mu_0 c^2}{4\pi}\frac{1}{\alpha}\frac{q}{|\boldsymbol{x}-\boldsymbol{X}(t_{\rm ret})|}, \tag{8.98}$$

[19] Equation (8.94) follows from the fact that we can change coordinates to $\boldsymbol{y}=\boldsymbol{f}(\boldsymbol{x})$. The volume element d^3y in the new coordinates is related to d^3x by $d^3y = |\mathcal{J}|d^3x$. The integral on the left side of eq. (8.94) thus becomes $\int g(\boldsymbol{y})\delta(\boldsymbol{y})(1/|\mathcal{J}(\boldsymbol{y})|)d^3y$.

$$A(t,\boldsymbol{x}) = \frac{\mu_0}{4\pi}\frac{1}{\alpha}\frac{q}{|\boldsymbol{x}-\boldsymbol{X}(t_{\text{ret}})|}\frac{d\boldsymbol{X}}{dt}(t_{\text{ret}}), \tag{8.99}$$

where

$$\alpha = 1 - \frac{1}{c}\hat{\boldsymbol{n}}\cdot\frac{d\boldsymbol{X}}{dt}(t_{\text{ret}}), \tag{8.100}$$

with

$$\hat{\boldsymbol{n}} = \frac{\boldsymbol{x}-\boldsymbol{X}(t_{\text{ret}})}{|\boldsymbol{x}-\boldsymbol{X}(t_{\text{ret}})|} \tag{8.101}$$

(i.e., $\hat{\boldsymbol{n}}$ is the unit vector that points from the position of the particle at t_{ret} to the observation point $\boldsymbol{x}$). Equations (8.98) and (8.99) are called the *Lienard-Wiechert potentials.*

The electric and magnetic fields can be computed from the Lienard-Wiechert potentials in the usual manner via eqs. (5.2) and (5.3). However, when taking derivatives of ϕ and $\boldsymbol{A}$ with respect to t and $\boldsymbol{x}$, it is important to recognize that in eqs. (8.98) and (8.99), the quantity t_{ret} denotes the time of intersection of the past light cone of $(t,\boldsymbol{x})$ with the worldline of the particle. Thus, for the given particle worldline $\boldsymbol{X}(t)$, t_{ret} is the function of $(t,\boldsymbol{x})$ that is implicitly defined by

$$t_{\text{ret}} = t - \frac{1}{c}|\boldsymbol{x}-\boldsymbol{X}(t_{\text{ret}})|\,. \tag{8.102}$$

Taking the partial derivative of this equation with respect to t, we obtain

$$\frac{\partial t_{\text{ret}}}{\partial t} = 1 + \frac{1}{c}\hat{\boldsymbol{n}}\cdot\frac{d\boldsymbol{X}}{dt}(t_{\text{ret}})\frac{\partial t_{\text{ret}}}{\partial t}, \tag{8.103}$$

and hence

$$\frac{\partial t_{\text{ret}}}{\partial t} = \frac{1}{1-\frac{1}{c}\hat{\boldsymbol{n}}\cdot\frac{d\boldsymbol{X}}{dt}(t_{\text{ret}})} = \frac{1}{\alpha}\,. \tag{8.104}$$

Similarly, we find

$$\frac{\partial t_{\text{ret}}}{\partial x^i} = -\frac{1}{c}\sum_j \hat{\boldsymbol{n}}_j\left(\delta_{ij} - \frac{dX^j}{dt}\frac{\partial t_{\text{ret}}}{\partial x^i}\right), \tag{8.105}$$

which yields

$$\nabla t_{\text{ret}} = -\frac{1}{\alpha}\hat{\boldsymbol{n}}\,. \tag{8.106}$$

Using these relations, we can take the derivatives of the potentials (8.98) and (8.99) with respect to t and $\boldsymbol{x}$ to obtain the following formulas for the electric and magnetic fields:

$$\boldsymbol{E}(t,\boldsymbol{x}) = q\frac{\mu_0 c^2}{4\pi}\frac{\left(\hat{\boldsymbol{n}}-\frac{1}{c}\frac{d\boldsymbol{X}}{dt}\right)\left(1-\frac{1}{c^2}\left|\frac{d\boldsymbol{X}}{dt}\right|^2\right)}{\alpha^3|\boldsymbol{x}-\boldsymbol{X}(t_{\text{ret}})|^2} + q\frac{\mu_0}{4\pi}\frac{\hat{\boldsymbol{n}}\times\left[\left(\hat{\boldsymbol{n}}-\frac{1}{c}\frac{d\boldsymbol{X}}{dt}\right)\times\frac{d^2\boldsymbol{X}}{dt^2}\right]}{\alpha^3|\boldsymbol{x}-\boldsymbol{X}(t_{\text{ret}})|}, \tag{8.107}$$

$$c\boldsymbol{B}(t,\boldsymbol{x}) = \hat{\boldsymbol{n}}\times\boldsymbol{E}(t,\boldsymbol{x}), \tag{8.108}$$

where it is understood that $d\boldsymbol{X}/dt$ and $d^2\boldsymbol{X}/dt^2$ are evaluated at $t_{\rm ret}$. It should be emphasized that eqs. (8.107) and (8.108) are exact solutions and hold at all distances from the worldline $\boldsymbol{X}(t)$ of the point charge.

It is of interest to calculate the electromagnetic energy radiated to infinity, as given by the Poynting flux as $|\boldsymbol{x}| \to \infty$. The first term in eq. (8.107) falls off as $1/|\boldsymbol{x}|^2$, so only the second term contributes to the energy flux at infinity. In addition, the difference between $\hat{\boldsymbol{n}}$ and $\hat{\boldsymbol{x}}$ can be neglected at this order, where $\hat{\boldsymbol{x}}$ is the unit outward radial vector. We find that the radiated power per unit solid angle at a given retarded time $t_{\rm ret}$ is given by

$$\frac{dP}{d\Omega} = \lim_{|\boldsymbol{x}|\to\infty} |\boldsymbol{x}|^2 \boldsymbol{\mathcal{S}}\cdot\hat{\boldsymbol{x}} = \frac{1}{\mu_0}\lim_{|\boldsymbol{x}|\to\infty} |\boldsymbol{x}|^2 (\boldsymbol{E}\times\boldsymbol{B})\cdot\hat{\boldsymbol{x}}$$

$$= \frac{q^2\mu_0}{16\pi^2 c\alpha^6}\left|\hat{\boldsymbol{x}}\times\left[\left(\hat{\boldsymbol{x}}-\frac{1}{c}\frac{d\boldsymbol{X}}{dt}\right)\times\frac{d^2\boldsymbol{X}}{dt^2}\right]\right|^2, \tag{8.109}$$

where $d\boldsymbol{X}/dt$ and $d^2\boldsymbol{X}/dt^2$ are evaluated at $t_{\rm ret}$. Note that the limit $|\boldsymbol{x}|\to\infty$ at fixed retarded time means that the limit is being taken on spheres of radius $|\boldsymbol{x}|$ lying on the future light cone of the event on the worldline of the particle at $t = t_{\rm ret}$.

Since the right side of eq. (8.109) is most naturally evaluated on spheres of constant retarded time at infinity it is natural to define the notion of energy flux per unit solid angle per unit *retarded* time by

$$\frac{dP'}{d\Omega} \equiv \frac{dP}{d\Omega}\frac{dt}{dt_{\rm ret}} = \frac{dP}{d\Omega}\alpha = \frac{q^2\mu_0}{16\pi^2 c\alpha^5}\left|\hat{\boldsymbol{x}}\times\left[\left(\hat{\boldsymbol{x}}-\frac{1}{c}\frac{d\boldsymbol{X}}{dt}\right)\times\frac{d^2\boldsymbol{X}}{dt^2}\right]\right|^2. \tag{8.110}$$

The total flux of energy between two retarded times would then be obtained by integrating the right side of eq. (8.110) over each sphere of constant retarded time and then integrating with respect to $t_{\rm ret}$. As we shall see below, a simple formula can be obtained for the integral of $dP'/d\Omega$ over a sphere of constant retarded time.

If the particle is at rest at some event on its worldline, then at the retarded time corresponding to that event, eq. (8.109) with $d\boldsymbol{X}/dt = 0$ and $\alpha = 1$ reduces to

$$\left(\frac{dP}{d\Omega}\right)_{\frac{d\boldsymbol{X}}{dt}=0} = \frac{q^2\mu_0}{16\pi^2 c}\left|\hat{\boldsymbol{x}}\times\left[\hat{\boldsymbol{x}}\times\frac{d^2\boldsymbol{X}}{dt^2}\right]\right|^2 = \frac{q^2\mu_0}{16\pi^2 c}\left[\left|\frac{d^2\boldsymbol{X}}{dt^2}\right|^2 - \left(\hat{\boldsymbol{x}}\cdot\frac{d^2\boldsymbol{X}}{dt^2}\right)^2\right]. \tag{8.111}$$

This agrees with eq. (5.79), since the dipole moment of the charged particle is $\boldsymbol{p} = q\boldsymbol{X}$. In particular, the angular distribution of the energy flux has a purely dipole pattern (i.e., it varies with angle as $\sin^2\theta$, where θ is the angle between $d^2\boldsymbol{X}/dt^2$ and the observation point). The effects of nonzero velocity, $d\boldsymbol{X}/dt \neq 0$, on the nature of the angular distribution of the radiated power at the corresponding retarded time can be illustrated by considering the following two special cases. We orient our spatial axes so that $d\boldsymbol{X}/dt$ at the given retarded time instantaneously points in the z-direction: $d\boldsymbol{X}/dt = v\hat{\boldsymbol{z}}$. Then we have

$$\alpha = 1 - \frac{v}{c}\cos\theta\,. \tag{8.112}$$

First consider the case where $d^2\boldsymbol{X}/dt^2$ points in the same direction as $d\boldsymbol{X}/dt$—as occurs for linear acceleration. Then eq. (8.109) becomes

$$\left(\frac{dP}{d\Omega}\right)_{\parallel} = \frac{q^2\mu_0}{16\pi^2 c}\frac{\sin^2\theta}{(1-\frac{v}{c}\cos\theta)^6}\left|\frac{d^2\mathbf{X}}{dt^2}\right|^2, \tag{8.113}$$

whereas $(dP'/d\Omega)_{\parallel}$ would be given by the same formula with one fewer power of $(1-\frac{v}{c}\cos\theta)$ in the denominator. As the second example, consider the case where, at a given retarded time, $d^2\mathbf{X}/dt^2$ points orthogonally to $d\mathbf{X}/dt$—as occurs for circular motion. For relativistic velocities, the radiation in this case in known as *synchrotron radiation*. We further orient our axes so that $d^2\mathbf{X}/dt^2$ instantaneously points in the x-direction. Equation (8.109) then yields

$$\left(\frac{dP}{d\Omega}\right)_{\perp} = \frac{q^2\mu_0}{16\pi^2 c}\frac{1}{(1-\frac{v}{c}\cos\theta)^6}\left[(1-\frac{v}{c}\cos\theta)^2-(1-\frac{v^2}{c^2})\sin^2\theta\cos^2\varphi\right]\left|\frac{d^2\mathbf{X}}{dt^2}\right|^2, \tag{8.114}$$

where (θ,φ) are the spherical coordinates of the observation point, with the axes chosen as above. Again, $(dP'/d\Omega)_{\perp}$ would be given by the same formula with one fewer power of $(1-\frac{v}{c}\cos\theta)$ in the denominator. As eq. (8.113) and eq. (8.114) illustrate, if $v\to c$, the radiation becomes highly beamed in the forward direction. This beaming of the radiation as compared with that seen by an observer who is at rest relative to the particle can be understood as resulting from the combined effects of the Lorentz boosting of the field strengths, eq. (8.68), and aberration (see problem 4).

The total power radiated at a fixed retarded time is obtained by integrating $dP/d\Omega$ over angles:

$$P = \int \frac{dP}{d\Omega}d\Omega = \int \frac{dP}{d\Omega}\sin\theta\, d\theta\, d\varphi\,. \tag{8.115}$$

For the case of a particle instantaneously at rest at a given retarded time, the angular integration of eq. (8.111) was already carried out in eq. (5.81). We obtain Larmor's formula

$$P_0 = \frac{q^2\mu_0}{6\pi c}\left|\frac{d^2\mathbf{X}}{dt^2}\right|^2, \tag{8.116}$$

where the subscript 0 on P indicates that this formula holds only when the particle is at rest at the given retarded time. For a particle at rest, we have

$$a^\mu \equiv \frac{du^\mu}{d\tau} = (0, \frac{d^2\mathbf{X}}{dt^2})\,. \tag{8.117}$$

Therefore, we may rewrite this formula as

$$P_0 = \frac{q^2\mu_0}{6\pi c}a^\mu a_\mu\,. \tag{8.118}$$

If the particle is not at rest, the angular integrals needed to calculate the radiated power are quite complicated. However, the total power per unit retarded time,

$$P' = \int \frac{dP'}{d\Omega}d\Omega, \tag{8.119}$$

can be obtained quite simply by the following argument. In the instantaneous rest frame of the particle, the energy radiated to infinity over an infinitesimally small retarded time interval Δt_0 as measured in the rest frame is given by

$$\Delta\mathscr{E}_0 = P_0 \Delta t_0. \tag{8.120}$$

In the instantaneous rest frame of the particle, the electromagnetic field carries no net momentum to infinity during this infinitesimal time interval, as can be seen from the fact that the radiation flux eq. (8.111) is parity invariant. Now, by arguments similar to that of part (d) of problem 5, the 4-momentum radiated to infinity between the given retarded times transforms as a 4-vector under Lorentz boosts. Thus, in an arbitrary Lorentz frame, we have

$$\Delta\mathscr{E} = \gamma\Delta\mathscr{E}_0 = \gamma P_0 \Delta t_0, \tag{8.121}$$

with $\gamma = (1 - v^2/c^2)^{-1/2}$. By definition of P', we have

$$\Delta\mathscr{E} = P' \Delta t_{\text{ret}}\,. \tag{8.122}$$

Finally, in the arbitrary Lorentz frame, we have $\Delta t_{\text{ret}} = \gamma \Delta t_0$, since the elapsed coordinate time along the worldline of the particle is time dilated relative to the rest frame. Putting all of this together, we see that in an arbitrary frame, we have

$$P' = P_0 = \frac{q^2\mu_0}{6\pi c} a^\mu a_\mu\,. \tag{8.123}$$

Thus, in any inertial frame, the total energy radiated to infinity in that frame between retarded times t_1 and t_2 is given by

$$\mathscr{E} = \frac{q^2\mu_0}{6\pi c}\int_{t_1}^{t_2} a^\mu(t_{\text{ret}}) a_\mu(t_{\text{ret}}) dt_{\text{ret}}\,, \tag{8.124}$$

where, in the integral, t_{ret} is the retarded time coordinate in that inertial frame.

Problems

1. (a) Consider the path in the x-y plane from $(x, y) = (0, 0)$ to $(x, y) = (0, 1)$ given by

$$x = \alpha y(1 - y),$$

where α is an arbitrary constant. Calculate the length of this path. [Hint: Use the analog of eq. (8.13) for Euclidean geometry.] The fact that the length is α-dependent is an example of what might be called the "distance paradox" of Euclidean geometry: two paths connecting the *same* points $(0, 0)$ and $(0, 1)$ can have *different* total lengths.

(b) Consider the path in spacetime from $(x, ct) = (0, 0)$ to $(x, ct) = (0, 1)$ given by

$$x = \alpha ct(1 - ct),$$

where α is a constant with $|\alpha| < 1$. Calculate the elapsed proper time along this path. The fact that the elapsed proper time depends on α is an example of the twin paradox of special relativity: If one twin moves inertially ($\alpha = 0$) between $(0, 0)$ and $(0, 1)$ whereas the other twin moves noninertially ($\alpha \neq 0$) between these events, then the noninertial twin will be younger than the inertial twin when they rejoin.

2. Show that the Lorentz transformation eq. (8.28) satisfies eq. (8.30). Conversely, show that any linear map $L^{\mu}{}_{\nu}$ that satisfies eq. (8.30) and leaves vectors in the y- and z-directions invariant must be of the form of a Lorentz transformation, eq. (8.28), possibly composed with the time reversal transformation $t \to -t$ and the spatial reflection $x \to -x$. (Note: If we drop the restriction that the y- and z-directions are left invariant, any linear map that satisfies eq. (8.30) must be a composition of a Lorentz transformation in some direction, a rotation about some axis, and possibly time reversal and parity transformations. However, you are only being asked to consider the case of transformations that leave y and z invariant in this problem.)

3. Let Λ_1 be a Lorentz transformation of velocity v_1 in the x-direction. Let Λ_2 be a Lorentz transformation of velocity v_2 in the y-direction (i.e., Λ_2 is of the form eq. (8.28) with x and y interchanged). Define the commutator of Λ_1 and Λ_2 by $C = \Lambda_2^{-1}\Lambda_1^{-1}\Lambda_2\Lambda_1$. Assume that $v_1, v_2 \ll c$, and compute C to lowest nontrivial (i.e., quadratic) order in v_1, v_2. Show that, at this order, C corresponds to a rotation about the z-axis.

4. Suppose an observer O sees a light ray moving in the x-y plane at an angle θ with respect to the x-axis (i.e., suppose that $k^{\mu} = (k, k\cos\theta, k\sin\theta, 0)$, where $k = \omega/c$). Another observer O' moves with respect to O with velocity v in the x-direction. Do *not* assume that $v \ll c$.
   (a) Show that the frequency ω' of the light ray as seen by the observer O' is
   $$\omega' = \gamma(1 - \frac{v}{c}\cos\theta)\omega.$$
   This change in frequency is referred to as the *Doppler shift*.
   (b) Show that the angle θ' at which observer O' sees the light ray move is given by
   $$\tan\theta' = \frac{\sin\theta}{\gamma(\cos\theta - \frac{v}{c})}.$$
   This change in angle is referred to as the *aberration of light.*

5. Consider a charge-current distribution J^{μ} in spacetime that is confined to a bounded region of space (or falls off sufficiently rapidly in spacelike directions). Define the total charge Q at time $t = t_0$ by
   $$Q = \frac{1}{c}\int_{t=t_0} J^0 d^3x = \int_{t=t_0} \rho d^3x.$$
   (a) Show that Q is independent of t_0 (i.e., total charge is conserved).
   (b) Show that Q is invariant under Lorentz transformations (i.e., the charge Q' defined by a Lorentz boosted observer is equal to Q). [Hint: You cannot solve this problem by applying a Lorentz transformation to J^{μ}

and the volume element, because the surfaces of constant t' are different surfaces in spacetime than the surfaces of constant t, that is, the integrals defining Q' and Q are taken over different spacetime regions. Instead, use the spacetime version of Gauss's theorem, which states that for any spacetime vector field v^μ and any bounded spacetime region $\mathcal{V}$, we have

$$\int_{\mathcal{V}} \partial_\mu v^\mu = \int_{\partial\mathcal{V}} n_\mu v^\mu,$$

where n^μ is the unit normal to the boundary $\partial\mathcal{V}$, which is taken to be "outward pointing" where $\partial\mathcal{V}$ is timelike and "inward pointing" where $\partial\mathcal{V}$ is spacelike.][20]

(c) Similarly, for a stress-energy tensor $T^{\mu\nu}$ that is confined to a bounded region of space (or falls off sufficiently rapidly in spacelike directions), define the total 4-momentum at time t_0 by

$$P^\mu = \frac{1}{c}\int_{t=t_0} T^{0\mu}\, d^3x.$$

Show that conservation of stress-energy, $\partial_\mu T^{\mu\nu} = 0$, implies that P^μ is conserved (i.e., it is independent of t_0).

(d) Show that a Lorentz-boosted observer obtains the 4-momentum $P'^\mu = \Lambda^\mu{}_\nu P^\nu$; that is, P^μ transforms as a 4-vector under Lorentz boosts. [Hint: Let s_μ be any dual vector field whose components are constant in spacetime. Apply the result of part (b) to the vector field $T^{\mu\nu}s_\nu$.]

6. Define

$$\epsilon_{\mu\nu\lambda\sigma} = \begin{cases} 1, & \text{if } \mu\nu\lambda\sigma \text{ is an even permutation of } 0,1,2,3, \\ -1, & \text{if } \mu\nu\lambda\sigma \text{ is an odd permutation of } 0,1,2,3, \\ 0, & \text{otherwise.} \end{cases}$$

(a) Show that $\epsilon_{\mu\nu\lambda\sigma}$ is a tensor of type $(0,4)$ (i.e., this formula for $\epsilon_{\mu\nu\lambda\sigma}$ is preserved under eq. (8.31)).

(b) The *dual*, ${}^*F_{\mu\nu}$, of the electromagnetic field tensor $F_{\mu\nu}$ is defined by

$${}^*F_{\mu\nu} = \frac{1}{2}\epsilon_{\mu\nu\lambda\sigma}F^{\lambda\sigma}.$$

Show that the duality rotation eq. (5.10) corresponds to

$$F_{\mu\nu} \to F_{\mu\nu}\cos\alpha - {}^*F_{\mu\nu}\sin\alpha\,.$$

(c) Show that the electromagnetic stress-energy-momentum tensor eq. (8.50) is invariant under a duality rotation; that is, $T^{\rm EM}_{\mu\nu}$ is unchanged if in eq. (8.50), $F_{\mu\nu}$ is replaced by $F_{\mu\nu}\cos\alpha - {}^*F_{\mu\nu}\sin\alpha$.

---

[20] These rules for Gauss's theorem come from translating the generalized Stokes's theorem (which takes the same form, independently of metrical structure) into the Gauss theorem form (see the boxed side comment on Gauss's theorem in section 2.1). The rules for the case where part of the boundary of $\mathcal{V}$ is a null surface require more explanation to define n_μ and the volume element, but this case is not needed here.

7. An infinitely long cylinder of cylindrical radius a has a uniform charge density $\rho = \alpha$ distributed inside it. The current density vanishes; $\boldsymbol{J} = 0$.

   (a) Find $\boldsymbol{E}$ and $\boldsymbol{B}$ outside the cylinder. (This is an elementary electrostatics problem and a trivial magnetostatics problem.)
   (b) An observer O' moves with velocity v in a direction parallel to the axis of the cylinder. What is the charge density ρ' and the current density $\boldsymbol{J}'$ as determined by O'? Do *not* assume $v \ll c$.
   (c) Use elementary electrostatics and magnetostatics to obtain the electric field $\boldsymbol{E}'$ and magnetic field $\boldsymbol{B}'$ associated with the ρ' and $\boldsymbol{J}'$ found in part (b).
   (d) Show that the results of part (c) agree with what would be obtained by Lorentz transforming the results of part (a).

8. Show that the quantities $|\boldsymbol{E}|^2 - c^2|\boldsymbol{B}|^2$ and $\boldsymbol{E} \cdot \boldsymbol{B}$ are Lorentz scalars; that is, their values do not change under a Lorentz transformation (8.67), (8.68), so all observers agree on the numerical values of these quantities at any event. (The Lorentz invariance of these quantities is actually an immediate consequence of $\frac{1}{c^2}|\boldsymbol{E}|^2 - |\boldsymbol{B}|^2 = -\frac{1}{2}F_{\mu\nu}F^{\mu\nu}$, and $\frac{1}{c}\boldsymbol{E} \cdot \boldsymbol{B} = \frac{1}{4}{}^*F_{\mu\nu}F^{\mu\nu}$, where ${}^*F_{\mu\nu}$ is defined in problem 6. However, you are being asked in this problem to verify the invariance by direct calculation using eqs. (8.67) and (8.68).)

9. A uniform electric field points in the z-direction, $\boldsymbol{E} = E_0\hat{\boldsymbol{z}}$, and a uniform magnetic field points in the y-direction, $\boldsymbol{B} = B_0\hat{\boldsymbol{y}}$, where $c|B_0| > |E_0|$. At $t = 0$, a particle of charge q and mass m is placed at rest at the origin. Show that the subsequent motion of the particle is nonrelativistic for all time if and only if $c|B_0| \gg |E_0|$. Obtain the motion of the particle in parametric form in the general case, and obtain it in explicit form in the nonrelativistic case, $c|B_0| \gg |E_0|$. [Hint: Show that there is a Lorentz frame in which $\boldsymbol{E}' = 0$. Obtain the motion in this frame, and Lorentz transform back to the original frame.]

10. Consider a particle of charge q that moves on the *spacelike* worldline $X(t) = Y(t) = 0$, $Z(t) = vt$, with $v > c$. Of course, such motion with $v > c$ is not possible if the particle corresponds to physically acceptable matter, but there is no mathematical inconsistency in considering such a charged particle source term in Maxwell's equations—and charged particle motion with $v > c/n$ is possible in a medium with $n > 1$ (see part (d) of this problem). An observer is at rest at $\boldsymbol{x}_0 = (x_0, 0, 0)$, where $x_0 > 0$.

    (a) Show that the past light cone of the observer at $(t, \boldsymbol{x}_0)$ does not intersect the worldline of the charged particle when

    $$t < t_0 \equiv \frac{x_0}{c}\sqrt{1 - (c/v)^2}.$$

    Thus, the retarded solution vanishes at the location of the observer for $t < t_0$.
    (b) Show that if $t > t_0$, the past light cone of the observer at $(t, \boldsymbol{x}_0)$ intersects the worldline of the charged particle at 2 events. Obtain the retarded solution for the potentials $\phi, \boldsymbol{A}$ at $(t, \boldsymbol{x}_0)$ for $t > t_0$ in parallel

with the derivation of the Lienard-Wiechert potentials for a charged particle moving on a timelike worldline.

(c) Show that the solution of part (b) is singular at the time $t = t_0$. Show that this time corresponds to when an observer would receive a light ray from the source propagating at angle $\cos\theta = c/v$ with respect to the z-axis.

(d) Using the correspondence obtained in section 6.1, reformulate the results of parts (b) and (c) for a charged particle moving with velocity $v > c/n$ in a medium[21] with index of refraction n. Particle motion in a medium with $v > c/n$ is commonly observed, and the resulting solution is known as *Cherenkov radiation*. The angle, $\sin\theta_C \equiv c/nv$, corresponding to the complement of the angle found in part (c) is known as the *Cherenkov angle*.

11. In the Lienard-Wiechert solution (8.107) and (8.108)—and, more generally, for radiation arising from any bounded source—at a fixed retarded time at order $1/|\boldsymbol{x}|$, $\boldsymbol{E}$ and $\boldsymbol{B}$ are orthogonal to $\hat{\boldsymbol{x}}$, orthogonal to each other, and satisfy $|\boldsymbol{E}| = c|\boldsymbol{B}|$. Show that this implies that the electromagnetic field tensor $F_{\mu\nu}$ takes the form

$$F_{\mu\nu} = \frac{1}{|\boldsymbol{x}|}[s_\mu k_\nu - s_\nu k_\mu] + O(1/|\boldsymbol{x}|^2),$$

where k^μ is the radially outward pointing null vector, $k^\mu = (1, \hat{\boldsymbol{x}})$, and $s^\mu(t, \theta, \varphi)$ is a spacelike vector orthogonal to k^μ, $s^\mu k_\mu = 0$. Show, further, that the stress-energy-momentum tensor of the electromagnetic field takes the form

$$T^{\rm EM}_{\mu\nu} = \frac{f(t, \theta, \varphi)}{|\boldsymbol{x}|^2} k_\mu k_\nu + O(1/|\boldsymbol{x}|^3)\,.$$

[21]The instantaneous response model of section 6.1 would not be realistic for describing phenomena involving charged particle motion with $v > c/n$. The model of section 6.2 would be more realistic, but you are only being asked to consider the model of section 6.1 here. The frequency dependence of the index of refraction in a more realistic model would smooth out the singularity found in part (c).

CHAPTER 9

Electromagnetism as a Gauge Theory

As I have emphasized throughout this book, the fundamental fields in electromagnetism should be taken to be the potentials $A_\mu = (-\phi/c, A_1, A_2, A_3)$, not the field strengths. This chapter elucidates this view by explaining how the electromagnetic field can naturally be viewed as a "gauge field." Section 9.1 gives a Lagrangian formulation of electromagnetism and discuss its fundamental couplings to charged matter. We will see that the fundamental couplings involve A_μ in an essential way and cannot be written in terms of the field strengths. Section 9.2 shows how the electromagnetic field can be viewed as a "connection" that provides a notion of differentiation of charged fields. In this way, electromagnetism can be viewed as a simple example of a more general class of theories known as Yang-Mills theories. Finally, section 9.3 shows that this new viewpoint naturally would allow for the presence of magnetic monopoles, provided that the monopole charge is suitably quantized.

9.1 Lagrangian for the Electromagnetic Field and Its Interactions

It is possible—and extremely useful—to give a Lagrangian formulation of electromagnetism. By a Lagrangian formulation of a field theory, we mean providing a local function $\mathcal{L}$ of the fields and their derivatives—called the *Lagrangian density*—such that the field equations are obtained by extremizing the action

$$S = \int \mathcal{L} dt d^3x = \frac{1}{c} \int \mathcal{L} d^4x \tag{9.1}$$

(where $d^4x = dx^0 dx^1 dx^2 dx^3 = c dt d^3x$) with respect to all field variations that vanish outside a bounded spacetime region. A Lagrangian formulation a theory provides the theory with additional structure,[1] which plays an essential role in quantization. It also implies a relationship between symmetries of the theory and conservation laws. However, we shall not consider these aspects of the Lagrangian formulation here, but will merely use the Lagrangian formulation to provide a simple characterization of properties of the electromagnetic field and its interactions.

[1] This additional structure comes mainly from the "boundary terms" that arise in the variation of $\mathcal{L}$.

We first consider electromagnetism in the absence of any charged matter and give the Lagrangian density for the source-free Maxwell equations. The field variables in electromagnetism are the potentials, most naturally taken to be the spacetime dual vector field A_μ, eq. (8.41). The Lagrangian density for the source-free Maxwell equations is

$$\mathcal{L}_{\rm EM} = -\frac{1}{4\mu_0} F^{\mu\nu} F_{\mu\nu}, \tag{9.2}$$

where $F_{\mu\nu}$ is defined by

$$F_{\mu\nu} = \partial_\mu A_\nu - \partial_\nu A_\mu \tag{9.3}$$

(see eq. (8.43)). Since $F_{\mu\nu}$ is gauge invariant (i.e., it is unchanged under $A_\mu \to A_\mu + \partial_\mu \chi$), $\mathcal{L}_{\rm EM}$ clearly is gauge invariant. In terms of $\boldsymbol{E}$ and $\boldsymbol{B}$, we have

$$\mathcal{L}_{\rm EM} = \frac{1}{2\mu_0}\left(\frac{1}{c^2}|\boldsymbol{E}|^2 - |\boldsymbol{B}|^2\right). \tag{9.4}$$

Under a variation δA_ν of the field A_ν, it is easily seen that

$$\delta\mathcal{L}_{\rm EM} = -\frac{1}{\mu_0} F^{\mu\nu}\partial_\mu(\delta A_\nu). \tag{9.5}$$

It follows that the action will be extremized for all variations δA_μ that vanish outside a bounded region of spacetime if and only if

$$\partial_\mu F^{\mu\nu} = 0, \tag{9.6}$$

which is Maxwell's equations in the form eq. (8.49) with $J^\mu = 0$. Equivalently, if one started with the Lagrangian in the form eq. (9.4) and varied it with respect to ϕ and $\boldsymbol{A}$, one would obtain the source-free Maxwell's equations in the form eqs. (5.4) and (5.5), respectively.

Before considering the coupling of the electromagnetic field to charged matter, we make a side comment on the stress-energy-momentum tensor of the electromagnetic field. In general relativity, the field equations arise from an action obtained by adding the Einstein-Hilbert action for the spacetime metric to a curved spacetime generalization of the action, $S_{\rm M} = \int \mathcal{L}_{\rm M} d^4x$, of the matter fields, where x^μ are now arbitrary coordinates, and in this paragraph, we set $c = 1$. For the electromagnetic field, the natural generalization of the Lagrangian density eq. (9.2) to a curved spacetime is

$$\mathcal{L}_{\rm EM} = -\frac{1}{4\mu_0} g^{\mu\alpha} g^{\nu\beta} F_{\mu\nu} F_{\alpha\beta}\sqrt{-g}, \tag{9.7}$$

where $g^{\mu\nu}$ is the inverse of the curved spacetime metric $g_{\mu\nu}$, and g denotes the determinant of the metric.[2] The Einstein field equation of general relativity is obtained by extremizing the action with respect to variations of the spacetime metric. Einstein's equation takes the form

$$G_{\mu\nu} = 8\pi G T_{\mu\nu}, \tag{9.8}$$

---

[2]The factor of $\sqrt{-g}$ is needed to convert the coordinate volume element d^4x to a proper volume element in the action.

where G is the gravitational constant, $G_{\mu\nu}$ is the Einstein tensor (constructed from the curvature of the metric), and $T_{\mu\nu}$ is the *stress-energy-momentum tensor* of the matter, defined by

$$T_{\mu\nu} \equiv -\frac{2}{\sqrt{-g}} \frac{\delta S_{\mathrm{M}}}{\delta g^{\mu\nu}} . \tag{9.9}$$

For the electromagnetic Lagrangian eq. (9.7), we obtain

$$T^{\mathrm{EM}}_{\mu\nu} = \frac{1}{\mu_0} g^{\alpha\beta} \left[F_{\mu\alpha} F_{\nu\beta} - \frac{1}{4} g_{\mu\nu} g^{\rho\sigma} F_{\alpha\rho} F_{\beta\sigma} \right], \tag{9.10}$$

which agrees with eq. (8.50) when $g_{\mu\nu} = \eta_{\mu\nu}$. My purpose in mentioning the above is not to attempt to provide an introduction to general relativity—which cannot be adequately done in one paragraph—but to elucidate the following two points previously made in section 1.2: (i) The formula for the stress-energy-momentum tensor of the electromagnetic field is determined by the electromagnetic Lagrangian density, eq. (9.7). (ii) The stress-energy-momentum tensor acts as a source term for gravity, and in principle, it can be measured by its gravitational effects.

We now return to the flat spacetime of special relativity and consider the coupling of the electromagnetic field to fundamental charged matter. As best as is known at present, at a fundamental level, matter consists of fields. The standard model of particle physics describes all presently known fields and their interactions. The charged fields in the standard model (i.e., the ones that interact with the electromagnetic field) are the electron, muon, and tau leptons, the six types of quarks, and the $W^{\pm}$ gauge bosons. The quarks and leptons are described by spin-1/2 fields, and the $W^{\pm}$ fields are of spin-1. The charged fields that dominantly contribute to the electromagnetic properties of ordinary matter are the electron field and the up and down quark fields (which, along with the gluon field, comprise the proton and neutron). These fields are spin-1/2 fields and hence are fermionic, so many of their physical properties can only be properly described in the context of quantum theory. To avoid dealing with any fundamentally quantum issues—as well as to avoid having to introduce the notion of spinor fields—we will instead consider the coupling of the electromagnetic field to a charged scalar field $\Phi(t, \boldsymbol{x})$. As far as is known, there are no charged scalar fields in nature—the Higgs is the only scalar field in the standard model, and it is uncharged. Nevertheless, the coupling of the electromagnetic field to a charged scalar field shares all the essential features of the coupling of the electromagnetic field to all other charged fields, so nothing will be lost with regard to the issues of interest here by restricting consideration to a charged scalar field.

To motivate the form of the coupling of a charged field to the electromagnetic field, we first consider the coupling of the electromagnetic field to a classical point particle from the Lagrangian and Hamiltonian points of view. We then consider the coupling of the electromagnetic field to a nonrelativistic quantum mechanical particle described by a Schrodinger wave function. Finally, we describe the coupling of the electromagnetic field to a charged scalar field.

The Lorentz force equation of motion of a charged particle in an electromagnetic field is given in eqs. (8.69) and (8.77). The charge-current source term J^{μ} of a charged particle is given by eq. (8.72). We now show that the equations of motion for the particle can be derived from a Lagrangian, and the source term it contributes to Maxwell's equations can be obtained by adding the particle action to the electromagnetic action and extremizing the total action with respect to A_μ.

We can parametrize the worldline of a particle using the time coordinate t of some inertial coordinate system, so that the worldline is given by $\boldsymbol{x}=\boldsymbol{X}(t)$. The dynamical variables of the particle may then be taken to be $\boldsymbol{X}(t)$. We take the Lagrangian L_{P} of a particle of mass m and charge q in an electromagnetic field A_μ to be

$$L_{\mathrm{P}}=-\frac{mc^2}{\gamma}+\frac{q}{\gamma}u^\mu A_\mu(t,\boldsymbol{X}(t)), \tag{9.11}$$

where $\gamma=1/\sqrt{1-|\dot{\boldsymbol{X}}|^2/c^2}$, and $u^\mu=(c\gamma,\gamma\dot{\boldsymbol{X}})$ is the 4-velocity of the particle. For this Lagrangian, for $i=1,2,3$, we have[3]

$$\frac{\partial L_{\mathrm{P}}}{\partial \dot{X}^i}=m\gamma\dot{X}_i+qA_i(t,\boldsymbol{X}), \tag{9.12}$$

$$\frac{\partial L_{\mathrm{P}}}{\partial X^i}=\frac{q}{\gamma}u^\mu\frac{\partial A_\mu}{\partial x^i}(t,\boldsymbol{X})=-q\frac{\partial\phi}{\partial x^i}(t,\boldsymbol{X})+q\sum_{j=1}^{3}\dot{X}^j\frac{\partial A_j}{\partial x^i}(t,\boldsymbol{X})\,. \tag{9.13}$$

Thus, the Euler-Lagrange equations are

$$\begin{aligned}0&=\frac{d}{dt}\left(\frac{\partial L_{\mathrm{P}}}{\partial \dot{X}^i}\right)-\frac{\partial L_{\mathrm{P}}}{\partial X^i}\\&=m\frac{d}{dt}(\gamma\dot{X}_i)+q\frac{\partial A_i}{\partial t}+q\sum_{j=1}^{3}\frac{\partial A_i}{\partial x^j}\dot{X}^j+q\frac{\partial\phi}{\partial x^i}-q\sum_{j=1}^{3}\dot{X}^j\frac{\partial A_j}{\partial x^i}\\&=m\frac{d}{dt}(\gamma\dot{X}_i)-qE_i-q\sum_{j=1}^{3}F_{ij}\dot{X}^j,\end{aligned} \tag{9.14}$$

which readily can be seen to be equivalent to eq. (8.79).

To obtain the source term in Maxwell's equations contributed by a particle with Lagrangian eq. (9.11), we add the Maxwell action, $S_{\mathrm{EM}}=\int\mathcal{L}_{\mathrm{EM}}dtd^3x$, to the particle action:

$$S_{\mathrm{P}}=\int L_{\mathrm{P}}dt=\int L_{\mathrm{P}}\delta(\boldsymbol{x}-\boldsymbol{X}(t))dtd^3x\,. \tag{9.15}$$

We then vary the total action with respect to A_ν. The resulting equation of motion for the electromagnetic field is

$$0=\frac{q}{\gamma}u^\nu\delta(\boldsymbol{x}-\boldsymbol{X}(t))+\frac{1}{\mu_0}\partial_\mu F^{\mu\nu}, \tag{9.16}$$

[3] Here $\dot{X}_i=\sum_j e_{ij}\dot{X}^j$, where e_{ij} is the Euclidean metric eq. (8.9), so in Cartesian coordinates, $\dot{X}_i=\dot{X}^i$. We maintain this notational distinction here, because we are now carefully distinguishing between vectors and dual vectors and it does not make sense to take the sum of a vector and a dual vector.

which is equivalent to Maxwell's equations in the form eq. (8.49) with J^μ given by eq. (8.72). Thus, we see that the Lagrangian eq. (9.11) correctly reproduces both the Lorentz force equation of motion of the particle and the source term in Maxwell's equations.[4]

To consider the coupling of the electromagnetic field to a quantum mechanical particle, it is useful to start with the Hamiltonian formulation of a classical charged particle in an electromagnetic field. For the Lagrangian eq. (9.11), we define the *canonical momenta* p_i by

$$p_i = \frac{\partial L_{\mathrm{P}}}{\partial \dot{X}^i} = m\gamma \dot{X}_i + qA_i(t, \boldsymbol{X}) \,. \tag{9.17}$$

The Hamiltonian H_{P} corresponding to L_{P} is then defined by

$$H_{\mathrm{P}}(\boldsymbol{p}, \boldsymbol{X}) = \boldsymbol{p} \cdot \dot{\boldsymbol{X}} - L_{\mathrm{P}}(\boldsymbol{X}, \dot{\boldsymbol{X}}), \tag{9.18}$$

where it is understood that $\dot{\boldsymbol{X}}$ is to be eliminated in favor of $\boldsymbol{p}$ using eq. (9.17), which yields

$$\dot{\boldsymbol{X}} = \frac{\boldsymbol{p} - q\boldsymbol{A}}{\left[m^2 + \frac{1}{c^2}|\boldsymbol{p} - q\boldsymbol{A}|^2\right]^{1/2}} \,. \tag{9.19}$$

Substituting eq. (9.19) in eq. (9.18), we obtain

$$H_{\mathrm{P}}(\boldsymbol{p}, \boldsymbol{X}) = \left[m^2 c^4 + c^2 |\boldsymbol{p} - q\boldsymbol{A}|^2\right]^{1/2} + q\phi \,. \tag{9.20}$$

For nonrelativistic motion, this can be approximated as

$$H_{\mathrm{P}}(\boldsymbol{p}, \boldsymbol{X}) \approx mc^2 + \frac{1}{2m}|\boldsymbol{p} - q\boldsymbol{A}|^2 + q\phi \,. \tag{9.21}$$

Although it will not be needed for the developments of this chapter, it is worth making a short excursion to obtain a Hamiltonian for the source-free electromagnetic field in a similar manner. We define the canonical 4-momentum density π^μ of the electromagnetic field by

$$\pi^\mu = \frac{\partial \mathcal{L}_{\mathrm{EM}}}{\partial \dot{A}_\mu} \,. \tag{9.22}$$

Thus, we have

$$\pi^0 = 0, \qquad \pi^i = -\frac{1}{\mu_0 c^2} E^i \,. \tag{9.23}$$

The Hamiltonian density is

$$\mathcal{H}_{\mathrm{EM}} = \pi^\mu \dot{A}_\mu - \mathcal{L}_{\mathrm{EM}} \,, \tag{9.24}$$

[4] However, note that the coupled system, eqs. (9.14) and (9.16), does not have anything that sensibly could be called "solutions" since (9.16) implies that the fields blow up on the worldine of the particle, in which case, (9.14) does not make sense. This issue is addressed in chapter 10.

and the Hamiltonian is

$$H_{\text{EM}} = \int \mathcal{H}_{\text{EM}}\, d^3x = \int \left[-\frac{1}{\mu_0 c^2} \boldsymbol{E} \cdot (-\boldsymbol{E} - \boldsymbol{\nabla}\phi) - \frac{1}{2\mu_0} \left(\frac{1}{c^2} |\boldsymbol{E}|^2 - |\boldsymbol{B}|^2 \right) \right] d^3x$$

$$= \frac{1}{2\mu_0} \int \left[\frac{1}{c^2} |\boldsymbol{E}|^2 + |\boldsymbol{B}|^2 - \frac{2}{c^2} \phi \boldsymbol{\nabla} \cdot \boldsymbol{E} \right] d^3x, \tag{9.25}$$

where we integrated by parts on ϕ in the last line. It should be understood that the dynamical variables in this Hamiltonian are $\boldsymbol{A}$ and its canonical momentum $\boldsymbol{\pi} = -\boldsymbol{E}/(\mu_0 c^2)$; that is in the Hamiltonian, $\boldsymbol{B} = \boldsymbol{\nabla} \times \boldsymbol{A}$ should be viewed as a function of $\boldsymbol{A}$, but $\boldsymbol{E}$ is a dynamical variable in its own right and is not a function of the potentials. The potential ϕ is also a dynamical variable, but since its canonical momentum vanishes, it effectively plays the role of a Lagrange multiplier that enforces the source-free initial value constraint $\boldsymbol{\nabla} \cdot \boldsymbol{E} = 0$ (see eq. (5.139)).

Returning to the dynamics of a charged particle, we now consider the coupling of a nonrelativistic, quantum mechanical charged particle to the electromagnetic field. In Schrodinger quantum mechanics, the state of the particle is described by a complex wave function $\Psi(t, \boldsymbol{x})$, which gives the amplitude for finding the particle at point $\boldsymbol{x}$ at time t. The wave function satisfies the Schrodinger equation

$$-\frac{\hbar}{i} \frac{\partial \Psi}{\partial t} = \hat{H} \Psi, \tag{9.26}$$

where $\hbar$ is the reduced Planck's constant, and $\hat{H}$ is an operator corresponding to the classical Hamiltonian. In Schrodinger quantum mechanics, the position operator is represented by multiplication by $\boldsymbol{x}$, and the momentum operator is represented by $\hat{\boldsymbol{p}} = \frac{\hbar}{i} \boldsymbol{\nabla}$. This suggests that the Hamiltonian operator for a nonrelativistic charged particle in an electromagnetic field corresponding to the classical Hamiltonian eq. (9.21) should be

$$\hat{H} = \frac{1}{2m} \left(\frac{\hbar}{i} \boldsymbol{\nabla} - q\boldsymbol{A} \right) \cdot \left(\frac{\hbar}{i} \boldsymbol{\nabla} - q\boldsymbol{A} \right) + q\phi, \tag{9.27}$$

where we have dropped the constant term mc^2. The Schrodinger equation therefore takes the form

$$-\left(\frac{\partial}{\partial t} + i\frac{q}{\hbar}\phi \right) \Psi = \frac{1}{2m} \left(\boldsymbol{\nabla} - i\frac{q}{\hbar}\boldsymbol{A} \right) \cdot \left(\boldsymbol{\nabla} - i\frac{q}{\hbar}\boldsymbol{A} \right) \Psi . \tag{9.28}$$

Note that for a classical charged particle, the Lagrangian eq. (9.11) has an essential dependence on A_μ that cannot be expressed in terms of the field strengths. Nevertheless, the resulting equation of motion (9.14) for a classical charged particle can be written purely in terms of the field strengths. However, potentials enter the equation of motion (9.28) of a quantum charged particle in an essential way (i.e., the Schrodinger equation for a charged particle cannot be written in terms of the field strengths).

The above discussion of the coupling of the electromagnetic field to classical and (nonrelativistic) quantum particles was given mainly to provide a motivation for the coupling of the electromagnetic field to fundamental charged matter: relativistic quantum fields. By a *field*, I simply mean a tensor quantity (or more generally, a spinor quantity—however, I will not attempt to define spinors here) that is defined at every event $(t, \boldsymbol{x})$ of spacetime. The simplest example of a field is a function $\Phi(t, \boldsymbol{x})$ on spacetime, which would be referred to as a "scalar field." An example of a (dual) vector field occurring in nature that should be very familiar by now to all readers is the electromagnetic field $A_\mu(t, \boldsymbol{x})$. Leptons and quarks are described by spinor fields. In a field theory, the dynamical variables are the fields themselves. States of a free field (or asymptotic states of an interacting field) can be given an interpretation in terms of "particles," but the dynamics are fundamentally described by the fields. Since the fields are observables, in quantum field theory, the fields are represented by operators. However, we shall be concerned only with classical fields here. As previously stated, we restrict our consideration here to describing the coupling of the electromagnetic field to a charged scalar field. Although no charged scalar field is known to exist in nature, the coupling of the electromagnetic field to a charged scalar field shares all essential features of the coupling of the electromagnetic field to all other charged fields.

In the absence of electromagnetic coupling, a *Klein-Gordon charged scalar field* is a complex function $\Phi(t, \boldsymbol{x})$ on spacetime whose Lagrangian density is given by

$$\mathcal{L}_{\mathrm{KG}} = -\frac{1}{2}\left[\eta^{\mu\nu}\partial_\mu \Phi^* \partial_\nu \Phi + \left(\frac{mc}{\hbar}\right)^2 |\Phi|^2\right], \tag{9.29}$$

where the $*$ denotes complex conjugation. The equation of motion arising from eq. (9.29) is

$$\Box \Phi - \left(\frac{mc}{\hbar}\right)^2 \Phi = 0\,. \tag{9.30}$$

Here, the presence of $\hbar$ in eq. (9.29) and eq. (9.30) does not indicate anything necessarily quantum about Φ—we will treat it as a classical field—but $\hbar$ is introduced so that in the quantum theory, the one-particle states will have mass m (in the sense that $\eta^{\mu\nu}p_\mu p_\nu = -m^2$, where p_μ denotes the 4-momentum of the one-particle states). We refer to m as the *mass* of the Klein-Gordon field, and we refer to the quantity $\hbar/mc$ as the *Compton wavelength* of the field. Note that if $m = 0$ and if the field were taken to be real, it would correspond to the field ψ used in many places in this book as a toy model for the electromagnetic field.

Nonrelativistic solutions to eq. (9.30) correspond to ones in which space derivatives of Φ are small compared with the inverse Compton wavelength. These solutions oscillate in time nearly as $\Phi \sim e^{\pm imc^2 t/\hbar}$. For the solutions that oscillate nearly as $e^{-imc^2 t/\hbar}$, we write

$$\Phi(t, \boldsymbol{x}) = e^{-imc^2 t/\hbar}\Upsilon(t, \boldsymbol{x}), \tag{9.31}$$

and substitute this expression in eq. (9.30). Neglecting the term where both time derivatives act on Υ, we obtain

$$0 = 2i\frac{m}{\hbar}\frac{\partial \Upsilon}{\partial t} + \nabla^2 \Upsilon, \tag{9.32}$$

which is identical to the Schrodinger equation for a free particle. In the quantum field theory of the Klein-Gordon field Φ, the positive frequency solutions to eq. (9.30) span a Hilbert space of one-particle states. The fact that nonrelativistic positive frequency solutions correspond to solutions to the Schrodinger equation allows us to make a correspondence between the quantum field theory of Φ and the quantum mechanics of particles.

We now turn to the coupling of Φ to an electromagnetic field. We want Φ to satisfy a relativistically covariant equation (i.e., we want the coupling to be compatible with the spacetime structure of special relativity). In view of the comments made in the previous paragraph, we also want the coupling to be such that in the nonrelativistic limit, the equation satisfied by Φ reduces to the Schrodinger equation (9.28) for a charged particle in an electromagnetic field. There is a simple and natural way to achieve both goals: In the Lagrangian density eq. (9.29), we make the replacement

$$\partial_\mu \Phi \to \partial_\mu \Phi - i\frac{q}{\hbar} A_\mu \Phi, \tag{9.33}$$

so that the Klein-Gordon Lagrangian density becomes

$$\mathcal{L}_{\rm KG}^{\rm EM} = -\frac{1}{2}\eta^{\mu\nu}(\partial_\mu \Phi^* + i\frac{q}{\hbar}A_\mu \Phi^*)(\partial_\nu \Phi - i\frac{q}{\hbar}A_\nu \Phi) - \frac{1}{2}\left(\frac{mc}{\hbar}\right)^2 |\Phi|^2 . \tag{9.34}$$

For this Lagrangian density, the equation of motion satisfied by Φ in the presence of an electromagnetic field is

$$\eta^{\mu\nu}(\partial_\mu - i\frac{q}{\hbar}A_\mu)(\partial_\nu - i\frac{q}{\hbar}A_\nu)\Phi - \left(\frac{mc}{\hbar}\right)^2 \Phi = 0 . \tag{9.35}$$

Taking the same sort of nonrelativistic limit as described in the previous paragraph, we obtain the Schrodinger equation (9.28), as desired. Again, there is nothing necessarily "quantum" about the presence of $\hbar$ in the coupling term $iq/\hbar$ (i.e., we may treat Φ as a classical field). The factor of $\hbar$ is introduced so that in the quantum field theory associated with Φ, the one-particle states will have charge q.

The total Lagrangian density for a Klein-Gordon field interacting with the electromagnetic field is obtained by adding the Klein-Gordon Lagrangian density eq. (9.34) to the electromagnetic Lagrangian density eq. (9.2). Extremization of the total action with respect to A_ν yields

$$0 = \frac{1}{\mu_0}\partial_\mu F^{\mu\nu} - \frac{iq}{2\hbar}\left[\Phi^*\left(\partial^\nu \Phi - i\frac{q}{\hbar}A^\nu \Phi\right) - \Phi\left(\partial^\nu \Phi^* + i\frac{q}{\hbar}A^\nu \Phi^*\right)\right] . \tag{9.36}$$

This corresponds to Maxwell's equations in the form eq. (8.49) with J^μ given by

$$J^\mu = \frac{q}{\hbar}{\rm Im}\left[\Phi^*\left(\partial^\nu \Phi - i\frac{q}{\hbar c}A^\nu \Phi\right)\right], \tag{9.37}$$

where "Im" stands for the imaginary part.

Equations (9.35) and (9.36) describe the coupled interactions of a charged Klein-Gordon field with the electromagnetic field. They are a well-posed system of equations. These equations involve the potentials A_μ in an essential way (i.e., the equations cannot be written in terms of the field strengths). In section 9.2, we consider the issue of the gauge invariance of these equations as well as further issues concerning their mathematical interpretation.

9.2 Gauge Invariance and the Reinterpretation of the Electromagnetic Field as a Connection

Up to this point in the book, by a gauge transformation, we have meant the transformation

$$A_\mu \to A'_\mu = A_\mu + \partial_\mu \chi, \tag{9.38}$$

where $\chi(t, \boldsymbol{x})$ is an arbitrary function on spacetime. In view of the manner in which fundamental charged matter couples to the electromagnetic field as seen in eqs. (9.35) and (9.36), we are about to make a major change to this notion. It is readily seen that these equations are not invariant under the transformation eq. (9.38) alone. Nevertheless, these equations—as well as the Lagrangian eq. (9.34)—are invariant under the combined transformation of both Φ and A_μ given by

$$\Phi \to \Phi' = e^{i\frac{q}{\hbar}\chi}\,\Phi\,, \qquad A_\mu \to A'_\mu = A_\mu + \partial_\mu \chi\,. \tag{9.39}$$

We now view eq. (9.39) *as defining the fundamental notion of gauge transformations for the coupled Maxwell-charged scalar field system.* If (Φ, A_μ) and (Φ', A'_μ) are related by eq. (9.39), then they are considered to be physically equivalent.

The gauge transformation eq. (9.39) acts on the charged field Φ by multiplying it by a spacetime dependent phase factor $e^{i\frac{q}{\hbar}\chi}$. Multiplication by a phase factor corresponds to a unitary map on complex numbers. More precisely, for any real number α, the map $U: \mathbb{C} \to \mathbb{C}$ defined by $Uz = e^{i\frac{q}{\hbar}\alpha} z$ for all $z \in \mathbb{C}$ is a unitary map (i.e., it is a linear map that has the property $|Uz|^2 = |z|^2$). Furthermore, all unitary maps on $\mathbb{C}$ are of this form. The unitary maps on $\mathbb{C}$ naturally comprise a group, with "multiplication" defined by composition of the maps. This group is denoted U(1)—where the "U" stands for "unitary" and the "1" stand for the 1-dimensional complex vector space $\mathbb{C}$. Note that since unitary maps have the form $Uz = e^{i\frac{q}{\hbar}\alpha} z$, the unitary maps are in one-to-one correspondence with real numbers α modulo $2\pi\hbar/q$. Thus, the group U(1) can be naturally identified with a circle S^1 of circumference $2\pi\hbar/q$. The key conclusion from these remarks is that *the gauge transformations eq.* (9.39) *correspond to the action of the group* U(1) *on charged fields*, where this action has arbitrary dependence on the spacetime point. The electromagnetic field A_μ transforms correspondingly, as given by eq. (9.39).

The quantity $\partial_\mu \Phi - i\frac{q}{\hbar} A_\mu \Phi$ appearing in the Lagrangian eq. (9.34) (as well as in eqs. (9.35) and (9.36)) can naturally be viewed as defining a notion of differentiation of the charged field Φ. Specifically, we define the *gauge covariant derivative operator* $\mathcal{D}_\mu$ by

$$\mathcal{D}_\mu = \partial_\mu - i\frac{q}{\hbar} A_\mu\,. \tag{9.40}$$

Under the gauge transformation eq. (9.39), we have

$$\mathcal{D}_\mu \Phi \to e^{i\frac{q}{\hbar}\chi}\,\mathcal{D}_\mu \Phi, \tag{9.41}$$

so $\mathcal{D}_\mu \Phi$ transforms in the same way as the charged field Φ itself. The Klein-Gordon Lagrangian density may be written in a manifestly gauge invariant form as

$$\mathcal{L}^{\rm EM}_{\rm KG} = -\frac{1}{2}\eta^{\mu\nu}(\mathcal{D}_\mu \Phi)^* \mathcal{D}_\nu \Phi - \frac{1}{2}\left(\frac{mc}{\hbar}\right)^2 |\Phi|^2\,. \tag{9.42}$$

The gauge covariant notion of differentiation defined by the operator $\mathcal{D}_\mu$ has a close mathematical similarity to a notion of covariant differentiation that can be defined in ordinary Riemannian (or Lorentzian) geometry. As mentioned below eq. (8.32), given a metric $g_{\mu\nu}$, there is an associated natural notion of covariant differentiation of tensor fields. In a flat geometry, this notion corresponds to taking the partial derivatives of the components of the tensor field in Cartesian/inertial coordinates, but in a non-flat geometry (or in non-Cartesian/inertial coordinates in a flat geometry), one must correct the partial derivative by adding "Christoffel terms" in a manner analogous to eq. (9.40). The covariant derivative of a tensor field then defines a coordinate-independent notion of differentiation of a tensor field in the same sense that eq. (9.40) defines a gauge-independent notion of differentiation of Φ.

In ordinary geometry, the notion of *curvature* is defined in terms the failure of successive covariant differentiations of a tensor field to commute.[5] It is natural to define a notion of the "curvature of the electromagnetic field" in terms of the failure of successive applications of the gauge covariant derivative operator $\mathcal{D}_\mu$ to commute; that is, we define the curvature $\mathcal{C}_{\mu\nu}$ of the electromagnetic field by

$$\left[\mathcal{D}_\mu, \mathcal{D}_\nu\right]\Phi = \mathcal{D}_\mu\mathcal{D}_\nu\Phi - \mathcal{D}_\nu\mathcal{D}_\mu\Phi = \mathcal{C}_{\mu\nu}\Phi\,. \tag{9.43}$$

A simple computation gives

$$\mathcal{C}_{\mu\nu} = -i\frac{q}{\hbar}\left(\partial_\mu A_\nu - \partial_\nu A_\mu\right) = -i\frac{q}{\hbar}F_{\mu\nu}\,. \tag{9.44}$$

Thus, we see that the electric and magnetic fields $\boldsymbol{E}$ and $\boldsymbol{B}$—as represented by the field strength tensor $F_{\mu\nu}$—have the interpretation of being the curvature associated with the gauge covariant derivative operator $\mathcal{D}_\mu$.

Up to this point, we have treated the 4-vector potential $A_\mu = (-\phi/c, A_1, A_2, A_3)$ as a dual vector field $A_\mu(t, \boldsymbol{x})$ on spacetime. However, since A_μ and $A'_\mu = A_\mu + \partial_\mu\chi$ represent the same physical electromagnetic field, an electromagnetic field really corresponds to an equivalence class of 4-vector potentials, where two 4-vector potentials are considered to be equivalent if they differ by a gauge transformation. It is somewhat awkward to work with gauge equivalence classes. For many purposes, this awkwardness can be avoided by working with the gauge invariant field strength $F_{\mu\nu}$. However, as we have seen above, the coupling of the electromagnetic field to charged fields cannot be expressed in terms of $F_{\mu\nu}$, so we must describe the electromagnetic field in terms of A_μ. Therefore, it is very useful to give an alternative characterization of A_μ as a uniquely defined quantity on a higher-dimensional space, as we shall now do.

Let $M = \mathbb{R}^4$ denote spacetime (i.e., the 4-dimensional set of events), labeled by $(t, \boldsymbol{x})$. Consider the 5-dimensional space $\mathcal{P} = M \times \mathrm{U}(1)$ obtained by taking the Cartesian product of spacetime M with the group $\mathrm{U}(1)$. As noted above, we can identify $\mathrm{U}(1)$ with the circle S^1 of circumference $2\pi\hbar/q$. Thus, points in $\mathcal{P}$ can be labeled as (x^μ, s), where x^μ is an event in spacetime, and s is a real number modulo $2\pi\hbar/q$. The group $\mathrm{U}(1)$ naturally acts on $\mathcal{P}$ via $U(x^\mu, s) = (x^\mu, U(s))$, where $U(s)$ denotes the composition of the unitary map U with the unitary map represented by s. This action gives $\mathcal{P}$ the structure of a *principal fiber bundle* with *structure group* $\mathrm{U}(1)$ and *base space* M.

[5]As previously mentioned, in a flat geometry, covariant differentiation corresponds to partial differentiation in Cartesian/inertial coordinates. Thus, successive covariant differentiations commute (i.e., the curvature vanishes).

The orbits of the U(1) action (i.e., the points in $\mathcal{P}$ corresponding to any given fixed x^μ) are referred to as the *fibers* of $\mathcal{P}$. A fiber bundle with the global structure of a Cartesian product as considered here is said to be a "trivial bundle." As we discuss further in section 9.3, a general principal fiber bundle $\mathcal{P}$ will have a Cartesian product structure only in a local sense, but we consider only $\mathcal{P} = M \times \mathrm{U}(1)$ for now.

A *connection*[6] $\mathscr{A}_\Lambda$ (with $\Lambda = 0, 1, 2, 3, 4$) is a dual vector field on $\mathcal{P}$ with the property that $\mathscr{A}_\Lambda(x^\mu, s)$ is independent of s and $\mathscr{A}_4 = 1$, where the 4-direction is the direction tangent to U(1). Since $\mathscr{A}_\Lambda(x^\mu, s)$ is independent of s, we may identify its $\Lambda = 0, 1, 2, 3$ components with a dual vector field $A_\nu(x^\mu)$ on M. However, on $\mathcal{P}$, we may perform the coordinate transformation

$$s \to s' = s - \chi(x^\mu) \tag{9.45}$$

for an arbitrary function $\chi(x^\mu)$, since this transformation preserves the action of U(1) on $\mathcal{P}$ and leaves the fundamental structure of $\mathcal{P}$ unchanged. Under this coordinate transformation, the components of the dual vector field $\mathscr{A}_\Lambda$ on $\mathcal{P}$ transform as

$$\mathscr{A}_4 \to \mathscr{A}'_4 = \mathscr{A}_4 = 1\,; \qquad \mathscr{A}_\Lambda \to \mathscr{A}'_\Lambda = \mathscr{A}_\Lambda + \partial_\Lambda \chi\,, \quad \text{for } \Lambda = 0, 1, 2, 3\,. \tag{9.46}$$

The dual vector field on M corresponding to $\mathscr{A}'_\Lambda$ is

$$A'_\mu = A_\mu + \partial_\mu \chi\,. \tag{9.47}$$

Thus, we see that the gauge equivalence class of A_μ on M corresponds to different coordinate representations under the coordinate transformation eq. (9.45) of the single object $\mathscr{A}_\Lambda$ on $\mathcal{P}$. It is natural to view this single object $\mathscr{A}_\Lambda$ on $\mathcal{P}$ as providing the fundamental description of the electromagnetic field.

The above considerations allow us to answer the following question that has been asked for many centuries: What are electricity and magnetism? *Electricity and magnetism are the phenomena arising from the electromagnetic field and its coupling to charged fields. The electromagnetic field itself is a connection on a principal fiber bundle over spacetime with structure group* U(1). This is undoubtedly a quite different answer than would have been provided by, say, Gilbert, Coulomb, Faraday, or Maxwell.

It is very much worth mentioning that an analogous mathematical construction to what has been described in this section can be performed with the group U(1) replaced by an arbitrary Lie group G. "Charged fields" then would be correspondingly replaced by fields valued in a vector space that carries a representation of G. However, the following aspects of the construction are somewhat modified and/or must be done with a bit more care: (i) If G is non-abelian, one must distinguish clearly between "right" and "left" actions of G. (ii) A connection on $\mathcal{P}$ must now be defined to be a dual vector field on $\mathcal{P}$ that is valued in the Lie algebra of G. (The Lie algebra of U(1) may be identified with the real numbers, so we were able to treat $\mathscr{A}_\Lambda$ above as an ordinary dual vector field on $\mathcal{P}$ in that case.) (iii) The curvature eq. (9.43) will pick up a term involving the Lie bracket of the connection must with itself. The curvature will thereby depend nonlinearly on

[6]The word "connection" arises as follows. The vectors V^Λ at any point in $\mathcal{P}$ that satisfy $\sum_\Lambda V^\Lambda \mathscr{A}_\Lambda = 0$ define a subspace of the same dimension as M. Using these "horizontal subspaces," we can uniquely "lift" any curve in M passing through a point $x \in M$ to a curve in $\mathcal{P}$ passing through a point p in the fiber over x. This gives us a way (dependent on a curve in M) to connect points on different fibers. The ability to do this is directly related to the notion of gauge covariant differentiation defined above. However, we shall not pursue these issues further here.

the connection. (iv) The above condition that $\mathscr{A}_4 = 1$ must be replaced by the condition that the connection must map vectors tangent to G-orbit directions in $\mathcal{P}$ into the corresponding Lie algebra vectors in G. (v) The above condition that $\mathscr{A}_\Lambda(x^\mu, s)$ is independent of s must be replaced by an equivariance condition. It would take me much too far afield to explain these differences in more detail (and define "Lie group," "Lie algebra," "equivariance," etc.), so I shall not attempt to provide any further elucidation of these points here.

For the case of a "compact, semi-simple" Lie group—terms that I also shall not attempt to explain here—a positive-definite metric can be defined on the Lie algebra that is invariant under the (left and right) Lie group action. Using this metric, one can define a natural generalization of the Maxwell Lagrangian eq. (9.2), where the dynamical field is taken to be the connection. The resulting generalization of electromagnetism is known as *Yang-Mills theory*. The Yang-Mills field interacts with fields carrying a Yang-Mills charge via a generalization of the gauge covariant differentiation given by eq. (9.40).

My reason for mentioning the Yang-Mills generalization of electromagnetism to other Lie groups is that nature appears to have made significant use of this generalization. The strong interactions are described by a Yang-Mills theory involving the gauge group SU(3). The electroweak interactions are described by a Yang-Mills theory associated with the gauge group SU(2) × U(1). As a consequence of its interaction with the Higgs field, the SU(2) × U(1) gauge symmetry is "spontaneously broken" to the U(1) subgroup[7] of electromagnetism. I will not attempt to explain the notion of spontaneous symmetry breaking here, but it gives rise to very different properties of the weak and electromagnetic interactions at the energy scales normally considered. However, fundamentally, both the weak and electromagnetic interactions are aspects of an SU(2) × U(1) gauge field interaction. It is possible that, similarly, the strong and electroweak interactions may be aspects of a single "grand unified" theory involving a Yang-Mills gauge group that contains SU(3) × SU(2) × U(1). Whether or not this is the case, it is clear that gauge fields of the Yang-Mills type underlie the fundamental interactions of nature. Electromagnetism is the simplest example of a Yang-Mills gauge theory.

9.3 Dirac Magnetic Monopoles

In Section 9.2, we defined the electromagnetic field as a connection on the space $\mathcal{P} = M \times \mathrm{U}(1)$. In fact, although it is essential that $\mathcal{P}$ have a Cartesian product structure of this sort locally, there is no mathematical necessity for $\mathcal{P}$ to have this structure globally. As we shall see in this section, the generalization of allowing the electromagnetic field to be a connection on a nontrivial (i.e., non-product) principal fiber bundle allows for the presence of magnetic charge, but this magnetic charge can only take on the values $n2\pi\hbar/(\mu_0 q)$, where n is an integer.

Until this point in this chapter, it was essential for most of our considerations to deal with the full vector potential A_μ on spacetime. However, for the issue of magnetic charge, the time component $A_0 = -\phi/c$ will play no role, and the possible time dependence of the electromagnetic field also will not be relevant. Thus, in this section, we focus attention on the vector potential $\boldsymbol{A}$ on space at a fixed t.

[7] This U(1) subgroup is *not* the U(1) factor of SU(2) × U(1).

We say that a *magnetic charge* is present if the electromagnetic field has the property that on some 2-sphere S, the quantity

$$g \equiv \frac{1}{\mu_0} \int_S \boldsymbol{B} \cdot \hat{\boldsymbol{n}} dS \tag{9.48}$$

is nonvanishing. A simple argument that $g \neq 0$ is not possible in electromagnetism can be given as follows. By Gauss's theorem (see section 2.1) and Maxwell's equation $\nabla \cdot \boldsymbol{B} = 0$, we have

$$g = \frac{1}{\mu_0} \int_S \boldsymbol{B} \cdot \hat{\boldsymbol{n}} dS = \frac{1}{\mu_0} \int_{\mathcal{V}} \nabla \cdot \boldsymbol{B} d^3x = 0, \tag{9.49}$$

where $\mathcal{V}$ is the volume bounded by S. However, this argument would break down if the topology of space were such that S did not bound a (compact) volume $\mathcal{V}$, as could happen, for example, in general relativity. It would also break down if $\nabla \cdot \boldsymbol{B} \neq 0$ due to nonlinear interactions of the electromagnetic field with other fields in some small region of space in the volume enclosed by S. This latter possibility is not at all far-fetched, since, as mentioned at the end section 9.2, the electromagnetic field is part of a larger Yang-Mills theory. This larger Yang-Mills theory could admit violations of $\nabla \cdot \boldsymbol{B} = 0$ in suitably strong field regions.

However, the following is an alternative argument that $g = 0$ that does not rely on the global topology of space nor on the validity of $\nabla \cdot \boldsymbol{B} = 0$ throughout the region enclosed by S. Using only $\boldsymbol{B} = \nabla \times \boldsymbol{A}$ on S together with Stokes's theorem on S, we obtain

$$g = \frac{1}{\mu_0} \int_S (\nabla \times \boldsymbol{A}) \cdot \hat{\boldsymbol{n}} dS = \frac{1}{\mu_0} \int_{\partial S} \boldsymbol{A} \cdot d\boldsymbol{l} = 0, \tag{9.50}$$

since S has no boundary (i.e., $\partial S = 0$). This second argument that $g = 0$ is impervious to the possible breakdowns of the first argument given in the previous paragraph that are caused by nontrivial topology of all of space or a violation of $\nabla \cdot \boldsymbol{B} = 0$ away from S. As long as $\boldsymbol{A}$ is nonsingular on S, we must have $g = 0$.

However, the second argument has a significant loophole that was exploited by Dirac to argue for the possible presence of magnetic monopoles. Let $\hat{\boldsymbol{\varphi}}$ denote the unit vector in the φ-direction. The vector potential

$$\boldsymbol{A} = \frac{\mu_0 g}{4\pi} \frac{1}{r \sin\theta} (1 - \cos\theta)\, \hat{\boldsymbol{\varphi}} \tag{9.51}$$

gives rise to the magnetic field

$$\boldsymbol{B} = \nabla \times \boldsymbol{A} = \frac{\mu_0 g}{4\pi r^2} \hat{\boldsymbol{r}}, \tag{9.52}$$

which corresponds to a magnetic charge g, eq. (9.48), on any sphere S that encloses the origin. However, this is not a counterexample to the second argument that $g = 0$, because $\boldsymbol{A}$ is singular[8] at $\theta = \pi$, which includes the south pole of S. Thus, $\boldsymbol{A}$ is singular on S and is not a legal vector potential by the rules that normally would be imposed.

[8]This singularity in $\boldsymbol{A}$ at $\theta = \pi$—which extends from the origin to infinity—is referred to as a *Dirac string*.

However, we could alternatively consider the vector potential

$$A' = -\frac{\mu_0 g}{4\pi}\frac{1}{r\sin\theta}(1+\cos\theta)\,\hat{\boldsymbol{\varphi}}, \tag{9.53}$$

which also yields the magnetic field eq. (9.52). In this case, $\boldsymbol{A}'$ is singular on the north pole, $\theta = 0$, so it also is not a legal vector potential.

Nevertheless, one could slightly modify the rules so that instead of requiring a single, globally defined vector potential that is smooth everywhere, one merely requires that space be covered in patches with smooth vector potentials in each patch. If these vector potentials are related by a gauge transformation in each overlap region of the patches—so that they represent the same physical electromagnetic field—then this should be "as good" as having a single, smooth vector potential that is globally defined on space. Under these modified rules, we could use the vector potential eq. (9.51) in a region that excludes $\theta = \pi$ and the vector potential eq. (9.53) in a region that excludes $\theta = 0$. Of course, both $\boldsymbol{A}$ and $\boldsymbol{A}'$ are singular at the origin, but we are concerned here with the second argument that only considers the vector potential in a neighborhood of a sphere S; as indicated above, the singularity at the origin potentially could be eliminated by a departure from Maxwell's equations near the origin.

The gauge transformation relating $\boldsymbol{A}$ and $\boldsymbol{A}'$ in the overlap region that excludes both $\theta = 0$ and $\theta = \pi$ is

$$\chi = -\frac{\mu_0 g}{2\pi}\varphi, \tag{9.54}$$

that is, with this choice of χ, we have $\boldsymbol{A}' = \boldsymbol{A} + \nabla\chi$. However, χ is not a legal gauge transformation, since φ is not single-valued (or it undergoes a discontinuous jump if one forces it to be single-valued). Nevertheless, since $\nabla\chi$ is single-valued, if there were no charged fields, one could modify the rules to allow gauge transformations of this sort. Under these modified rules, magnetic charges with any value of g would be allowed.

However, if a charged field Φ is present, then a major difficulty arises with allowing the gauge transformation eq. (9.54), because according to eq. (9.39), Φ transforms as

$$\Phi \to \Phi' = e^{i\frac{q}{\hbar}\chi}\,\Phi = e^{-i\frac{\mu_0 g q}{2\pi\hbar}\varphi}\,\Phi\,. \tag{9.55}$$

Since φ is not single valued, this transformation is ill defined in general. Consequently, a gauge transformation of the form eq. (9.54) is not mathematically acceptable in general. However, an important exception occurs if g satisfies the *Dirac quantization condition*

$$g = n\frac{2\pi\hbar}{\mu_0 q}, \tag{9.56}$$

where n is any integer. If g satisfies this condition, then the gauge transformation eq. (9.55) becomes

$$\Phi \to e^{-in\varphi}\,\Phi, \tag{9.57}$$

which is well defined despite the multivalued nature of φ, since $e^{-in\varphi}$ is unchanged under $\varphi \to \varphi + 2\pi$. *Thus, magnetic charge is mathematically acceptable, provided that the magnetic charge g satisfies eq. (9.56).*

The new rules that allow the vector potentials (9.51) and (9.53) to cover S only in patches—provided that g satisfies eq. (9.56)—can be formulated in an elegant mathematical manner as follows. First, consider the case where we have a vector potential that is everywhere smooth on S (so in particular, $g=0$) and also is everywhere tangent to S. Then, by the same construction as discussed in section 9.2—but replacing spacetime M with S and replacing A_μ with $\boldsymbol{A}$—we can represent the gauge equivalence class of this vector potential on S by a connection on $S \times \mathrm{U}(1)$. Now consider the case that we have just encountered in eqs. (9.51) and (9.53), where we have a vector potential $\boldsymbol{A}$ that is smooth in a region containing the northern hemisphere, $S_+ = \{0 \leq \theta \leq \pi/2\}$, and a vector potential $\boldsymbol{A}'$ that is smooth in a region containing the southern hemisphere, $S_- = \{\pi/2 \leq \theta \leq \pi\}$, with $\boldsymbol{A}$ and $\boldsymbol{A}'$ related by a gauge transformation in the overlap region of the form eq. (9.54) with g satisfying eq. (9.56). We can represent the gauge equivalence class of the electromagnetic field represented by $\boldsymbol{A}$ and $\boldsymbol{A}'$ as a globally well-defined, smooth quantity on a higher-dimensional space as follows:

We start with the spaces $S_+ \times \mathrm{U}(1)$ and $S_- \times \mathrm{U}(1)$. If we were to "glue" S_+ to S_- along the equator, $\theta = \pi/2$, and if we were to correspondingly "glue" the $\mathrm{U}(1)$ factors to each other at each point on the equator, we would produce the space $\mathcal{P}_0 = S \times \mathrm{U}(1)$, which would have the structure of a (trivial) principal bundle. The connections on $\mathcal{P}_0$ would correspond to the gauge equivalence classes of smooth vector potentials on S, and these would automatically have vanishing magnetic charge, $g=0$. But now suppose we again glue S_+ to S_- along the equator, but at each point φ on the equator of S_+, we apply a rotation of $-n\varphi$ for some integer n to the $\mathrm{U}(1)$ factor before we glue it to the corresponding $\mathrm{U}(1)$ factor of S_-. This gives a consistent "gluing scheme," because the $\mathrm{U}(1)$ factor is rotated by $-2\pi n$ when one has gone all the way around the equator, which is equivalent to no rotation at all. The resulting space $\mathcal{P}_n$ will no longer have the product structure[9] of $S \times \mathrm{U}(1)$. Nevertheless, the group $\mathrm{U}(1)$ acts naturally on $\mathcal{P}_n$ and gives $\mathcal{P}_n$ the structure required to be a principal fiber bundle with structure group $\mathrm{U}(1)$ and base space S. We shall not give a general definition of "principal fiber bundle" here, but the key property to note is that $\mathcal{P}_n$ has the local—but not global—structure of a Cartesian product with $\mathrm{U}(1)$.

For the nontrivial principal bundle $\mathcal{P}_n$, a connection $\mathscr{A}_\Lambda$ can defined to be a dual vector field on $\mathcal{P}_n$ that is globally well defined and smooth on $\mathcal{P}_n$ and, in each local Cartesian product region, satisfies the properties of a connection as described in section 9.2. In each local Cartesian product region, $\mathscr{A}_\Lambda$ corresponds to a gauge equivalence class of smooth vector potentials $\boldsymbol{A}$ on that region of S, as described in section 9.2. However, since one can make this correspondence only in local Cartesian product regions, even though $\mathscr{A}_\Lambda$ is globally well defined on $\mathcal{P}_n$, there is no globally well-defined vector potential $\boldsymbol{A}$ that can be defined on S. This is exactly the situation we encountered in eqs. (9.51) and (9.53). Thus, we now see that eqs. (9.51) and (9.53) for the magnetic charge values eq. (9.56) correspond to a globally well-defined, smooth connection on $\mathcal{P}_n$.

In the above discussion, we have restricted attention to how the vector potential eqs. (9.51) and (9.53) can be represented as a connection on the principal $\mathrm{U}(1)$-bundle $\mathcal{P}_n$ over a single 2-dimensional sphere S. Of course, what we really want to do is construct a corresponding principal $\mathrm{U}(1)$-bundle over spacetime and represent the full vector potential A_μ as a connection on this bundle. This would be straightforward to do

[9] For $n=1$, the resulting space has the topology of a 3-dimensional sphere. The expression of the 3-sphere in this manner as a principal $\mathrm{U}(1)$ bundle over a 2-sphere is called a *Hopf fibration* of the 3-sphere.

if space had the topology of $S^2 \times \mathbb{R}$. However, if space has the topology $\mathbb{R}^3$ (and, correspondingly, spacetime has topology $\mathbb{R}^4$) as is normally assumed, it is impossible to construct a nontrivial principal U(1)-bundle over spacetime. Thus, if the electromagnetic field is represented by a smooth connection on a principal U(1)-bundle defined over all spacetime, magnetic charges would not be possible. This is in accord with the first argument against magnetic monopoles that was given near the beginning of this section based on $\nabla \cdot \boldsymbol{B} = 0$ holding everywhere. However, as already mentioned, since the electromagnetic field is part of a larger Yang-Mills theory, it is possible for violations of $\nabla \cdot \boldsymbol{B} = 0$ to occur in small regions of space due to nonlinear interactions. Only outside such regions would the electromagnetic field be described by a connection in a principal U(1)-bundle. But space with such "holes" present admits nontrivial principal U(1)-bundles. Thus, magnetic monopoles of the sort we have described above may well exist in nature.

No magnetic monopoles have yet been observed in nature. Thus, nature may not have availed itself of the opportunity to provide us with magnetic monopoles—at least not in the abundance needed for us to find them. Nevertheless, the possible existence and properties of magnetic monopoles provides considerable insight into the features of electromagnetism when viewed as a gauge theory.

CHAPTER 10

Point Charges and Self-Force

In this chapter, we first show in section 10.1 how the notion of a point charge can be obtained in a mathematically rigorous manner by taking a suitable limit of a body whose size, charge, and mass go to zero. We show in section 10.2 that the limiting worldline of such a body satisfies the Lorentz force equation of motion. We then consider the leading-order corrections to this motion due to finite size and charge. The corrections due to finite charge involve the "self-field" of the body (i.e., the retarded solution associated with its charge-current). In electrostatics, we found that the electrostatic force on a body exerted by its own self-field vanishes (see eq. (2.41)). However, as we shall see, in electrodynamics, there are nontrivial self-force effects (often referred to as "radiation reaction"). We obtain a formula for the leading-order self-force in section 10.3. Finally, in section 10.4, we show how to obtain self-consistent equations of motion for a small charged body that take into account the leading-order self-force effects but do not admit spurious "runaway" solutions.

Many of the results of this chapter are based on the work of Gralla et al.,[1] and we occasionally refer to this reference for further details.

10.1 The Point Particle Limit

As discussed previously (see eq. (8.79)), the Lorentz force equation of motion for a charged point particle in an electromagnetic field is

$$m\frac{d(\gamma \boldsymbol{v})}{dt} = q\left(\boldsymbol{E} + \boldsymbol{v} \times \boldsymbol{B}\right) . \tag{10.1}$$

As also previously discussed, the charge density and current density of a charged point particle are given by eq. (8.73), so Maxwell's equations with a charged point particle source moving on the worldline $\boldsymbol{x} = \boldsymbol{X}(t)$ take the form

$$\nabla \cdot \boldsymbol{E} = \mu_0 c^2 q \delta(\boldsymbol{x} - \boldsymbol{X}(t)) , \qquad \nabla \times \boldsymbol{B} - \frac{1}{c^2}\frac{\partial \boldsymbol{E}}{\partial t} = \mu_0 q \delta(\boldsymbol{x} - \boldsymbol{X}(t))\boldsymbol{v} . \tag{10.2}$$

[1] S. Gralla, A. Harte, and R. M. Wald, "A Rigorous Derivation of Electromagnetic Self-force," *Physical Review* **D80**, 024031 (2009); arXiv:0905.2391.

For a given smooth $\boldsymbol{E}$ and $\boldsymbol{B}$, eq. (10.1) is well posed and has a unique solution for the motion of the particle, given the initial position and velocity of the particle. For a given particle motion, eq. (10.2) is well posed and has a unique retarded solution, which we obtained explicitly in section 8.3.2. However, if we wish to take point particles seriously in electrodynamics, we should be solving eqs. (10.1) and (10.2) simultaneously, so that the $\boldsymbol{E}$ and $\boldsymbol{B}$ in eq. (10.1) are the full electromagnetic field, with the source as in eq. (10.2). In other words, we should consistently take self-field and corresponding self-force effects into account. However, it is clear that the joint system of eqs. (10.1) and (10.2) does not have anything that reasonably could be called a "solution." Any solution to (10.2) will necessarily have an infinite self-field on the worldline of the particle, and consequently, eq. (10.1) will not make sense.

If, in this book, we had treated point charges as fundamental objects on which classical electrodynamics is based, the inconsistency of eqs. (10.1) and (10.2) would require some serious reckoning, wherein the mathematical consistency of classical electrodynamics would be thrown into doubt. At the very least, some "regularization scheme" would need to be applied to eqs. (10.1) and (10.2) to eliminate the singular behavior of the electromagnetic field on the worldline of the particle. At best, the rules for such a regularization would be ad hoc, and the foundations of classical electrodynamics would become correspondingly murky. However, as we have emphasized in this book, at a fundamental level, the charge density ρ and current density $\boldsymbol{J}$ should be taken to be smoothly distributed in spacetime. If the charged matter satisfies suitable well-posed equations, then there will be no difficulties with the coupled system describing the electromagnetic field and charged matter. In particular, the coupled charged scalar field and electromagnetic system described by eqs. (9.35) and (9.36) is well posed, and the coupled evolution of Φ and A_μ is completely well defined. The "self-force" effects occurring in this system are nonsingular, and eqs. (9.35) and (9.36) automatically take these effects fully into account in a self-consistent manner.

Nevertheless, even though ideal point charges satisfying both eq. (10.1) and eq. (10.2) do not make mathematical sense, one can consider a charged body of finite size that is extremely small. It is of considerable interest to know how such a body would move in an electromagnetic field, taking self-force effects into account.

One way to investigate this issue would be to consider a specific model of the body—such as a charged fluid or a charged elastic solid—wherein the charged matter comprising the body satisfies well-posed equations, and the dynamics of the coupled matter-electromagnetic system is well defined. However, if one gives the charged matter its own independent dynamical degrees of freedom, a very complicated dynamics will result, since the motion of the body in an electromagnetic field will produce nonuniform forces on the body, which should induce internal oscillations of the body. These oscillations and the complicated dynamics that arise from them will be highly model dependent for a body of finite size. Thus, this approach does not appear to be a fruitful way to proceed to analyze how a small charged body moves in an electromagnetic field, taking self-force effects into account.

In nonrelativistic mechanics, the complications that would arise from the internal degrees of freedom of the body can be avoided by considering a rigid body, defined as a body for which the spatial relationships between the points of the body do not change with time. A rigid body of finite size will have only a finite number of degrees of freedom—namely, its position and orientation—so the dynamical complications that would arise from a more realistic model do not occur. However, in special relativity, it would be unnatural/unphysical to require a body to be rigid in this sense in some fixed inertial frame, since in essence, the body would have to physically expand to compensate for the effects of Lorentz contraction. In special relativity, the natural notion of rigid body motion would be "Born rigidity," which requires that in the rest frame of any worldline of the matter composing the body, the spatial relationships between the infinitesimally nearby matter worldlines do not change with time. However, only very a limited class of Born rigid motions are possible, so this notion is too stringent to impose on the motion of a body if one wishes to study general motion. Thus, rigid body models would not appear to be useful for the study of the motion of a body of finite size in a general electromagnetic field.

A more promising approach would be to consider a body of finite size but then take a limit as the size of the body goes to zero. In the limit of zero size, one might expect that any complications caused by any internal dynamics of the body would become negligible, so that one could get a well-defined dynamical evolution. However, if one tries to take a limit of zero size of a body holding the total charge q of the body fixed, the electromagnetic energy will diverge to infinity. Consequently, one would need the charged matter to have a negative mass that diverges to infinity in this limit to keep the total energy finite. One also would need the stress of the charged matter to diverge to infinity in this limit to keep the body from flying apart due to electrostatic repulsion as it shrinks to zero size. Therefore, it is clear that such a limit of vanishing size of a body at fixed charge requires highly unphysical matter and reintroduces the singular behavior of eqs. (10.1) and (10.2).

However, as we shall now show, there is another way of taking a point particle limit of a charged body that is mathematically well defined and leads to entirely nonsingular results. At first sight, this limit may appear vacuous, but as we shall see, we can use it to rigorously derive Lorentz force motion and self-force corrections to the motion. The key idea is that, as we take a limit as the size R of the body goes to zero, we simultaneously let the charge q and mass m of the body go to zero in a manner proportional to the size. Since the self-energy of the body will scale as q^2/R, the self-energy will go to zero in this limit. Thus, this limit does not encounter the mathematical difficulties described in the previous paragraph that would arise if one tried to take a limit keeping q fixed at some nonzero value. However, it might seem absurd to take a limit of this sort, because the body itself also goes away completely in the limit! Since there is no longer any charge or mass present, what could be learned about charged particle motion by taking such a limit?

The answer to this question is that the shrinking down of the distribution of charge-current and matter stress-energy to a worldline is consistent only for certain worldlines. As we shall see in section 10.2, this worldline must satisfy the Lorentz force equation of motion (10.1), where q/m is the limiting value of the charge-to-mass ratio of the body (which is well defined and nonzero). One nontrivial feature that emerges from the analysis at this stage is that the m in the Lorentz force equation of motion includes the electromagnetic self-energy of the body. However, the main benefit of the analysis is that it allows one to do a well-defined perturbation theory to obtain the corrections to Lorentz force motion due to self-force as well as due to finite size effects. We will obtain these leading-order corrections in section 10.3.

We now state in a precise manner what we mean by "taking a limit of a charged body wherein its size, charge, and mass go to zero in a proportional way." A charged body is represented by its charge-current distribution $J^\mu(t, \boldsymbol{x})$ and its stress-energy-momentum tensor $T^{\mathrm{M}}_{\mu\nu}(t, \boldsymbol{x})$ describing all the (charged and uncharged) matter composing the body apart from the electromagnetic field. Thus, $T^{\mathrm{M}}_{\mu\nu}$ includes all "mechanical" energy, momentum, and stress of the body but does not include the energy, momentum, or stress of its electromagnetic self-field or the cross-term contributions of its self-field with an external electromagnetic field, which will be accounted for separately. The idea of shrinking a body down to vanishing size is represented by a one-parameter family of bodies, for which $J^{\mu}(\lambda;t,\boldsymbol{x})$ and $T^{\mathrm{M}}_{\mu\nu}(\lambda;t,\boldsymbol{x})$ have the property that as $\lambda \to 0$, both J^{μ} and $T^{\mathrm{M}}_{\mu\nu}$ vanish outside an arbitrarily small neighborhood of some timelike worldline $\boldsymbol{X}(t)$. We can implement this idea together with a proportional scaling to zero of the charge and mechanical mass by taking the one-parameter families, $J^{\mu}(\lambda;t,\boldsymbol{x})$ and $T^{\mathrm{M}}_{\mu\nu}(\lambda;t,\boldsymbol{x})$, to be of the functional form

$$J^{\mu}(\lambda;t,\boldsymbol{x}) = \frac{1}{\lambda^2}\mathscr{J}^{\mu}\left(\lambda;t,\left[\boldsymbol{x}-\boldsymbol{X}(t)\right]/\lambda\right), \tag{10.3}$$

$$T^{\mathrm{M}}_{\mu\nu}(\lambda;t,\boldsymbol{x}) = \frac{1}{\lambda^2}\mathscr{T}^{\mathrm{M}}_{\mu\nu}\left(\lambda;t,\left[\boldsymbol{x}-\boldsymbol{X}(t)\right]/\lambda\right). \tag{10.4}$$

Here $\mathscr{J}^{\mu}(\lambda;t,\boldsymbol{z})$ and $\mathscr{T}^{\mathrm{M}}_{\mu\nu}(\lambda;t,\boldsymbol{z})$ are smooth functions of $(\lambda;t,\boldsymbol{z})$ with the property that they vanish when $|\boldsymbol{z}| > R$ for some R. We now explain the meaning and interpretation of this one-parameter family.

For any function $h(\boldsymbol{z})$, the function $H(\lambda;\boldsymbol{z}) \equiv h(\boldsymbol{z}/\lambda)$ at any fixed λ is a function with the same shape as h but is compressed in $\boldsymbol{z}$ by a factor of $1/\lambda$. If we require h to vanish for $|\boldsymbol{z}| > R$, then $H(\lambda;\boldsymbol{z})$ vanishes for $|\boldsymbol{z}| > \lambda R$. Thus, the one-parameter family $H(\lambda;\boldsymbol{z})$ can be interpreted as rigidly scaling $h(\boldsymbol{z})$ to vanishing size as $\lambda \to 0$. An exactly rigid scaling of the charge-current and stress-energy would be too strong a requirement for our purposes (since, in particular, it would be a coordinate dependent notion), so we allow additional smooth λ-dependence in the families we consider. According to what we have just said, the quantities $\mathscr{J}^{\mu}(\lambda;t,[\boldsymbol{x}-\boldsymbol{X}(t)]/\lambda)$ and $\mathscr{T}^{\mathrm{M}}_{\mu\nu}(\lambda;t,[\boldsymbol{x}-\boldsymbol{X}(t)]/\lambda)$ can be interpreted as representing the charge-current and stress-energy distributions of a one-parameter family of bodies that shrinks down to the worldline $\boldsymbol{X}(t)$ as $\lambda \to 0$ with the size of the body being proportional to λ. For such a family, the peak magnitude of the charge-current and stress-energy would remain bounded as $\lambda \to 0$. Since the spatial volume occupied by the body scales as λ^3, the charge of $\mathscr{J}^{\mu}(\lambda;t,\boldsymbol{x})$ and mechanical mass of $\mathscr{T}^{\mathrm{M}}_{\mu\nu}(\lambda;t,\boldsymbol{x})$ would scale to zero as λ^3 as $\lambda \to 0$. However, we wish to have the charge and mechanical mass scale as λ, so that self-force corrections will appear at the same order as finite size corrections, and we will not have to work to high order in perturbation theory to calculate them. Therefore, we have inserted the factor of $1/\lambda^2$ in eqs. (10.3) and (10.4) to ensure that the one-parameter families we consider have the desired scaling of charge and mechanical mass as $\lambda \to 0$.

We further assume that the body is placed in a smooth external electromagnetic field $F^{\mathrm{ext}}_{\mu\nu}$ that satisfies the source-free Maxwell equations, and that the total electromagnetic field is thus[2]

$$F_{\mu\nu}(\lambda;t,\boldsymbol{x}) = F^{\mathrm{ext}}_{\mu\nu}(t,\boldsymbol{x}) + F^{\mathrm{self}}_{\mu\nu}(\lambda;t,\boldsymbol{x}), \tag{10.5}$$

where the self-field $F^{\mathrm{self}}_{\mu\nu}$ is taken to be the retarded solution with source (10.3).

For any $\lambda > 0$, the body is of finite size, so its dynamical behavior depends on the details of the charged matter composing the body. However, whatever the body is

[2]We take $F^{\mathrm{ext}}_{\mu\nu}$ to be independent of λ, but there would be no harm done in allowing $F^{\mathrm{ext}}_{\mu\nu}$ to have a smooth dependence on λ.

composed of, its charge current $J^\mu(\lambda; t, \boldsymbol{x})$ must be conserved:

$$\partial_\mu J^\mu = 0\,. \tag{10.6}$$

Furthermore, whatever the body is composed of, the total stress-energy

$$T_{\mu\nu}(\lambda; t, \boldsymbol{x}) = T^{\mathrm{M}}_{\mu\nu}(\lambda; t, \boldsymbol{x}) + T^{\mathrm{EM}}_{\mu\nu}(\lambda; t, \boldsymbol{x}) \tag{10.7}$$

must be conserved:

$$\partial^\mu T_{\mu\nu} = 0\,. \tag{10.8}$$

In eq. (10.7), $T^{\mathrm{EM}}_{\mu\nu}$ is the stress-energy tensor (eq. (8.50)) for the electromagnetic field given by eq. (10.5). In particular, $T^{\mathrm{EM}}_{\mu\nu}$ includes the electromagnetic self-energy of the body as well as its interaction energy with the external field. As we shall now show, Lorentz force motion and the leading-order self-force correction can be derived using only eqs. (10.6) and (10.8). Thus, our results hold for any form of the charged matter composing the body.

10.2 Lorentz Force

In this section, we show that the conservation equations (10.6) and (10.8) can be satisfied only if the worldline $\boldsymbol{X}(t)$ in eqs. (10.3) and (10.4) satisfies the Lorentz force equation of motion eq. (10.1). This shows that, in a precise sense, any sufficiently small body of sufficiently small charge must behave like an ideal point charge. In section 10.3, we obtain the leading-order corrections to the motion due to finite size and charge.

For the one-parameter family given by eqs. (10.3) and (10.4), the pointwise limit as $\lambda \to 0$ of $J^\mu(\lambda; t, \boldsymbol{x})$ and $T^{\mathrm{M}}_{\mu\nu}(\lambda; t, \boldsymbol{x})$ vanishes if the event $(t, \boldsymbol{x})$ does not lie on the worldline $\boldsymbol{X}(t)$ and diverges if this event does lie on this worldline. Thus, the pointwise limits as $\lambda \to 0$ of $J^\mu(\lambda; t, \boldsymbol{x})$ and $T^{\mathrm{M}}_{\mu\nu}(\lambda; t, \boldsymbol{x})$ are highly singular, as is the behavior of their λ-derivatives. Nevertheless, as we shall now describe, if we view $J^\mu(\lambda; t, \boldsymbol{x})$ and $T^{\mathrm{M}}_{\mu\nu}(\lambda; t, \boldsymbol{x})$ as *distributions*, then their behavior as $\lambda \to 0$ can be described in a manner that is completely well defined mathematically. As we shall see, $J^\mu(\lambda; t, \boldsymbol{x})$ and $T^{\mathrm{M}}_{\mu\nu}(\lambda; t, \boldsymbol{x})$ will go to zero as $\lambda \to 0$, but their first derivative with respect to λ will limit to a δ-function.

To explain how to describe the limiting behavior of $J^\mu(\lambda; t, \boldsymbol{x})$ and $T^{\mathrm{M}}_{\mu\nu}(\lambda; t, \boldsymbol{x})$ as distributions, let us first consider a function $G(\lambda; t, \boldsymbol{x})$ of the form

$$G(\lambda; t, \boldsymbol{x}) = \lambda^{-2} g(\lambda; t, [\boldsymbol{x} - \boldsymbol{X}(t)]/\lambda), \tag{10.9}$$

where $g(\lambda; t, \boldsymbol{z})$ is a smooth function of its arguments and vanishes for $|\boldsymbol{z}| > R$. Each component of $J^\mu(\lambda; t, \boldsymbol{x})$ and $T^{\mathrm{M}}_{\mu\nu}(\lambda; t, \boldsymbol{x})$ is of this form, so any results we obtain for G will apply directly to J^μ and $T^{\mathrm{M}}_{\mu\nu}$. Instead of considering the limiting behavior as $\lambda \to 0$ of G as a function, we can consider the limiting behavior of the corresponding λ-dependent distribution:

$$\mathscr{G}(\lambda; [f]) \equiv \int G(\lambda; t, \boldsymbol{x}) f(t, \boldsymbol{x}) dt d^3x\,. \tag{10.10}$$

Here $f(t, \boldsymbol{x})$ is an arbitrary "test function" (i.e., a smooth function that is nonvanishing only in a bounded spacetime region). Equation (10.10) defines $\mathscr{G}$ as a distribution for all $\lambda > 0$; that is, $\mathscr{G}$ is a linear map taking test functions into numbers that is suitably continuous in f. We wish to consider the limiting behavior of $\mathscr{G}$ and its λ-derivatives as distributions as $\lambda \to 0$.

In terms of the smooth function g, we have

$$\mathscr{G}(\lambda;[f]) = \frac{1}{\lambda^2}\int g\left(\lambda;t,[\boldsymbol{x}-\boldsymbol{X}(t)]/\lambda\right) f(t,\boldsymbol{x})dtd^3x\,. \tag{10.11}$$

We make the change of variables

$$\boldsymbol{x} \to \boldsymbol{x}' = \frac{\boldsymbol{x}-\boldsymbol{X}(t)}{\lambda}\,. \tag{10.12}$$

Since $dtd^3x = \lambda^3 dtd^3x'$, we obtain

$$\mathscr{G}(\lambda;[f]) = \lambda\int g(\lambda;t,\boldsymbol{x}')f(t,\boldsymbol{X}(t)+\lambda\boldsymbol{x}')dtd^3x'\,. \tag{10.13}$$

Since g is smooth in all its variables and vanishes for $|\boldsymbol{x}'| > R$, it is manifest from this equation that $\mathscr{G}(\lambda;[f])$ depends smoothly on λ at $\lambda = 0$. In other words, despite the singular pointwise behavior of $G(\lambda;t,\boldsymbol{x})$ as $\lambda \to 0$, when viewed as a distribution $\mathscr{G}$, the behavior of $\mathscr{G}$ and its λ-derivatives is entirely nonsingular at $\lambda = 0$. In particular, it follows immediately from eq. (10.13) that as $\lambda \to 0$, we have

$$\mathscr{G}(\lambda;[f]) \to 0, \tag{10.14}$$

whereas

$$\frac{\partial\mathscr{G}}{\partial\lambda}(\lambda,[f]) \to \int g(0;t,\boldsymbol{x}')f(t,\boldsymbol{X}(t))dtd^3x' = \int dtf(t,\boldsymbol{X}(t))\int d^3x' g(0;t,\boldsymbol{x}')\,. \tag{10.15}$$

Section 10.2 can be re-expressed as saying that, as a distribution, we have

$$\frac{\partial\mathscr{G}}{\partial\lambda} \to C(t)\delta\left(\boldsymbol{x}-\boldsymbol{X}(t)\right), \tag{10.16}$$

where $C(t) = \int g(0;t,\boldsymbol{x}')d^3x'$.

We now apply the above results to our one-parameter family given by eqs. (10.3) and (10.4). Viewing $J^\mu(\lambda;t,\boldsymbol{x})$ and $T^{\mathrm{M}}_{\mu\nu}(\lambda;t,\boldsymbol{x})$ as distributions, we see that as $\lambda \to 0$, we have

$$J^\mu \to 0\,, \qquad T^{\mathrm{M}}_{\mu\nu} \to 0, \tag{10.17}$$

whereas their first λ-derivatives limit to

$$J^{\mu(1)} \equiv \lim_{\lambda\to 0}\frac{\partial J^\mu}{\partial\lambda} = \mathcal{J}^\mu(t)\delta\left(\boldsymbol{x}-\boldsymbol{X}(t)\right), \tag{10.18}$$

$$T^{\mathrm{M}(1)}_{\mu\nu} \equiv \lim_{\lambda\to 0}\frac{\partial T^{\mathrm{M}}_{\mu\nu}}{\partial\lambda} = \mathcal{T}^{\mathrm{M}}_{\mu\nu}(t)\delta\left(\boldsymbol{x}-\boldsymbol{X}(t)\right), \tag{10.19}$$

where $\mathcal{J}^\mu(t) = \int \mathscr{J}^\mu(0;t,\boldsymbol{x})d^3x$ and $\mathcal{T}^{\mathrm{M}}_{\mu\nu}(t) = \int \mathscr{T}^{\mathrm{M}}_{\mu\nu}(0;t,\boldsymbol{x})d^3x$.

Next, we impose the restrictions on $J^{\mu(1)}$ arising from charge-current conservation, eq. (10.6). For $\lambda > 0$, we have for any test function f,

$$0 = \int \left[\partial_\mu J^\mu(\lambda; t, \boldsymbol{x})\right] f(t, \boldsymbol{x}) dt d^3x = -\int J^\mu(\lambda; t, \boldsymbol{x}) \partial_\mu f(t, \boldsymbol{x}) dt d^3x . \tag{10.20}$$

Taking the λ-derivative of this equation and letting $\lambda \to 0$, we have for any test function f

$$\lim_{\lambda \to 0} \int \frac{\partial J^\mu}{\partial \lambda}(\lambda; t, \boldsymbol{x}) \partial_\mu f(t, \boldsymbol{x}) dt d^3x = 0 . \tag{10.21}$$

If the limiting behavior of $\partial J^\mu / \partial \lambda$ were such that $\mathcal{J}^\mu(t)$ in (10.18) failed to be tangent to the worldline $\boldsymbol{X}(t)$, we would obtain a contradiction with eq. (10.21) by choosing f to vanish on the worldline $\boldsymbol{X}(t)$ but have a nonvanishing gradient orthogonal to this worldline. Thus, eq. (10.21) requires $\mathcal{J}^\mu(t)$ to be of the form

$$\mathcal{J}^\mu(t) = j(t) u^\mu , \tag{10.22}$$

where u^μ denotes the 4-velocity of the worldline $\boldsymbol{X}(t)$. The factor of proportionality $j(t)$ is related to the charge:

$$cq^{(1)} = \lim_{\lambda \to 0} \int \frac{\partial J^0}{\partial \lambda}(\lambda; t, \boldsymbol{x}) d^3x = c\gamma j(t) . \tag{10.23}$$

Furthermore, eq. (10.21) implies that $q^{(1)}$ is independent of t. Thus, we find that charge conservation implies that $J^{\mu(1)}$ takes the form

$$J^{\mu(1)} = \frac{q^{(1)}}{\gamma} \delta\left(\boldsymbol{x} - \boldsymbol{X}(t)\right) u^\mu , \tag{10.24}$$

which is exactly the form, eq. (8.72), of the charge-current of a point charge of charge $q^{(1)}$ moving along the wordline $\boldsymbol{X}(t)$. However, no restrictions on the worldline $\boldsymbol{X}(t)$ arise from charge-current conservation.

Next, we consider the restrictions that arise from stress-energy conservation, eq. (10.8). This equation involves the electromagnetic stress-energy tensor $T^{\text{EM}}_{\mu\nu}(\lambda; t, \boldsymbol{x})$, so we must examine its behavior as $\lambda \to 0$. Since the electromagnetic stress-energy tensor eq. (8.50) is quadratic in $F_{\mu\nu}$, and since $F_{\mu\nu}$ is given by eq. (10.5), we can write $T^{\text{EM}}_{\mu\nu}$ as

$$T^{\text{EM}}_{\mu\nu} = T^{\text{ext}}_{\mu\nu} + T^{\text{cross}}_{\mu\nu} + T^{\text{self}}_{\mu\nu} . \tag{10.25}$$

Here $T^{\text{ext}}_{\mu\nu}$ is the stress-energy tensor of $F^{\text{ext}}_{\mu\nu}$, $T^{\text{self}}_{\mu\nu}$ is the stress-energy tensor of $F^{\text{self}}_{\mu\nu}$, and $T^{\text{cross}}_{\mu\nu}$ represents the cross-terms in $T^{\text{EM}}_{\mu\nu}$ arising from one factor of $F^{\text{ext}}_{\mu\nu}$ and one factor of $F^{\text{self}}_{\mu\nu}$. Since $F^{\text{ext}}_{\mu\nu}$ satisfies the source-free Maxwell equations, we have $\partial^\mu T^{\text{ext}}_{\mu\nu} = 0$, so this term does not contribute to the conservation equation (10.8). Again, using the fact that $F^{\text{ext}}_{\mu\nu}$ satisfies the source-free Maxwell equations, we find in parallel with eq. (8.54) that for $\lambda > 0$, we have

$$\partial^\mu T^{\text{cross}}_{\mu\nu}(\lambda; t, \boldsymbol{x}) = F^{\text{ext}}_{\mu\nu}(t, \boldsymbol{x}) J^\mu(\lambda; t, \boldsymbol{x}) . \tag{10.26}$$

Since $J^\mu \to 0$ as a distribution when $\lambda \to 0$, it follows that—viewing $T^{\rm cross}_{\mu\nu}$ as a distribution—we have

$$\partial^\mu T^{\rm cross}_{\mu\nu}(\lambda) \to 0 \text{ as } \lambda \to 0, \tag{10.27}$$

whereas we have

$$\partial^\mu T^{\rm cross(1)}_{\mu\nu} \equiv \lim_{\lambda\to 0}\frac{\partial}{\partial\lambda}\left(\partial^\mu T^{\rm cross}_{\mu\nu}\right) = F^{\rm ext}_{\mu\nu} J^{\mu(1)} = \frac{q^{(1)}}{\gamma} F^{\rm ext}_{\mu\nu} u^\mu \delta\left(\boldsymbol{x} - \boldsymbol{X}(t)\right). \tag{10.28}$$

The behavior of $T^{\rm self}_{\mu\nu}$ as $\lambda \to 0$ is much more delicate to analyze. The self-field $F^{\rm self}_{\mu\nu}$ is the retarded solution with source J^μ. We have seen above that, as a distribution, when $\lambda \to 0$, we have $J^\mu(\lambda) \to 0$ whereas $\partial J^\mu/\partial\lambda$ approaches the point charge distribution eq. (10.24). It follows that, as a distribution, we have $F^{\rm self}_{\mu\nu}(\lambda) \to 0$, as $\lambda \to 0$, whereas

$$F^{\rm self(1)}_{\mu\nu} \equiv \lim_{\lambda\to 0}\frac{\partial F^{\rm self}_{\mu\nu}}{\partial\lambda} = F^{\rm LW}_{\mu\nu}, \tag{10.29}$$

where $F^{\rm LW}_{\mu\nu}$ denotes the Lienard-Wiechert field (8.107) and (8.108) of a point charge of charge $q^{(1)}$ moving on the worldline $\boldsymbol{X}(t)$. However, knowing these limits of $F^{\rm self}_{\mu\nu}$ is not directly helpful for calculating the limit as $\lambda \to 0$ of $T^{\rm self}_{\mu\nu}(\lambda)$, since we must first calculate[3] the stress energy of $F^{\rm self}_{\mu\nu}(\lambda)$ before taking the limit as $\lambda \to 0$. To do so, we start with the fact that the vector potential $A^{\rm self}_\mu$ is the retarded solution with source J^μ and thus is given by

$$\begin{aligned} A^{\rm self}_\mu(\lambda; t, \boldsymbol{x}) &= \frac{\mu_0}{4\pi}\int \frac{[J_\mu(\lambda; t', \boldsymbol{x}')]_{\rm ret}}{|\boldsymbol{x}-\boldsymbol{x}'|} d^3x' \\ &= \frac{\mu_0}{4\pi\lambda^2}\int \frac{\mathscr{J}_\mu(\lambda; t - |\boldsymbol{x}-\boldsymbol{x}'|/c, [\boldsymbol{x}' - \boldsymbol{X}(t)]/\lambda)}{|\boldsymbol{x}-\boldsymbol{x}'|} d^3x', \end{aligned} \tag{10.30}$$

where eq. (10.3) was used. By making the change of variables $\boldsymbol{x}' \to \boldsymbol{y} = [\boldsymbol{x}' - \boldsymbol{X}(t)]/\lambda$, it can be seen that $A^{\rm self}_\mu$ is of the form

$$A^{\rm self}_\mu(\lambda; t, \boldsymbol{x}) = \mathscr{A}^{\rm self}_\mu\left(\lambda; t, [\boldsymbol{x} - \boldsymbol{X}(t)]/\lambda\right), \tag{10.31}$$

where $\mathscr{A}^{\rm self}_\mu(\lambda; t, \boldsymbol{z})$ is smooth in all of its arguments. It follows that $F^{\rm self}_{\mu\nu}$ is of the form

$$F^{\rm self}_{\mu\nu}(\lambda; t, \boldsymbol{x}) = \partial_\mu A^{\rm self}_\nu - \partial_\nu A^{\rm self}_\mu = \frac{1}{\lambda}\mathscr{F}^{\rm self}_{\mu\nu}\left(\lambda; t, [\boldsymbol{x} - \boldsymbol{X}(t)]/\lambda\right), \tag{10.32}$$

where $\mathscr{F}^{\rm self}_{\mu\nu}(\lambda; t, \boldsymbol{z})$ is smooth in all of its arguments. It follows immediately that $T^{\rm self}_{\mu\nu}$ is of the form

$$T^{\rm self}_{\mu\nu}(\lambda; t, \boldsymbol{x}) = \frac{1}{\lambda^2}\mathscr{T}^{\rm self}_{\mu\nu}\left(\lambda; t, [\boldsymbol{x} - \boldsymbol{X}(t)]/\lambda\right), \tag{10.33}$$

[3] If $F^{\rm self}_{\mu\nu}(\lambda; t, \boldsymbol{x})$ were a continuous function of its variables at $\lambda = 0$, then taking products of $F^{\rm self}_{\mu\nu}(\lambda)$ would commute with taking the limit as $\lambda \to 0$, so we could obtain the limiting behavior of $T^{\rm self}_{\mu\nu}(\lambda)$ from the limiting behavior of $F^{\rm self}_{\mu\nu}(\lambda)$. However, this is not the case here.

where $\mathscr{T}^{\text{self}}_{\mu\nu}(\lambda;t,\boldsymbol{z})$ is a smooth function of all of its arguments. Note that the exact functional form of $\mathscr{T}^{\text{self}}_{\mu\nu}$ depends on the functional form of $\mathscr{J}^{\mu}$; that is, $\mathscr{T}^{\text{self}}_{\mu\nu}$ depends on the structure of the body being considered.

Remarkably, $T^{\text{self}}_{\mu\nu}(\lambda;t,\boldsymbol{x})$ has the same behavior as $\lambda \to 0$ as the matter stress-energy tensor, eq. (10.4), except that $\mathscr{T}^{\text{M}}_{\mu\nu}(\lambda;t,\boldsymbol{z})$ must vanish for $|\boldsymbol{z}| > R$, whereas $\mathscr{T}^{\text{self}}_{\mu\nu}(\lambda;t,\boldsymbol{z})$ has no such restriction. Nevertheless, by analyzing the behavior of $F^{\text{self}}_{\mu\nu}(\lambda;t,\boldsymbol{x})$ for λ near 0, one can show that the arguments leading to eq. (10.14) and eq. (10.16) continue to apply to $T^{\text{self}}_{\mu\nu}$. Consequently, when viewed as a distribution, as $\lambda \to 0$, we have

$$T^{\text{self}}_{\mu\nu} \to 0\,, \qquad T^{\text{self}(1)}_{\mu\nu} \equiv \lim_{\lambda\to 0} \frac{\partial T^{\text{self}}_{\mu\nu}}{\partial\lambda} = \mathcal{T}^{\text{self}}_{\mu\nu}(t)\delta\left(\boldsymbol{x} - \boldsymbol{X}(t)\right), \tag{10.34}$$

where $\mathcal{T}^{\text{self}}_{\mu\nu}(t)$ depends smoothly on t. Thus, at first order in λ, the electromagnetic self-stress-energy tensor of the body is localized on the worldline of the body.

We define the total stress-energy of the body $T^{\text{B}}_{\mu\nu}(\lambda;t,\boldsymbol{x})$ by adding the matter stress-energy $T^{\text{M}}_{\mu\nu}(\lambda;t,\boldsymbol{x})$ of the body to the electromagnetic stress energy $T^{\text{self}}_{\mu\nu}(\lambda;t,\boldsymbol{x})$ of the self-field of the body:

$$T^{\text{B}}_{\mu\nu} \equiv T^{\text{M}}_{\mu\nu} + T^{\text{self}}_{\mu\nu}\,. \tag{10.35}$$

By eqs. (10.19) and (10.34), we have

$$T^{\text{B}}_{\mu\nu} \to 0\,, \qquad T^{\text{B}(1)}_{\mu\nu} \equiv \lim_{\lambda\to 0} \frac{\partial T^{\text{B}}_{\mu\nu}}{\partial\lambda} = \mathcal{T}^{\text{B}}_{\mu\nu}(t)\delta\left(\boldsymbol{x} - \boldsymbol{X}(t)\right), \tag{10.36}$$

where $\mathcal{T}^{\text{B}}_{\mu\nu}(t)$ depends smoothly on t.

We are finally ready to impose the restrictions that arise from conservation of total stress-energy, eq. (10.8). For $\lambda > 0$, we have

$$\partial^{\mu} T^{\text{B}}_{\mu\nu} = -\partial^{\mu} T^{\text{cross}}_{\mu\nu} = -F^{\text{ext}}_{\mu\nu} J^{\mu}, \tag{10.37}$$

where eq. (10.26) was used for the last equality. Thus, for any smooth vector field $f^{\nu}(t,\boldsymbol{x})$ that vanishes outside a bounded region, we have for all $\lambda > 0$:

$$\int T^{\text{B}}_{\mu\nu}(\lambda;t,\boldsymbol{x})\partial^{\mu} f^{\nu}(t,\boldsymbol{x})dtd^3x = \int F^{\text{ext}}_{\mu\nu}(t,\boldsymbol{x})J^{\mu}(\lambda;t,\boldsymbol{x})f^{\nu}(t,\boldsymbol{x})dtd^3x\,. \tag{10.38}$$

By paralleling the steps that led from eq. (10.20) to eq. (10.22)—where we choose f^{ν} to vanish on the worldline $\boldsymbol{X}(t)$ but have nonvanishing gradient there—we find that $\mathcal{T}^{\text{B}}_{\mu\nu}$ must be of the form

$$\mathcal{T}^{\text{B}}_{\mu\nu}(t) = s(t)u_{\mu}u_{\nu}\,. \tag{10.39}$$

We define the *rest mass* m of the body to be $1/c^2$ times the total energy associated with $T^{\text{B}}_{\mu\nu}$ as measured in the rest frame of the worldline $\boldsymbol{X}(t)$. The rest mass vanishes at $\lambda = 0$, but at first order in λ at any time t, we have in the instantaneous rest frame of $\boldsymbol{X}(t)$:

$$m^{(1)}c^2 = \lim_{\lambda\to 0} \int \frac{\partial T^{\text{B}}_{00}}{\partial\lambda}(\lambda;t,\boldsymbol{x})d^3x\,. \tag{10.40}$$

Since the 3-momentum vanishes in the instantaneous rest frame and the 4-momentum transforms as a 4-vector (see part (d) of problem 5 of chapter 8), in an arbitrary fixed inertial frame, the integrated energy density at time t is $\gamma m^{(1)}c^2$. Thus, we have

$$\gamma m^{(1)}c^2 = s(t)(u_0)^2 = s(t)\gamma^2 c^2, \tag{10.41}$$

so $s(t) = m^{(1)}/\gamma$. Thus, in arbitrary inertial coordinates, we have

$$T^{\mathrm{B}(1)}_{\mu\nu} = \frac{m^{(1)}}{\gamma} u_\mu u_\nu \delta\left(\boldsymbol{x} - \boldsymbol{X}(t)\right). \tag{10.42}$$

Note that both $T^{\mathrm{M}}_{\mu\nu}$ and $T^{\mathrm{self}}_{\mu\nu}$ contribute to $m^{(1)}$.

Equation (10.42) provides all the information that arises from taking the limit as $\lambda \to 0$ of the λ-derivative of eq. (10.38) when one takes f^ν to vanish on the worldline $\boldsymbol{X}(t)$. The remaining information arising from conservation of the stress-energy tensor is obtained by taking f^ν to be nonvanishing on the worldline. Choosing f^ν to be proportional to u^ν on the worldline, we obtain

$$\frac{dm^{(1)}}{dt} = 0 \tag{10.43}$$

(i.e., the mass of the body is constant to first order in λ). Choosing f^ν to be such that it is of the form $f^\nu = Z^\nu(t)$ on the worldline, where Z^ν is orthogonal to u^μ (i.e., $\eta_{\mu\nu}u^\mu Z^\nu = 0$), we obtain

$$\int \frac{m^{(1)}}{\gamma} u^\mu u^\nu \partial_\mu Z_\nu(t) dt = \int F^{\mathrm{ext}}_{\mu\nu}(t) \frac{q^{(1)}}{\gamma} u^\mu Z^\nu(t) dt. \tag{10.44}$$

Since we have

$$u^\mu u^\nu \partial_\mu Z_\nu = u^\mu \partial_\mu (u^\nu Z_\nu) - Z_\nu u^\mu \partial_\mu u^\nu = -Z_\nu u^\mu \partial_\mu u^\nu = -Z^\nu u^\mu \partial_\mu u_\nu, \tag{10.45}$$

it readily can be seen that eq. (10.44) can hold for all $Z^\nu(t)$ only if

$$m^{(1)} u^\mu \partial_\mu u_\nu = -q^{(1)} F^{\mathrm{ext}}_{\mu\nu} u^\mu. \tag{10.46}$$

This is just the Lorentz force equation (10.1) written in the form of eqs. (8.69) and (8.77).

10.3 Corrections to Lorentz Force Motion

We have just gone through a great deal of effort, requiring careful mathematical analysis, to derive the Lorentz force equation of motion (10.46)—a result that was previously obtained in section 8.3.1 with very little effort by simply ignoring the self-field of the charge. One benefit we have already obtained from our more careful analysis is that we have seen explicitly how the electromagnetic self-energy of the body contributes to the rest mass $m^{(1)}$ of the particle that appears in the Lorentz force equation (10.46). An even more significant benefit is that we are now in a position to do perturbation theory, wherein the corrections to the Lorentz force motion can be systematically calculated, including both self-force corrections and finite size corrections.

Equations (10.3) and (10.4) describe a one-parameter family of bodies that approach the worldline $\boldsymbol{X}(t)$ as $\lambda \to 0$. As we have shown in section 10.2, the Lorentz force equation (10.46) arises from examining the $O(\lambda)$ behavior of this family. One might expect that the leading-order corrections could be obtained by repeating the same analysis but keeping terms second order in λ. However, this cannot work straightforwardly, because the worldline $\boldsymbol{X}(t)$ appearing in eqs. (10.3) and (10.4) is independent of λ and is therefore fixed, once and for all, by eq. (10.46). Thus, there can be no corrections to $\boldsymbol{X}(t)$. Deviations from Lorentz force motion arise in our approach because for small but finite λ, the worldine $\boldsymbol{X}(t)$ no longer represents the motion of the center of mass of the body. However, to analyze the center-of-mass behavior, it is very fruitful to take the limit as $\lambda \to 0$ in a different way.

The key idea is to "zoom in" on the body and suitably rescale quantities as we take the limit as $\lambda \to 0$ in such a way that we get a smooth, finite limit of all quantities. To see how the zooming in works, choose a point $\boldsymbol{X}(t_0)$ on the limiting worldline, and define scaled coordinates

$$\bar{t} = \frac{t - t_0}{\lambda}, \qquad \bar{x}^i = \frac{x^i - X^i(t_0)}{\lambda}. \tag{10.47}$$

In these new λ-dependent coordinates, the body will remain of finite coordinate size as $\lambda \to 0$. Of course, its physical size is determined by the spacetime metric. The spacetime interval between events separated by $\Delta\bar{x}^\mu$ is $\lambda^2 \eta_{\mu\nu}\Delta x^\mu \Delta x^\nu$. Thus, in $\bar{x}^\mu$-coordinates, we have

$$\eta_{\bar{\mu}\bar{\nu}} = \lambda^2 \eta_{\mu\nu}. \tag{10.48}$$

Here the bars over indices denote components in $\bar{x}^\mu$-coordinates, and in equalities such as eq. (10.48), it is understood that $\bar{\mu}$ stands for the component in $\bar{x}^\mu$-coordinates corresponding to μ in x^μ-coordinates. Thus, although the coordinate size of the body remains finite, its physical size goes to zero—as, of course, it must, since its physical size does not depend on the choice of coordinates. However, suppose we define a new, scaled metric $\bar{\eta}_{\mu\nu}$ by

$$\bar{\eta}_{\mu\nu} = \lambda^{-2} \eta_{\mu\nu}. \tag{10.49}$$

Then we have

$$\bar{\eta}_{\bar{\mu}\bar{\nu}} = \eta_{\mu\nu}. \tag{10.50}$$

Thus, $\bar{x}^\mu$ are inertial coordinates for the metric $\bar{\eta}_{\bar{\mu}\bar{\nu}}$, and the physical size of the body in the metric $\bar{\eta}_{\bar{\mu}\bar{\nu}}$ will remain finite as $\lambda \to 0$. However, J^μ, $T^{\rm M}_{\mu\nu}$, and $F_{\mu\nu}$ satisfy Maxwell's equations with respect to the physical spacetime metric $\eta_{\mu\nu}$, not with respect to the scaled metric eq. (10.49). Nevertheless, we can define a new, scaled current density $\bar{J}^\mu$, matter stress-energy $\bar{T}^{\rm M}_{\mu\nu}$, and electromagnetic field tensor $\bar{F}_{\mu\nu}$ by

$$\bar{J}^\mu = \lambda^3 J^\mu, \qquad \bar{T}^{\rm M}_{\mu\nu} = T^{\rm M}_{\mu\nu}, \qquad \bar{F}_{\mu\nu} = \lambda^{-1} F_{\mu\nu}. \tag{10.51}$$

In the $\bar{x}^\mu$-coordinates, we have

$$\bar{J}^{\bar{\mu}} = \lambda^{-1} \bar{J}^\mu = \lambda^2 J^\mu, \qquad \bar{T}^{\rm M}_{\bar{\mu}\bar{\nu}} = \lambda^2 \bar{T}^{\rm M}_{\mu\nu} = \lambda^2 T^{\rm M}_{\mu\nu}, \qquad \bar{F}_{\bar{\mu}\bar{\nu}} = \lambda^2 \bar{F}_{\mu\nu} = \lambda F_{\mu\nu}. \tag{10.52}$$

With the scalings eq. (10.51), we have

$$\partial^{\bar{\mu}}\bar{F}_{\bar{\mu}\bar{\nu}} \equiv \bar{\eta}^{\bar{\mu}\bar{\alpha}}\partial_{\bar{\alpha}}\bar{F}_{\bar{\mu}\bar{\nu}} = \lambda^2\eta^{\mu\alpha}\partial_\alpha F_{\mu\nu} = -\mu_0\lambda^2 J_\nu = -\mu_0\lambda^2\eta_{\nu\beta}J^\beta$$
$$= -\mu_0\bar{\eta}_{\bar{\nu}\bar{\beta}}\bar{J}^{\bar{\beta}} = -\mu_0\bar{J}_{\bar{\nu}}, \qquad (10.53)$$

$$\partial^{\bar{\mu}}\left(\bar{T}^{\mathrm{M}}_{\bar{\mu}\bar{\nu}} + \bar{T}^{\mathrm{EM}}_{\bar{\mu}\bar{\nu}}\right) \equiv \bar{\eta}^{\bar{\mu}\bar{\alpha}}\partial_{\bar{\alpha}}\left(\bar{T}^{\mathrm{M}}_{\bar{\mu}\bar{\nu}} + \bar{T}^{\mathrm{EM}}_{\bar{\mu}\bar{\nu}}\right) = 0, \qquad (10.54)$$

so Maxwell's equations and conservation of total stress-energy hold for our scaled fields with respect to the scaled metric eq. (10.49). For $\lambda > 0$, we can equally well work with the original system $(\eta_{\mu\nu}, J^\mu, T^{\mathrm{M}}_{\mu\nu}, F_{\mu\nu})$ or the scaled system $(\bar{\eta}_{\bar{\mu}\bar{\nu}}, \bar{J}^{\bar{\mu}}, \bar{T}^{\mathrm{M}}_{\bar{\mu}\bar{\nu}}, \bar{F}_{\bar{\mu}\bar{\nu}})$ in scaled coordinates. Any result we obtain for one system can be translated to the other using the above scalings. As we shall see, the limit as $\lambda \to 0$ of the barred system is very useful for studying deviations from Lorentz force motion.

However, the barred coordinates $\bar{x}^\mu$ that we have just introduced have the following significant deficiency. Since we wish to follow the motion of the body, we will want to vary the time t_0 in eq. (10.47) with respect to which these coordinates are defined. However, since the inertial coordinate position $\boldsymbol{X}(t_0)$ of the worldline varies with t_0, the $\bar{x}^i$ coordinates of a fixed event will vary with t_0 by an amount that diverges as $\lambda \to 0$, making these coordinates unsuitable for studying behavior in the limit $\lambda \to 0$. This deficiency can be overcome if, before scaling the coordinates, we start with coordinates x'^μ adapted to the worldline $\boldsymbol{X}(t)$ instead of with inertial coordinates x^μ. Specifically, we take the time coordinate t' to agree with proper time along the worldline $\boldsymbol{X}(t)$, and we take the surfaces of constant t' to coincide with the constant time surfaces in the instantaneous rest frame of the worldline at time t'. We take the spatial coordinates x'^i to be Cartesian coordinates in the surfaces of constant t', with origin on the worldline and with the Cartesian axes "Fermi transported" along the worldline.[4] The resulting coordinates x'^μ are known as *Fermi normal coordinates*. For inertial motion, Fermi normal coordinates agree with inertial coordinates in which the worldline is at the origin. For noninertial motion, different surfaces of constant t' will intersect, resulting in a breakdown of the coordinates. However, the Fermi normal coordinates are always well defined and smooth in a neighborhood of the worldline $\boldsymbol{X}(t)$. Since we will be concerned only with an arbitrarily small neighborhood of $\boldsymbol{X}(t)$ as we take the limit $\lambda \to 0$, the breakdown of Fermi normal coordinates away from $\boldsymbol{X}(t)$ is of no concern.

For any time t'_0, we define the scaled coordinates $\bar{x}'^\mu$ by $\bar{t}' = (t' - t'_0)/\lambda$ and $\bar{x}'^i = x'^i/\lambda$. We again define the scaled metric, charge-current, matter stress-energy tensor, and electromagnetic field by eqs. (10.49) and (10.51), but we express these quantities in Fermi normal coordinates[5] rather than inertial coordinates. From eqs. (10.3), (10.4), and (10.32), it is not difficult to show that when expressed in scaled Fermi normal coordinates, these scaled quantities are of the form

$$\bar{J}'^{\bar{\mu}'}(\lambda; \bar{t}', \bar{x}'^i) = \mathscr{J}'^{\mu'}\left(\lambda; t'_0 + \lambda\bar{t}', \bar{x}'^i\right), \qquad (10.55)$$

$$\bar{T}'^{\mathrm{M}}_{\bar{\mu}'\bar{\nu}'}(\lambda; \bar{t}', \bar{x}'^i) = \mathscr{T}'^{\mathrm{M}}_{\mu'\nu'}\left(\lambda; t'_0 + \lambda\bar{t}', \bar{x}'^i\right), \qquad (10.56)$$

[4] Fermi transport corresponds to keeping the directions of the spatial axes fixed except for the projection necessary to make them orthogonal to the 4-velocity of the worldline. If the direction of acceleration changes with time, these projections will result in a rotation of the Fermi transported axes relative to inertial axes. This effect is known as *Thomas precession* (see eq. (10.69)) and is related to the fact that the commutator of two infinitesimal Lorentz transformations is a rotation (see problem 3 of chapter 8).

[5] Note that, since Fermi normal coordinates are noninertial if the worldline undergoes acceleration, the spacetime metric components $\eta_{\mu'\nu'}$ in Fermi normal coordinates will *not* take the simple form eq. (8.7).

$$F'_{\bar{\mu}'\bar{\nu}'}(\lambda;\bar{t}',\bar{x}'^{i}) = \mathscr{F}'^{\text{self}}_{\mu'\nu'}\left(\lambda;t'_0+\lambda\bar{t}',\bar{x}'^{i}\right) + \lambda F'^{\text{ext}}_{\mu'\nu'}\left(\lambda;t'_0+\lambda\bar{t}',\lambda\bar{x}'^{i}\right), \quad (10.57)$$

where $\mathscr{J}'^{\mu'}$, $\mathscr{T}'^{\text{M}}_{\mu'\nu'}$, $\mathscr{F}'^{\text{self}}_{\mu'\nu'}$, and $F'^{\text{ext}}_{\mu'\nu'}$ are smooth functions of their arguments, which are related, respectively, to the smooth quantities $\mathscr{J}^{\mu}$, $\mathscr{T}^{\text{M}}_{\mu\nu}$, $\mathscr{F}^{\text{self}}_{\mu\nu}$, and $F^{\text{ext}}_{\mu\nu}$ (see eqs.(10.3), (10.4), (10.32), and (10.5)) by the smooth coordinate transformation taking inertial coordinates x^{μ} to Fermi normal coordinates x'^{μ}. The key point is that the scaled metric, charge-current, matter stress-energy tensor, and electromagnetic field expressed in the scaled coordinates $\bar{x}'^{\mu}$ behave smoothly in λ as $\lambda \to 0$.

Since we have introduced many new quantities, it may be helpful to summarize our notation for the case of the current density:

> $J^{\mu}(\lambda;t,\boldsymbol{x})$ denotes the current density of our one-parameter family, with the components taken in some inertial coordinates of the spacetime metric $\eta_{\mu\nu}$;
> $\mathscr{J}^{\mu}$ is a smooth function of its arguments that is related to J^{μ} by eq. (10.3);
> $\bar{J}^{\mu}$ is the scaling of J^{μ} given by eq. (10.51);
> $\bar{J}^{\bar{\mu}}$ is the scaled current density expressed in the scaled inertial coordinates $\bar{x}^{\mu}$ of eq. (10.47);
> $\bar{J}'^{\bar{\mu}'}$ is the scaled current density expressed in scaled Fermi normal coordinates; and
> $\mathscr{J}'^{\mu'}$ is the smooth function of its arguments related to $\bar{J}'^{\bar{\mu}'}$ by eq. (10.55).

For $\lambda > 0$, the description of the system in terms of the scaled quantites $(\bar{\eta}'_{\bar{\mu}'\bar{\nu}'}(\lambda)$, $\bar{J}'^{\bar{\mu}'}(\lambda), \bar{T}'^{\text{M}}_{\bar{\mu}'\bar{\nu}'}(\lambda), \bar{F}'_{\bar{\mu}'\bar{\nu}'}(\lambda))$ in scaled Fermi normal coordinates $\bar{x}'^{\mu}$ is, of course, equivalent to our original description in terms of $(\eta_{\mu\nu}, J^{\mu}(\lambda), T^{\text{M}}_{\mu\nu}(\lambda), F_{\mu\nu}(\lambda))$ in inertial coordinates x^{μ}. That is, we can go from one description to the other by performing the scalings and coordinate transformations described above. However, the expansion about $\lambda = 0$ of these families is quite different, both because of the field scalings and because of the λ-dependent scaling of coordinates, which results in a "mixing of orders" when comparing the two expansions. In particular, as we saw in section 10.2, in the original description, the body itself "disappears" at $\lambda = 0$ while the external field $F^{\text{ext}}_{\mu\nu}$ remains present. In the barred description, the body remains present at $\lambda = 0$, but the external field goes to zero. The expansion about $\lambda = 0$ obtained in the original unbarred description corresponds to what is called the "far-zone approximation" in the theory of matched asymptotic expansions, whereas the expansion about $\lambda = 0$ arising from the barred description corresponds to the "near-zone approximation." In the theory of matched asymptotic expansions, one normally assumes that there is a "buffer zone" where both expansions apply. In our case, there is no need to make any such assumption, since we have set up the problem so that the two descriptions of the systems agree exactly for $\lambda > 0$. Nevertheless, we will use the terminology "far-zone" and "near-zone" below to describe the expansion in λ for the unbarred and barred systems, respectively.

In the near-zone description, we define the total stress-energy of the body by applying the same scaling and coordinate change to $T^{\text{B}}_{\mu\nu} = T^{\text{M}}_{\mu\nu} + T^{\text{self}}_{\mu\nu}$ (see eq. (10.35)) as we have done to $T^{\text{M}}_{\mu\nu}$. In the near-zone picture, the mass of the body at time t_0 at $\lambda = 0$ is given by

$$m(t_0)c^2 = \int \bar{T}'^{\text{B}}_{\bar{0}'\bar{0}'}(\lambda = 0;\bar{t}' = 0,\bar{x}'^{i})d^3\bar{x}'. \quad (10.58)$$

It is not difficult to see that this zeroth order mass in the near-zone picture corresponds to the first order mass $m^{(1)}$ in the far-zone picture, as given by eq. (10.40). In the near-zone picture, we define the *center of mass* $\bar{X}^i_{\rm CM}$ to zeroth order in λ at time t_0 by[6]

$$\bar{X}^i_{\rm CM} = \frac{1}{mc^2} \int \bar{T}'^{\rm B}_{\bar{0}'\bar{0}'}(0; 0, \bar{x}'^j)\bar{x}'^i d^3\bar{x}' \,. \tag{10.59}$$

We take the center of mass as describing the "true position" of the body. A finite displacement at $\lambda = 0$ of the center of mass from the origin in the scaled Fermi normal coordinates $\bar{x}'^i$ corresponds to a displacement from the origin that is $O(\lambda)$ in the unscaled Fermi normal coordinates x'^i. This in turn corresponds to a displacement of first order in λ from the worldline $\boldsymbol{X}(t)$ in the original inertial coordinates x^μ. Thus, if we can find the motion of the center of mass at zeroth order in λ in the near-zone picture, we can obtain the leading-order corrections to Lorentz force motion in the far-zone picture.

The center-of-mass motion is determined by conservation of stress-energy. At zeroth order in λ in the near-zone picture, conservation of stress-energy does not place any constraints on $\bar{X}^i_{\rm CM}$. At first order in λ, conservation of stress-energy also does not place any constraints on $\bar{X}^i_{\rm CM}$, but it does require a relationship between the worldline acceleration associated with the Fermi normal coordinates and the external electromagnetic field. This relationship is equivalent to the Lorentz force equation of motion eq. (10.46). Finally, conservation of stress-energy at second order in λ yields equations governing the time evolution of $\bar{X}^i_{\rm CM}$ as well as an evolution equation for the perturbed mass.

The calculations leading to these evolution equations are somewhat lengthy and require some details (such as expressing the spacetime metric in Fermi normal coordinates) that would not be appropriate to present here. Therefore, we refer the reader to Gralla et al. (see footnote 1 in section 10.1) for the details of the calculations and merely describe the final results here. These results are obtained by working in the near-zone picture and then translating the final results to the far-zone picture. The quantities that enter these results and our notation for them are as follows:

- The leading-order mass and charge of the body, which arise at zeroth order in λ in the near-zone picture and first order in λ in the far-zone picture. We denote these quantities as m and q. They are the same quantities denoted as $m^{(1)}$ and $q^{(1)}$ in section 10.2.
- The leading-order correction to the mass,[7] which arises at first order in λ in the near-zone picture and second order in λ in the far-zone picture. We denote this quantity as δm.
- The 4-velocity u^μ and 4-acceleration, $a^\mu \equiv u^\nu \partial_\nu u^\mu$, of the worldline $\boldsymbol{X}(t)$ appearing in eqs. (10.3) and (10.4). As we have seen in eq. (10.46), this worldline satisfies the Lorentz force equation $ma_\mu = -qF^{\rm ext}_{\nu\mu}u^\nu$.

---

[6]It is not difficult to show that the fall-off of $\bar{T}'^{\rm self}_{\bar{0}'\bar{0}'}$ at $\lambda = 0$ at large $|\bar{\boldsymbol{x}}'|$ is sufficient for the integral in eq. (10.59) to be well defined. However, it would not be obvious how to define the center of mass including self-field stress-energy when $\lambda > 0$.

[7]We could also allow our one-parameter family to have a leading-order modification, δq, to the charge. However, by conservation of charge, δq does not dynamically evolve, so nothing is lost by requiring $\delta q = 0$, as we shall do. By contrast, δm does dynamically evolve (see eq. (10.68)), so it is not consistent to set $\delta m = 0$.

- The perturbed acceleration δa^μ of the worldline $\boldsymbol{X}(t)$. This is obtained by calculating how $\bar{X}^i_{\rm CM}$ varies with t'_0 at $\lambda = 0$ in the near-zone picture and translating the result to the far-zone picture.
- The spin $\boldsymbol{S}$ of the body. This quantity is defined in the near-zone picture at $\lambda = 0$ as the angular momentum of $\bar{T}'^{\rm B}_{\bar{\mu}'\bar{\nu}'}$ about the center of mass of the body. It corresponds to a quantity that is second order in λ in the far-zone picture.
- The electric dipole moment $\boldsymbol{p}$ and magnetic dipole moment $\boldsymbol{\mu}$ of the body, computed in the near-zone picture at $\lambda = 0$ in terms of $\bar{J}'^{\bar{\mu}'}$, with origin chosen at the center of mass. These quantities are second order in λ in the far-zone picture.

It is useful to break up the corrections to Lorentz force motion into two separate cases, as we shall now do in sections 10.3.1 and 10.3.2.

10.3.1 SELF-FORCE CORRECTIONS

To isolate the effects of self-force, we set the spin $\boldsymbol{S}$, electric dipole moment $\boldsymbol{p}$, and magnetic dipole moment $\boldsymbol{\mu}$ to zero. We define the leading-order correction δf^μ to the 4-force by

$$\delta f^\mu \equiv \delta(ma^\mu) = m\delta a^\mu + a^\mu \delta m. \tag{10.60}$$

The results of Gralla et al. (see footnote 1 in section 10.1) yield the following expression for δf^μ when $\boldsymbol{S} = \boldsymbol{p} = \boldsymbol{\mu} = 0$:

$$\delta f_\mu = -\delta\left[qF^{\rm ext}_{\nu\mu}u^\nu\right] + \frac{\mu_0 q^2}{6\pi c}\left[u^\nu \partial_\nu a_\mu - a^\nu a_\nu u_\mu\right]. \tag{10.61}$$

Furthermore, we have

$$u^\mu \partial_\mu \delta m = \frac{d\delta m}{d\tau} = 0, \tag{10.62}$$

that is, δm is constant along the worldline.

The first term on the right side of eq. (10.61) is merely the modification to the Lorentz force due to the body actually being in a slightly different position and/or having a slightly different 4-velocity than assigned to it by the original worldline $\boldsymbol{X}(t)$. The second term on the right side is the term of interest. It is quadratic in q and represents the effects of the body's self-field $F^{\rm self}_{\mu\nu}$ on its motion; that is, it represents the self-force

$$\delta f^{\rm SF}_\mu = \frac{\mu_0 q^2}{6\pi c}\left[u^\nu \partial_\nu a_\mu - a^\nu a_\nu u_\mu\right], \tag{10.63}$$

which is usually referred to as the *Abraham-Lorentz-Dirac force.*[8] For nonrelativistic motion, we may replace $u^\mu \partial_\mu$ with d/dt and drop the second term in square brackets, so the self-force reduces to

$$\delta \boldsymbol{f}^{\rm SF} = \frac{\mu_0 q^2}{6\pi c}\frac{d\boldsymbol{a}}{dt}. \tag{10.64}$$

---

[8]The nonrelativistic version eq. (10.64) of this force was obtained by Lorentz and Abraham (prior to special relativity) from extended body models. It was then generalized to the relativistic case by Abraham and others. Dirac obtained eq. (10.63) by applying a regularization scheme to the field of a point charge.

The self-force eq. (10.63) leads to an apparent paradox with regard to energy conservation. For simplicity, we restrict consideration to the case of nonrelativistic motion, so that we can use the nonrelativistic formula (10.64) for self-force and the nonrelativistic formula (5.81) for energy radiated to infinity. During a period of uniform acceleration (such would occur if the body were placed in a uniform electric field), we have $\boldsymbol{da}/dt = 0$, so by eq. (10.64), the self-force vanishes.[9] Thus, there is no additional energy cost arising from self-force needed to keep a charged body in uniform acceleration. However, during a period of uniform acceleration, by eq. (5.81), electromagnetic energy is radiated to infinity. Thus, it might appear that energy is not conserved.

To pose this issue more clearly and precisely, suppose that one applies an additional external force to a body (e.g., literally, by hand) to exactly compensate for the self-force eq. (10.64), so that the body moves exactly under Lorentz force motion. (We apply such an additional external force to avoid dealing with the issue of self-consistent motion under self-force, which will be addressed in section 10.4.) Suppose, further, that the external electromagnetic field $F^{\text{ext}}_{\mu\nu}$ vanishes at early and late times, so that the body starts in inertial motion and then returns to inertial motion. Then the total work done in overcoming the self-force eq. (10.64) is

$$\begin{aligned} W &= -\int_{-\infty}^{\infty} \delta \boldsymbol{f}^{\text{SF}} \cdot \boldsymbol{v}\, dt = -\frac{\mu_0 q^2}{6\pi c}\int_{-\infty}^{\infty} \frac{d^2\boldsymbol{v}}{dt^2}\cdot \boldsymbol{v}\, dt \\ &= -\frac{\mu_0 q^2}{6\pi c}\int_{-\infty}^{\infty}\left[\frac{d}{dt}\left(\frac{d\boldsymbol{v}}{dt}\cdot \boldsymbol{v}\right) - \left|\frac{d\boldsymbol{v}}{dt}\right|^2\right] dt \\ &= \frac{\mu_0 q^2}{6\pi c}\int_{-\infty}^{\infty}\left|\frac{d\boldsymbol{v}}{dt}\right|^2 dt\,, \end{aligned} \tag{10.65}$$

where, in the last step, we used the fact that $d\boldsymbol{v}/dt \to 0$ at early and late times. On the other hand, by eq. (5.81), the total energy radiated to infinity to leading order in v/c is

$$\mathscr{E} = \int_{-\infty}^{\infty} \frac{\mu_0}{6\pi c}\left|\frac{d^2\boldsymbol{p}}{dt^2}\right|^2_{\text{ret}} dt = \frac{\mu_0 q^2}{6\pi c}\int_{-\infty}^{\infty}\left|\frac{d\boldsymbol{v}}{dt}\right|^2_{\text{ret}} dt = \frac{\mu_0 q^2}{6\pi c}\int_{-\infty}^{\infty}\left|\frac{d\boldsymbol{v}}{dt}\right|^2 dt\,. \tag{10.66}$$

Thus, the total work done in overcoming the self-force balances the total radiated energy, so total energy is conserved (i.e., the work done by an external agent in overcoming the effects of self-force goes into the electromagnetic radiation emitted by the body). It is comforting that we get equality of W and $\mathscr{E}$ when we integrate over all time. However, we do *not* get equality of these quantities if we integrate over a finite time. In particular, as already mentioned, during a period of uniform acceleration, no work is being done by the external agent, whereas a flux of energy is radiated to infinity.

Thus, although one has a conservation of energy over all time, it might appear that energy is not locally conserved. It would not be acceptable behavior for the system to radiate energy with only a promise that it will be paid back eventually by the work done on the particle; the laws of physics require that energy must be exactly conserved at all times. However, the failure of W and $\mathscr{E}$ to balance over finite times is not inconsistent

[9] For the relativistic motion (eq. (8.83)) in a uniform electric field, the self-force eq. (10.63) also vanishes.

with local energy conservation, which *does* hold exactly at all times for the self-force eq. (10.63). Indeed, as we have already described, the self-force eq. (10.63) was *derived* by imposing the conservation equation (10.8) for the total stress-energy tensor, which implies that energy and momentum conservation hold exactly at all times. It follows that any discrepancy between work done and energy radiated must be compensated by the energy stored in $T^{\rm B}_{\mu\nu} = T^{\rm M}_{\mu\nu} + T^{\rm self}_{\mu\nu}$. By the nature of the derivation, there cannot be a problem with exact energy conservation holding at all times.

10.3.2 SPIN AND DIPOLE EFFECTS

We now consider the effect on the motion of the body of its spin, electric dipole moment, and magnetic dipole moment. We omit the self-force term, since we have already discussed its properties, but we can simply add the self-force term back into our expression if we wish to include it. Again, we identify the leading-order correction to the 4-force as $\delta f_\mu = m\delta a_\mu + a_\mu \delta m$. And again, we obtain the term $-\delta[\frac{q}{c} F^{\rm ext}_{\nu\mu} u^\nu]$ arising from the modification to the Lorentz force due to the body actually being in a slightly different position and/or having a slightly different 4-velocity than assigned to it by the original worldline $\boldsymbol{X}(t)$. The remaining terms yield the dipole force $\delta \boldsymbol{f}^{\rm D}$ on the body due to its spin and electric and magnetic dipole moments. The full relativistic expression is given in Gralla et al., but for simplicity, we give only the nonrelativistic expression here, dropping all terms proportional to the velocity of the dipole:

$$\delta \boldsymbol{f}^{\rm D} = (\boldsymbol{p}\cdot\nabla)\boldsymbol{E}^{\rm ext} + (\boldsymbol{\mu}\cdot\nabla)\boldsymbol{B}^{\rm ext} + \frac{1}{c^2}(\boldsymbol{p}\cdot\boldsymbol{E}^{\rm ext})\boldsymbol{a} + \frac{d}{dt}\left(\frac{1}{c^2}\boldsymbol{S}\times\boldsymbol{a}\right) - \frac{1}{c^2}\frac{d\boldsymbol{\mu}}{dt}\times\boldsymbol{E}^{\rm ext} + \frac{d\boldsymbol{p}}{dt}\times\boldsymbol{B}^{\rm ext}, \tag{10.67}$$

where $\boldsymbol{E}^{\rm ext}$ and $\boldsymbol{B}^{\rm ext}$ are evaluated on the worldline of the particle, and d/dt denotes the time derivative along the worldline. In addition, we obtain a nontrivial evolution equation for the leading-order correction δm to the rest mass of the body:

$$c^2\frac{d}{dt}(\delta m) = -\boldsymbol{\mu}\cdot\frac{d\boldsymbol{B}^{\rm ext}}{dt} + \frac{d\boldsymbol{p}}{dt}\cdot\boldsymbol{E}^{\rm ext}. \tag{10.68}$$

Finally, we also obtain an evolution equation for the spin of the body:

$$\frac{d\boldsymbol{S}}{dt} = \boldsymbol{p}\times\boldsymbol{E}^{\rm ext} + \boldsymbol{\mu}\times\boldsymbol{B}^{\rm ext} + \frac{1}{2c^2}(\boldsymbol{v}\times\boldsymbol{a})\times\boldsymbol{S}. \tag{10.69}$$

There are no evolution equations for $\boldsymbol{p}$ and $\boldsymbol{\mu}$ (i.e., the manner in which they evolve depends on the nature of the body and is not determined by conservation of charge-current and stress-energy alone).

Let us discuss the meaning of the terms in the above equations. We first consider the force expression, eq. (10.67). The terms $(\boldsymbol{p}\cdot\nabla)\boldsymbol{E}^{\rm ext}$ and $(\boldsymbol{\mu}\cdot\nabla)\boldsymbol{B}^{\rm ext}$ should be expected to appear, since we previously encountered them in our studies of electrostatics and magnetostatics (see eq. (2.44) and eq. (4.68)). In the case of a stationary charge-current distribution in a stationary external field, these are the only terms that arise. However, if the charged body is accelerating, one has an additional term $(\boldsymbol{p}\cdot\boldsymbol{E}^{\rm ext})\boldsymbol{a}/c^2$. Since this term is proportional to $\boldsymbol{a}$, one could absorb it in the definition of perturbed mass δm.[10] However, we prefer not to do this, so that m and δm have the meaning of the energy of $T^{\mathrm{B}}_{\mu\nu} = T^{\mathrm{M}}_{\mu\nu} + T^{\mathrm{self}}_{\mu\nu}$ in the center-of-mass rest frame. The fourth term on the right side of eq. (10.67) involving the spin and acceleration of the body is entirely kinematic in origin (i.e., it would similarly arise if the acceleration were due to a nonelectromagnetic force). The fifth term involving the time derivative of the magnetic dipole moment can be understood as arising from the change in the "hidden momentum" of the body (see problem 7 of chapter 4 for the corresponding hidden momentum of the electromagnetic field). Finally, the last term involving the time derivative of the electric dipole moment is simply the Lorentz force on the current associated with the changing dipole moment (see eq. (10.70) below).

We turn now to the evolution equation for the rest mass, eq. (10.68). That the rest mass must change with time should not be surprising, since we already saw a manifestation of this in section 4.3. In that section, we quasi-statically moved a body (labeled as "body 2" in the discussion there) in a magnetic field $\boldsymbol{B}_1$ of another body ("body 1") and found that its self-energy must change by $c^2\delta m_2 = -\boldsymbol{\mu}_2 \cdot \delta \boldsymbol{B}_1$ (see eq. (4.75)). We also found that, in this process, the self-energy of body 1 must change by $c^2\delta m_1 = -\boldsymbol{\mu}_1 \cdot \delta \boldsymbol{B}_2$ (see eq. (4.80)). These changes for each body correspond to the first term in eq. (10.68). The second term in eq. (10.68) does not arise in magnetostatics, since $d\boldsymbol{p}/dt = 0$. However, it is natural that it would be present in electrodynamics, since, for any body of finite size, by conservation of charge-current, we have

$$\frac{d\boldsymbol{p}}{dt} = \int \boldsymbol{J}\, d^3x \tag{10.70}$$

(see eq. (5.73)). Thus, for a body of very small size, the second term in eq. (10.68) corresponds to

$$\frac{d\boldsymbol{p}}{dt} \cdot \boldsymbol{E}^{\mathrm{ext}} = \int \boldsymbol{J} \cdot \boldsymbol{E}^{\mathrm{ext}} d^3x . \tag{10.71}$$

By eq. (5.19), the right side is just the rate at which energy is transferred from the electromagnetic field to the body. It should be expected that this transfer of energy contributes to the rest mass of the body.

In eq. (10.69) for the spin evolution, the first two terms correspond, respectively, to the torque (eq. (2.45)) on an electric dipole $\boldsymbol{p}$ in electrostatics and the torque (eq. (4.69)) on a magnetic dipole in magnetostatics. The last term in eq. (10.69) is the Thomas precession (see footnote 4 in section 10.3), which is entirely kinematic in origin.

10.4 Self-Consistent Motion

We now turn to a discussion of the motion of a charged body in an external electromagnetic field $F^{\mathrm{ext}}_{\mu\nu}$, taking into account the corrections to Lorentz force. We are primarily interested in the self-force correction, so, for simplicity, we set $\boldsymbol{S} = \boldsymbol{p} = \boldsymbol{\mu} = 0$. However, there would be no essential modification to the discussion if we included

[10] In other words, we could define $\widehat{\delta m} = \delta m - \boldsymbol{p} \cdot \boldsymbol{E}^{\mathrm{ext}}/c^2$ and redefine the force as $m\delta\boldsymbol{a} + \boldsymbol{a}\widehat{\delta m}$, thereby eliminating the term $(\boldsymbol{p} \cdot \boldsymbol{E}^{\mathrm{ext}})\boldsymbol{a}/c^2$ from the force expression (10.67). Note that $-\boldsymbol{p} \cdot \boldsymbol{E}^{\mathrm{ext}}$ is the electrostatic interaction energy of $\boldsymbol{E}^{\mathrm{self}}$ with $\boldsymbol{E}^{\mathrm{ext}}$ (see eq. (2.34)).

the effects of $\boldsymbol{S}$, $\boldsymbol{p}$, and $\boldsymbol{\mu}$. In addition, for simplicity, we restrict consideration to nonrelativistic motion, so that we can use various nonrelativistic approximations, such as eq. (10.64) for the self-force. However, no essential modification to the discussion would be necessary if we treated the motion in a fully relativistic manner. Finally, since δm is constant by eq. (10.62), nothing will be lost by setting $\delta m = 0$, (i.e., we assume that our one-parameter family has been chosen, so that $\delta m = 0$).

As we have seen, at leading order and in the nonrelativistic approximation, the worldline $\boldsymbol{X}(t)$ of the body satisfies the Lorentz force equation:

$$m\frac{d^2\boldsymbol{X}}{dt^2}(t) = q\boldsymbol{E}^{\text{ext}}[t, \boldsymbol{X}(t)] + q\frac{d\boldsymbol{X}}{dt} \times \boldsymbol{B}^{\text{ext}}[t, \boldsymbol{X}(t)]. \tag{10.72}$$

This is a second order ordinary differential equation for $\boldsymbol{X}(t)$, which has a unique solution for any given initial position $\boldsymbol{X}(t_0)$ and velocity $\frac{d\boldsymbol{X}}{dt}(t_0)$ at time $t = t_0$. By eq. (10.61), in the nonrelativistic approximation the leading-order deviation $\delta\boldsymbol{X}(t)$ from Lorentz force motion satisfies

$$m\frac{d^2\delta\boldsymbol{X}}{dt^2} = q\delta\boldsymbol{E}^{\text{ext}} + q\delta\left[\frac{d\boldsymbol{X}}{dt} \times \boldsymbol{B}^{\text{ext}}\right] + \frac{\mu_0 q^2}{6\pi c}\frac{d^3\boldsymbol{X}}{dt^3}. \tag{10.73}$$

Here, the explicit meaning of $\delta\boldsymbol{E}^{\text{ext}}$ is

$$\delta\boldsymbol{E}^{\text{ext}} = (\delta\boldsymbol{X} \cdot \nabla)\,\boldsymbol{E}^{\text{ext}}, \tag{10.74}$$

where the derivative of $\boldsymbol{E}^{\text{ext}}$ is evaluated at $(t, \boldsymbol{X}(t))$. Similarly, we have

$$\delta\left[\frac{d\boldsymbol{X}}{dt} \times \boldsymbol{B}^{\text{ext}}\right] = \frac{d\delta\boldsymbol{X}}{dt} \times \boldsymbol{B}^{\text{ext}} + \frac{d\boldsymbol{X}}{dt} \times \left[(\delta\boldsymbol{X} \cdot \nabla)\,\boldsymbol{B}^{\text{ext}}\right], \tag{10.75}$$

where again, evaluation of all quantities is done on the unperturbed worldline. Equation (10.73) is a second order ordinary differential equation for $\delta\boldsymbol{X}(t)$, which has a unique solution for any given initial values of $\delta\boldsymbol{X}$ and $d(\delta\boldsymbol{X})/dt$ at time $t = t_0$. Note, in particular, that the appearance of third time derivatives in the self-force term in eq. (10.73) is not in any way problematic, since they act on the unperturbed worldline $\boldsymbol{X}(t)$, which has already been obtained by solving eq. (10.72). The self-force term merely acts as a known source term in eq. (10.73).

However, solving eq. (10.72) and eq. (10.73) is not a good way of determining the motion of a charged body over long intervals of time. The leading-order correction, $\delta\boldsymbol{X}$, to the position of the body will be a good approximation only when $\delta\boldsymbol{X}$ is very small compared with the scales set by the external fields. However, even if $\delta\boldsymbol{X} = d(\delta\boldsymbol{X})/dt = 0$ initially and even if the self-force term $(\mu_0 q^2/6\pi c)d^3\boldsymbol{X}/dt^3$ in eq. (10.73) is very small, the solution $\delta\boldsymbol{X}(t)$ to eq. (10.73) will generically become large at large t. When this happens, it will be a very poor approximation to use the unperturbed worldline $\boldsymbol{X}(t)$ to calculate the right side of eq. (10.73). This difficulty will not be overcome by including higher-order corrections to the motion, since the same difficulty will arise if one stops at any finite order in perturbation theory.

The above difficultly arises because, although the self-force correction to Lorentz motion is very small locally, its cumulative effects can be large. However, if this is the case, then even though the unperturbed solution $\boldsymbol{X}(t)$ to eq. (10.72) is no longer a good approximation to the position of the body at late times, the motion at all times should be well described locally by Newton's second law, with the force corresponding to the sum of the Lorentz force and the self-force. In other words, taking account of both the Lorentz force $\boldsymbol{f}^{\,\mathrm{LF}}$ and the self-force $\boldsymbol{f}^{\,\mathrm{SF}}$, the position $\boldsymbol{\mathcal{X}}(t)$ of the center of mass of the body should satisfy

$$m\frac{d^2\boldsymbol{\mathcal{X}}}{dt^2}(t)=\boldsymbol{f}^{\,\mathrm{LF}}+\boldsymbol{f}^{\,\mathrm{SF}}=q\boldsymbol{E}^{\mathrm{ext}}[t,\boldsymbol{\mathcal{X}}(t)]+q\frac{d\boldsymbol{\mathcal{X}}}{dt}\times\boldsymbol{B}^{\mathrm{ext}}[t,\boldsymbol{\mathcal{X}}(t)]+\frac{\mu_0 q^2}{6\pi c}\frac{d^3\boldsymbol{\mathcal{X}}}{dt^3}. \tag{10.76}$$

This equation automatically self-consistently corrects the unperturbed motion $\boldsymbol{X}(t)$ that would be obtained by solving eq. (10.72) alone. We will refer to an equation of motion obtained by adding a perturbative correction to the force to the original equation of motion in this manner as a *self-consistent perturbative equation*. The idea of using a self-consistent perturbative equation is sufficiently straightforward and obvious that it is normally done in most contexts without any comment.

However, in the present case, the self-consistent perturbative equation (10.76) is disastrous. The supposedly small self-force term $(\mu_0 q^2/6\pi c)d^3\boldsymbol{\mathcal{X}}/dt^3$ involves third time derivatives of $\boldsymbol{\mathcal{X}}$ and thus is higher order than the other terms in this equation. Consequently, this term drives the evolution of the system, and there is no reason it should remain small under the evolution of eq. (10.76), even if the self-force would be very small under Lorentz force motion. Indeed, it is easy to see that even with $\boldsymbol{E}^{\mathrm{ext}}=\boldsymbol{B}^{\mathrm{ext}}=0$, eq. (10.76) admits solutions of the form

$$\mathcal{X}(t)=Ce^{(6\pi mc/\mu_0 q^2)t}, \tag{10.77}$$

where C is a constant. Solutions with exponential growth of this sort are referred to as *runaway solutions*. Note that if q is made smaller (so that the self-force effects should become smaller relative to Lorentz force effects), the exponential growth rate of the runaway solution eq. (10.77) becomes larger.

Since eq. (10.76) is an ordinary differential equation for $\boldsymbol{\mathcal{X}}(t)$ that is third order in time, the initial conditions that can be freely specified at an initial time t_0 are $\boldsymbol{\mathcal{X}}(t_0)$, $(d\boldsymbol{\mathcal{X}}/dt)(t_0)$, and $(d^2\boldsymbol{\mathcal{X}}/dt^2)(t_0)$. If $\boldsymbol{E}^{\mathrm{ext}}=\boldsymbol{B}^{\mathrm{ext}}=0$, we could eliminate the runaway solution (10.77) by demanding that the initial conditions be chosen so that $(d^2\boldsymbol{\mathcal{X}}/dt^2)(t_0)=0$. However, if $\boldsymbol{E}^{\mathrm{ext}}$ and/or $\boldsymbol{B}^{\mathrm{ext}}$ become nonzero at some later time, then one would have to carefully adjust the initial value of the acceleration $(d^2\boldsymbol{\mathcal{X}}/dt^2)(t_0)$ to a value depending on the entire future history of $\boldsymbol{E}^{\mathrm{ext}}$ and $\boldsymbol{B}^{\mathrm{ext}}$ to avoid having runaway solutions.[11] Such a choice of initial acceleration that depends on the future history of $\boldsymbol{E}^{\mathrm{ext}}$ and $\boldsymbol{B}^{\mathrm{ext}}$ is referred to as *pre-acceleration*. Quite surprisingly, many references advocate pre-acceleration as a means to eliminate the runaway solutions to eq. (10.76), despite the obvious violation of causality that this proposal entails.

If we return to our original one-parameter family of bodies (10.3) and (10.4) and further assume that the matter stress energy $T^{\mathrm{M}}_{\mu\nu}(\lambda;t,\boldsymbol{x})$ satisfies physically reasonable positive energy conditions, then it is easy to see that runaway solutions like eq. (10.77) are blatantly inconsistent with conservation of energy. There is also no

[11] If the external fields are not turned off at late times, it may not be possible to choose any value of the initial acceleration that avoids runaway solutions.

runaway behavior in the perturbative description of motion, eq. (10.73). Since runaway solutions do not correspond to the physical behavior of bodies, it is clear that these solutions are entirely an artifact of the poor choice of self-consistent perturbative equation made in eq. (10.76). As we now discuss, the runaway behavior can be eliminated by simply making a better choice of self-consistent perturbative equation.

In eq. (10.73), the quantity $\boldsymbol{X}(t)$ is a solution to the unperturbed equation of motion (10.72). For a solution to eq. (10.72), we have

$$\frac{d^3\boldsymbol{X}}{dt^3} = \frac{d}{dt}\left(\frac{d^2\boldsymbol{X}}{dt^2}\right) = \frac{q}{m}\frac{d\boldsymbol{E}^{\text{ext}}}{dt} + \frac{q}{m}\frac{d}{dt}\left(\frac{d\boldsymbol{X}}{dt} \times \boldsymbol{B}^{\text{ext}}\right), \tag{10.78}$$

where d/dt is the derivative along the worldline $\boldsymbol{X}(t)$, that is,

$$\frac{d\boldsymbol{E}^{\text{ext}}}{dt} = \frac{\partial \boldsymbol{E}^{\text{ext}}}{\partial t} + \left(\frac{d\boldsymbol{X}}{dt} \cdot \nabla\right)\boldsymbol{E}^{\text{ext}}. \tag{10.79}$$

Furthermore, in the term

$$\frac{d}{dt}\left(\frac{d\boldsymbol{X}}{dt} \times \boldsymbol{B}^{\text{ext}}\right) = \frac{d^2\boldsymbol{X}}{dt^2} \times \boldsymbol{B}^{\text{ext}} + \frac{d\boldsymbol{X}}{dt} \times \frac{d\boldsymbol{B}^{\text{ext}}}{dt}, \tag{10.80}$$

we may replace $d^2\boldsymbol{X}/dt^2$ with the expression given by eq. (10.72). By doing the substitution eq. (10.78), followed by this substitution for $d^2\boldsymbol{X}/dt^2$, we may write the self-force term in eq. (10.73) in a form that involves only $\boldsymbol{X}(t)$ and its first time derivative but no higher derivatives. The procedure of making such substitutions that use the unperturbed equation to decrease the number of derivatives of unperturbed quantities appearing in the perturbed equation is referred to as *reduction of order*.[12] The reduced order equation for $\delta\boldsymbol{X}$ is entirely equivalent to the original equation of motion eq. (10.73).

However, the self-consistent perturbative equation corresponding to the reduced order perturbed equation of motion for $\delta\boldsymbol{X}$ is *not* equivalent to eq. (10.76). The reduced order self-consistent perturbative equation of motion for $\boldsymbol{\mathcal{X}}$ is

$$\begin{aligned} m\frac{d^2\boldsymbol{\mathcal{X}}}{dt^2}(t) = {} & q\boldsymbol{E}^{\text{ext}} + q\left(\frac{d\boldsymbol{\mathcal{X}}}{dt} \times \boldsymbol{B}^{\text{ext}}\right) + \frac{\mu_0 q^3}{6\pi mc}\left[\frac{\partial \boldsymbol{E}^{\text{ext}}}{\partial t} + \left(\frac{d\boldsymbol{\mathcal{X}}}{dt} \cdot \nabla\right)\boldsymbol{E}^{\text{ext}}\right] \\ & + \frac{\mu_0 q^4}{6\pi m^2 c}\left[\boldsymbol{E}^{\text{ext}} + \frac{d\boldsymbol{\mathcal{X}}}{dt} \times \boldsymbol{B}^{\text{ext}}\right] \times \boldsymbol{B}^{\text{ext}} \\ & + \frac{\mu_0 q^3}{6\pi mc}\frac{d\boldsymbol{\mathcal{X}}}{dt} \times \left[\frac{\partial \boldsymbol{B}^{\text{ext}}}{\partial t} + \left(\frac{d\boldsymbol{\mathcal{X}}}{dt} \cdot \nabla\right)\boldsymbol{B}^{\text{ext}}\right], \end{aligned} \tag{10.81}$$

[12] In the case of any system where the dynamics is described by ordinary differential equations in time, a similar reduction of order can be done for any perturbed equation that contains higher time derivatives of the unperturbed dynamical variables. However, if the dynamics of a system is described by partial differential equations, the reduction-of-order procedure would require selecting a time direction with respect to which the reduction is being done (thereby breaking Lorentz covariance if the original equations were Lorentz covariant), and in general, the procedure will increase the number of spatial derivatives in the equation (thereby possibly making the equations ill behaved). Thus, in general, there does not appear to be a satisfactory analog of reduction of order for partial differential equations.

where $\boldsymbol{E}^{\text{ext}}$ and $\boldsymbol{B}^{\text{ext}}$ as well as their derivatives are evaluated at $(t, \boldsymbol{\mathcal{X}}(t))$. Equation (10.81) is a second order ordinary differential equation for $\boldsymbol{\mathcal{X}}$, where the only second derivative term is the one on the left side of the equation. The initial data for this equation consist of $\boldsymbol{\mathcal{X}}(t_0)$ and $(d\boldsymbol{\mathcal{X}}/dt)(t_0)$, and there are no spurious runaway solutions analogous to eq. (10.77). Equation (10.81) thus provides a fully satisfactory description of nonrelativistic motion of a charged body, taking into account the leading-order effects of self-force.

Index